INTRODUCTION TO TRANSPORT PHENOMENA

William J. Thomson

Dept. of Chemical Engineering
Washington State University
Pullman, WA 99164-2710

ISBN 0-13-454828-0

Prentice Hall PTR
Upper Saddle River, New Jersey 07458
http://www.phptr.com

90000

9 780134 548289

Library of Congress Cataloging-in-Publication Data

Thomson, William J.
    Introduction to transport phenomena / William J. Thomson.
      p.   cm.
    Includes bibliographical references and index.
    ISBN 0-13-454828-0
    1. Transport theory. 2. Chemical engineering. I. Title.
TP156. T7T48 1999
660′.284–dc21   99-383330
                                         CIP

Acquisitions editor: *Bernard M. Goodwin*
Editorial assistant: *Diane Spina*
Cover designer: *Talar Agasyan*
Cover design director: *Jerry Votta*
Manufacturing manager: *Alan Fischer*
Marketing manager: *Lisa Konzelmann*
Project coordinator: *Anne Trowbridge*

© 2000 by Prentice Hall PTR
Prentice-Hall, Inc.
Upper Saddle River, New Jersey 07458

Prentice Hall books are widely used by corporations and government agencies for training, marketing, and resale.

The publisher offers discounts on this book when ordered in bulk quantities. For more information contact:

    Corporate Sales Department
    Phone: 800-382-3419
    Fax: 201-236-7141
    E-mail: corpsales@prenhall.com

    Or write:

    Prentice Hall PTR
    Corp. Sales Dept.
    One Lake Street
    Upper Saddle River, New Jersey 07458

Printed in the United States of America
10 9 8 7 6 5 4 3 2 1

ISBN: 0-13-454828-0

Prentice-Hall International (UK) Limited, *London*
Prentice-Hall of Australia Pty. Limited, *Sydney*
Prentice-Hall Canada Inc., *Toronto*
Prentice-Hall Hispanoamericana, S.A., *Mexico*
Prentice-Hall of India Private Limited, *New Delhi*
Prentice-Hall of Japan, Inc., *Tokyo*
Prentice-Hall (Singapore) Pte Ltd., *Singapore*
Editora Prentice-Hall do Brasil, Ltda., *Rio de Janeiro*

CONTENTS

Contents

Contents

PREFACE

When Bird, Stewart and Lightfoot first introduced their landmark text (TRANSPORT PHENOMENA, John Wiley & Sons, N.Y., 1960) there was an immediate adoption of this rather mathematical approach of describing fluid mechanics and heat and mass transfer; primarily at the graduate level. It is a testimony to the universality of this text that it still remains in its first edition almost 40 years after its introduction. However, despite many attempts at introducing the transport phenomena approach to the undergraduate curriculum, it is still considered to be a rather esoteric applied mathematical approach to the "practical" business of sizing piping systems, heat exchangers and continuous separation processes. Typically the course has been intended as a "primer" for those students interested in pursuing a graduate degree.

Times have changed. The advent of powerful, numerical differential equation solvers has greatly expanded the ability to solve problems that used to require a multitude of simplifying assumptions. Granted, complex turbulent mixing phenomena are still beyond the grasp of practical mathematical solutions, but who knows for how long! Increasingly, the transport phenomena approach to these problems is becoming a pre-requisite as opposed to an "elective". This statement is based on two premises. One, it is important that students have a basic understanding of the calculations they will do in describing the rates of momentum, energy and mass transfer and WHY they are doing it. Two, they should be capable of describing these rates from a differential point of view, so that they are prepared for the inevitable advances that will be forthcoming in computational techniques. In fact, it is not too difficult to imagine a time in the near future where empirical transfer coefficient correlations will no longer be needed to do sizing calculations.

This text is targeted to third year undergraduate chemical engineering students and thus the emphasis is placed on only the simplest of transport processes; i.e., one-dimensional and, for the most part, steady state systems (CHAPTERS 1-5). The "Equations of Change" are minimally derived and presented in an appendix in the event that an instructor might wish to approach transport problems from that perspective. Furthermore, in order to achieve the dual objectives of understanding and simplicity, the derivations of the applicable differential equations are approached only from a fixed control volume point of view. The goals are straight-forward: to understand the fundamental concepts of differential conservation balances, to be able to translate physical phenomena into mathematical terms and to appreciate the reasons for having to resort to semi-empiricism in order to quantify complex problems in heat, mass and momentum transfer.

The text is divided into three sections: (I) Molecular Transport, (II) Convective Transport and (III) Macroscopic Calculations. The basics of transport phenomena are covered in Section I (Chapters 1-6), using an "inductive" approach. That is, specific problems are analyzed and solved in order to illustrate the transport phenomena approach to physical problems. A separate chapter on similarity analyses (Chapter 7) shows how the mathematical solutions can be generalized and prepares the students for analysis of convective transport (Section II, Chapters 8-10). These three chapters cover convective transport problems that are mathematically tractable (Chap 8), the basics of turbulent flow (Chapter 9) and the concept of transfer coefficients to solve problems that are not mathematically tractable (Chapter 10). Section III of the text (Chapters 11-13) is intended to illustrate how transfer coefficients can be combined with the differential balance approach of transport phenomena to do more extensive calculations than is typical in the traditional "unit operations" approach to momentum, energy and mass transfer problems. A good example of this is the complexity of the gas absorber problems which are analyzed in Chapter 13. Personal experience with teaching this subject in this manner has demonstrated that students have only minimal difficulty with this approach. While the types of problems in Chapters 11-13 are not extensive, this experience might suggest that it is time to reconsider the traditional methods of teaching this material.

Again, this text is intended only as an introduction to the subject of transport phenomena and for students at the mid-point of their undergraduate program. Hopefully this approach to transport phenomena can contribute to the skill level of the typical BS engineer as well as serve as a preliminary exposure to those students who will pursue the subject at an advanced level of their studies.

NOMENCLATURE

GENERALIZED UNITS:
E = energy, **L** = length, **m** = mass, **M** = moles, **t** = time, **T** = temperature

a_I = interfacial surface area/volume (L^{-1})

a_p = surface area/volume of particle

A_i = area normal to i (L^2)

A_P = surface area of particle (L^2)

A_s = surface area (L^2)

A_{sP} = total surface area of packing (L^2)

$\hat{c}_P$ = heat capacity at constant pressure (E/m-T)

$\hat{c}_s$ = humid heat (E/m-T)

$\hat{c}v$ = heat capacity at constant volume (E/m-T)

C = total molar concentration (M/L^3)

C_i = molar concentration, species i (M/L^3)

d_p = particle diameter

D = diameter (L)

D_{eq} = equivalent diameter (L^2)

D_{ij} = diffusivity of i in j (L^2/t)

D_{K_i} = Knudsen diffusivity of "I" (L^2/t)

D_{eff} = "effective diffusivity (L^2/t)

D^t = turbulent diffusivity (L^2/t)

D_L = dispersion coefficient (L^2/t)

E = energy (E)

f = friction factor

F = force $(m\, L/t^2)$

F_P = Packing Factor

$\hat{F}$ = free energy (E/m)

F_Λ = generalized flux (Λ units/t-L^2)

g = gravitational acceleration (L/t^2)

G	= mass velocity)m/L^2-t
Gz	= Graetz Number, (dimensionless)
$G_{(L,G)f}$	= loading factor, Eq. (11.????)
h	= heat transfer coefficient (E/t-L^2-T
h_f	= fouling factor (E/t-L^2-T)
$h_{(c,e,f,s)}$	= friction losses due to contraction,expansion,fittings,skin (??)
H	= enthalpy (E)
H	= humidity
H_R	= relative humidity
$H_{(OG,OL)}$	= height of a transfer unit based on overall mass transfer coefficients (L)
$H^{(G,L)}$	= height of a gas or liquid Transfer Unit (L)
$\Delta \tilde{H}_s$	= heat of solution (E/M)
$\Delta \hat{H}_v$	= heat of vaporization (E/m)
$\Delta \tilde{H}_r$	= heat of reaction, (E/M)
H	= Henry's Law Constant (pressure/mole fraction)
J	= molar diffusional flux of mass (M/t-L^2)
j	= mass diffusional flux of mass (m/t-L^2)
$j_{(H,M)}$	= j-factor for heat (H) or mass (M) transfer (dimensionless)
k	= thermal conductivity (E/t-L-T)
$k_{m(x,c,p)}$	= mass transfer coefficient (M/t-L^2-Concentration)
$k_{(v,s)}$	= reaction rate constant for nth order reaction (M/L^3-Concentration^n, M/L^2-Concentration^n
K	= Boltzman constant (E/T)
$K_{OG(x,c,P)}$	= overall mass transfer coefficient, gas concentrations (M/L^2-t- concentration)
$K_{OL(x,c,P)}$	= overall mass transfer coefficient, liquid concentrations (M/L^2-t- concentration)
L_C	= characteristic dimension (L)
L_e	= Lewis Number, $\dfrac{\alpha}{D_{Am}}$, (dimensionless)
m	= mass (m)
m_i	= linear equilibrium parameter, $\dfrac{(X_i^G)*}{(X_i^L)*}$
M	= molecular weight (m/M)
n	= moles (M)
N_i	= molar flux of species i (M/t-L^2)
$(N_i)'$	= molar flux of species i on a "solute-free" basis (M/t-L^2)
$\dot{N}_i$	= molar flow rate of species i (M/t)
$\tilde{N}$	= Avogadro's Number (molecules/mole)

Nu = Nusselt Number (dimensionless heat transfer coefficient)

NTU = number of transfer units, Eq (13.45)

P = total pressure $(m/t^2\text{-}L)$

P_i = partial pressure of i $(m/t^2\text{-}L)$

Pe = Peclet Number, product of Re and Pr (or Sc), dimensionless)

Pr = Prandtl Number, $\dfrac{\nu}{\alpha}$, (dimensionless)

P_{Mi} = permeability of "i" in membrane, Eq (13.97) ("barrers")

P_t = tube "pitch"

PW = pump power (E-t)

$\wp$ = combined pressure - gravity force/volume (Eq. 5.19)

$\wp_w$ = wetted perimeter (L)

q = molecular energy flux $(E/t\text{-}L^2)$

$\dot{Q}$ = energy flow rate (E/t)

$(Q_s)_V$ = energy SOURCE or SINK per unit volume $(E/t\text{-}L^3)$

r = radial position (L)

$r_{(ad,d)}$ = rate of adsorption, desorption $(M/t\text{-}L^2)$

R = geometric radius (L)

R = parameter to correct $(\Delta T)_{LM}$, Eq(12.51)

R = resistance term in heat or mass transfer $(L^2\text{-}t/(E,M))$

R_g = gas law constant $(mL^2/t^2\text{-mole-}T)$

R_H = hydraulic radius (L)

R_{iV} = homogeneous reaction rate of species i $(M/t\text{-}L^3)$

R_{iS} = surface reaction rate of species i $(M/t\text{-}L^2)$

Re = Reynolds Number (dimensionless)

s = distance from wall, R - r, (L)

S = parameter to correct $(\Delta T)_{LM}$, Eq(12.51)

S_i = solubility of "i" in membrane, Eq (13.97)

$S_{(H,V)}$ = horizontal or vertical tube spacing (L)

$\hat{S}$ = surface area per mass (L^2/m)

Sc = Schmidt Number, $\dfrac{\nu}{D_{Am}}$, (dimensionless)

Sh = Sherwood Number (dimensionless mass transfer coefficient)

t = time (t)

T = temperature (T)

T_{DP} = dew point temperature (T)

$(\Delta T)_{LM}$ = "log mean" temperature driving force, Eq (12.46) (T)

u	= molecular velocity (L/t)
U	= internal energy (E)
$U_{(i,o)}$	- overall heat transfer coefficient based on inside, outside area (E/t-L^2-T)
V	= volume (L^3)
v	= fluid velocity (L/t)
[v]	= time-averaged velocity (L/t)
v′	= fluctuating velocity (L/t)
v_o	= superficial velocity (L/t)
v_*	= friction velocity defined by Equation (9.7)
WP	= pump input energy per mass of fluid (E/m)
x	= cartesian position (L)
X_i	= mole fraction of species i
y	= cartesian position (L)
z	= cartesian position (L)
Z	= molecular flux, defined by Equation (6.2)
Z_T	= Tower height (L)

GREEK LETTERS

α	= thermal diffusivity (L^2/t)
α	= membrane selectivity, Eq(13.99)
β	= volumetric expansion coefficient (T^{-1})
χ_i	= fraction conversion, species 'i'
δ	= differential distance, film thickness (L)
Δ	= difference
η	= dimensionless coordinate, combined variable (Eq. 7.16)
η	= pump efficiency
ε	= eddy viscosity (L^2/t)
ε	= void fraction
Γ	= hydraulic loading, Eq (12.35) (m/L)
λ	= mean free path (L)
μ	= viscosity (m/L-t)
μ^t	= turbulent viscosity (m/L-t)
μ_i	= chemical potential of species i
ν	= kinematic viscosity or momentum diffusivity (L^2/t)
Φ	= viscosity correction factor in heat transfer coefficient correlations
ξ	= enhancement factor, Eq(13.65)

ρ = density (m/L^3)

σ = surface tension (m/t^2)

σ_c = critical surface tension of packings (m/t^2)

τ_{ij} = momentum flux [j-momentum in the i-direction] or shear stress [force acting in j-direction on area perpendicular to i-direction] (m/L-t^2)

$\tau^{tot,t}$ = total, turbulent momentum flux (m/L-t^2)

τ_i = residence time of "i<u>th</u>" eddy

$\overline{\tau}$ = average eddy residence time

ϖ_i = mass fraction of species i

OVERBARS

($\tilde{\ }$)   = "per unit mole"

($\wedge$)   = "per unit mass"

($\bar{\ }$)   = "average value"

[ ]   = "time averaged value"

[ ]*   = "psuedo-equilibrium" concentration

(*)   = per unit time (t^{-1})

SUBSCRIPTS

G   = gas phase

L   = liquid phase

i,o   = referring to "inside", "outside"

I   = interface

f   = fluid

mf   = "at" minimum fluidizing velocity

p   = based on or pertaining to particle

s   = solid

V   = per unit volume

( )$_B$ = conditions at bottom of tower

( )$_T$ = conditions at top of tower

SUPERSCRIPTS

G   = gas phase

L   = liquid phase

n   = reaction order

( )*   = equilibrium condition

PART I

Molecular Transport

The first part of this text addresses the mechanisms by which energy, mass, and momentum are transported by molecular mechanisms. This mechanism of transport, termed *molecular transport*, is inherently a property of the molecules themselves and therefore, depends on molecule-molecule interactions. Because molecular transport lends itself to mathematical analysis, it is possible to describe the conservation of energy, mass and momentum transport *at a point*. That is, differential equations can be derived and, often solved, to predict the variation of temperature, concentration and velocity as they change with position and time. Consequently, this part of the text also introduces the concept of differential balances and describes techniques that can be used to derive the differential equations appropriate to specific physical problems. This is followed with separate chapters illustrating how these techniques can be applied to obtain useful solutions for a variety of problems in the molecular transport of energy, mass, and momentum. The last two chapters of this part of the text describe the relationship of molecular interactions to the phenomenological transport coefficients and the use of similarity analysis to transform the specific solutions so that they are more universally applicable.

CHAPTER 1

The Nature of Transport Phenomena

1-1 WHY TRANSPORT PHENOMENA?

The goal of engineering disciplines is generally stated to be a design function but can be more universally extended to include analysis functions as well. Chemical engineering is more uniquely focused on process engineering; whether it be the processes associated with petroleum refining or those related to microchip fabrication. In any process we are concerned with either "grass-roots" design or with analyzing data from existing processes in an attempt to discover what is wrong or to make improvements; i.e., to increase efficiency. Because of the incredible increase in the complexity of modern chemical engineering it is more important than ever to grasp the fundamentals which govern the kinds of things that engineers do on a routine basis. With respect to the design function, we wish to understand and quantify the <u>rates</u> at which matter and energy are moved from place-to-place. Hopefully it is intuitive that when rates are higher, equipment size (and cost) are lower or, for a given equipment size, production rates are higher. It is here where transport phenomena has its place since it deals with phenomena associated with energy transport (to predict heat exchange performance), mass transport (to describe separations processes), and momentum transport (to yield predictions of pressure losses in process piping systems). Notice also, that when we use the term <u>mass transport</u>, we are actually referring to the transport of individual species, <u>not</u> the transport of total mass since, in the absence of nuclear reactions, total mass is always constant. Of course, long before transport phenomena arrived on the scene, chemical engineers had always dealt with these same issues; so, what is the reason for this particular emphasis?

When Bird, Stewart, and Lightfoot first introduced their landmark text on transport phenomena over 30 years ago [1], it was viewed primarily as a rather mathematical approach of describing fluid mechanics and heat and mass transfer. Even today it is often considered to be a rather esoteric applied-mathematics approach, with little application to the "practical" business of sizing piping systems, heat exchangers and continuous separation processes.

At the current state of development, transport phenomena is certainly able to qualitatively explain all of the macroscopic phenomena associated with the transport of energy, mass, and momentum in complex mixing processes (such as is found in turbulent flow) but is still unable to produce quantitative predictions in these systems. Given the amazing developments in computational methods over the past ten years, practical computational predictions of turbulent processes may be on the horizon. If so, knowledge of the transport phenomena will be of great value to the practicing chemical engineer. Even if this does not come to pass very rapidly, an understanding of its principles will make for a better engineer; one who can understand the tools at hand and is able to apply them to new situations.

Keeping in mind that we wish to describe and quantify the rates at which energy, mass, and momentum are transported, it is useful to compare the approaches taken by the transport phenomena in contrast to that taken by the conventional "rate processes" approach [2-4]. First of all, the transport phenomena approach is a point-to-point description of the rates and is based on our understanding of how matter goes about transporting energy, mass, and momentum on a molecular scale. Consequently the mathematical description of the phenomena results in differential equations which model the conservation of energy, mass, and momentum transport at a point and utilizes the inherent properties of matter to calculate the pertinent transport rates. In contrast, the conventional modeling of heat, mass, and momentum transfer is based on macroscopic (or "overall") conservation balances and relies heavily on empirical correlations to estimate the transfer coefficients needed to calculate the transfer rates. However, because it is macroscopic, this approach can only predict overall rates and gives no information on the point-to-point variations (the profiles) of temperature, concentration, and velocity. Predictions of these profiles can often be critical as in the case of chemical reactor design where reaction rates usually have a high dependence on temperature and thus vary according to the existing temperature profiles within the reactor. The attractiveness of the transport approach is that, not only can it yield the same results as the conventional approach (by integration of the differential equations), but it also allows for prediction of the temperature, concentration, and velocity profiles.

So, why should we even use the conventional approach if the transport approach yields the same results as well as information on the profiles? The difficulty lies in our inability to solve many of the differential equations which arise from applying the transport approach. Although we can now solve many more of these problems due to the increasing power of computers and numerical methods, we are still unable to do a good job of predicting transport rates in systems with complex mixing phenomena; i.e., in turbulent flows. On the other hand, if there is no fluid flow or if the fluid flows in a "well-behaved" manner (e.g., in <u>laminar</u> flow where the fluid travels along "stream" lines), then we can generally solve the resulting differential equations.[#] However, if fluid elements are moving in a complex manner then we have <u>turbulent</u> flow and, although we can still predict transport rates within a fluid element, we are not able to consistently predict the movement of the elements themselves. In this latter case we have no alternative except to use empirical correlations to predict transfer rates and sometimes we can also use empirical correlations to predict the profiles. Table 1-1 summarizes the comparisons between the transport and rate processes approaches to the prediction of rates in transport processes.

Table 1-1 Transport vs. Rate Processes Approach

Transport Approach	Rate Process Approach
1. Fundamental Approach	1. Relies on correlations.
2. Results in <u>differential equations</u> capable of predicting temperature, concentration and velocity profiles.	2. Uses averages and integrated equations resulting in algebraic equations which can only predict overall rates.
3. Sometimes useful for design purposes (e.g., chemical reactors).	3. Primarily used for design purposes.
4. Not generally useful in turbulent flow systems.	4. Can be successfully used in turbulent flow systems as long as empirical correlations are applicable.
5. Good analytical tool for interpreting process data.	5. Not very useful for analytical purposes.
6. Useful for analyzing simultaneous transport processes.	6. Assumptions and iterations required to handle simultaneous transport processes.
7. Firm basis for analyzing and designing new technologies (e.g., membranes, chemical vapor deposition, biotechnology).	7. In the absence of suitable correlations, generally restricted to conventional technologies

---

[#]   Depending on the complexity of the problem, the computational effort required for a solution may not be justifiable.

1-2 MECHANISMS OF TRANSPORT PROCESSES

Throughout this text we will refer to two types of transport processes: <u>convective</u> transport and <u>molecular</u> transport. The former is transport due to bulk fluid motion while the latter is transport between aggregate groups of molecules and is <u>independent</u> of bulk fluid motion. Let us now examine these concepts in greater detail.

The term "convection" is familiar to most individuals and connotes a fluid movement. Convection currents are readily associated with oceans, rivers, and the phenomena of hot air rising (so called "natural" convection). If we think upon this for a moment, we can see that the aggregate of fluid molecules is capable of <u>carrying</u> with it any physical or chemical properties it may have. Thus the moving fluid carries its inherent internal energy, its momentum, and its own mass. A moving fluid will most probably encounter other fluid aggregates during its journeys; thus there is the potential of transporting these properties to the fluid it encounters. This bulk transport mechanism is a very efficient means of transporting energy, mass, and momentum although it can often be extremely difficult to analyze (turbulent flow for example). On the other hand, if you think about it, there must be some dimensional scale where convective motion no longer has much meaning. Thus, as two fluid elements approach one another, the mechanism of transport is eventually on a molecular scale.[#] The same is true in regions where fluid comes into contact with a pipe wall (as in a heat exchanger). Thus, even though we may have a very efficient means of convective transport, ultimately all of the transport must take place by a molecular mechanism. The natural question is then "If all transport ultimately takes place via molecular mechanisms, what is the advantage of convection?"

Experience tells us that a hot spoon is cooled more rapidly if it is waved it about in the air (convective energy transport). The link between molecular and convective transport lies in the empirical observations of Fourier, Fick, and Newton who observed the inherent rates of the transport of energy (heat), mass, and momentum, respectively, and then formulated simple mathematical expressions to describe their observations. The discussion of these "phenomenological" laws will be postponed until later, but for now consider Fourier's observations of energy transport in a solid. First of all, there is no fluid motion in a solid, so we know that we are not dealing with convective transport but rather with the inherent ability of the material to transport energy. Fourier noted that the rate of energy transport (energy/time) was proportional to the surface area of the solid as well as to the magnitude of the temperature difference between its hot and cold ends but inversely proportional to the distance between the hot and cold ends. The proportionality constant is called the <u>thermal conductivity</u> of the material and is a measure of the material's inherent capability of transporting energy. Since this

---

[#]   Actually, the dimensional scale cannot be too small or else we will lose the ability to use statistical averaging to describe the behavior of the molecules. When this happens we have a situation called "free molecule" flow and different mathematical approaches are necessary.

ability is inherent, it is really a "molecular" transport of energy and, indeed, all materials (solids, liquids, gases) have this ability to one degree or another.

These observations also give us the answer to the apparent dilemma mentioned above. If a material is moving in a direction which differs from the direction in which the transport of energy (or mass or momentum) is occurring, then its only means of transport is via molecular mechanisms. On the other hand, if there is bulk fluid motion in the direction of transport, then the transport rates will be higher. The reason convective transport is more efficient in this case is due to the fact that the convective fluid motion (or "mixing," if you prefer) brings fluid elements of different energy contents closer to one another. Since Fourier's observations indicate that transport rates are higher when the distance between them is smaller, we have an explanation for the phenomena; namely, convection serves to shorten the distance over which molecular transport takes place. It is important to keep these two mechanisms of transport in mind because molecular transport is fairly easy to describe mathematically, but convective transport is often intractable from a mathematical point of view. That is, it is very difficult to predict just how close convection can bring two fluid elements,

1-3 DRIVING FORCES FOR TRANSPORT PROCESSES

As pointed out earlier, it is the rate at which transport takes place which determines the size of equipment or the production capacity for a given sized equipment. However, a complementary part of all engineering problems is the question of how much can be transported if we can afford the luxury of infinite time. This question is the subject of thermodynamics and is a necessary one to pose since it is important to know whether conditions are right for transport to take place at all; and if so, what is the best we can do? For example, our experience tells us that energy transport will cease once temperatures have equilibrated (thermal equilibrium). Similarly, the mass of a particular species will no longer be transported when its chemical potential[#] is the same everywhere and there will be no net transport of momentum once momentum is equally distributed. When a given situation is not at these equilibrium conditions then there will be "driving forces" which will tend to establish equilibrium. In fact, the rates at which the system attempts to reach equilibrium are directly proportional to the magnitude of these driving forces.

The simplest driving force to consider is one consistent with our everyday experiences; viz., temperature differences which result in energy transport. The facts that energy transport ceases when temperatures are equal and that matter heats or cools more rapidly when there are larger temperature differences between the matter and the hot or cold surroundings are all

---

[#]   Chemical potential, μ_i, is another way of describing the free energy, $\hat{F}_i$, (per unit mass) which is assigned to a particular species. Equilibrium exists in a system whenever the free energy is minimized (or $\Delta\hat{F} = 0$).

consistent with our day-to-day observations. Not only is it is obvious that the appropriate driving force for energy transport is a temperature difference, but temperatures are easily measured. The temperature difference driving force can be applied between two macroscopic systems (such as is done in the rate processes approach) or, more typical of the transport approach, we can look at point-to-point differences. In the former case the driving force is usually referred to as an overall "delta T," or ΔT, and, in the latter case as an infinitesimal δT. As we shall see in the next chapter, when we use the transport approach we will be concerned, not only with the magnitude of δT, but also with its variation over infinitesimal distances, δx.

The fact that mass transport of a particular species ceases when its chemical potential is equilibrated, has already been mentioned. In analogy with energy transport it is tempting to define differences in chemical potential as the appropriate driving force for mass transport. Indeed, this is often done in complex systems where mass transport can occur by a variety of driving forces [5]. The difficulty here is that chemical potential cannot be measured directly but can only be calculated from other types of measurements. Since in this introductory text we will only be concerned with the more conventional aspects of mass transport, the driving forces will be expressed in terms of concentrations; parameters which are easily measured and which, for many processes, are equivalent to chemical potential. Now the difficulty becomes one of defining concentrations. For example, we can express concentration in <u>volumetric</u> terms such as mass or molar density (ρ_i = mass/vol or C_i = moles/vol) or in relative terms such as mass or mole fractions (ω_i or X_i) where the subscript, i, represents a particular species. There are other methods as well, but we will restrict ourselves to these four methods. As we shall see in the next chapter, depending on how we define the concentration driving forces, some of the numerical values associated with mass transport rates can be drastically different.

One of the more formidable concepts to grasp, when studying transport phenomena, is the transport of momentum. The reason for the difficulty is due to the fact that momentum is a vector quantity; i.e, *momentum* = $m\,\vec{v}$, and it is not easy to visualize a vector quantity, which has a specific direction, being transported in some other direction. This will be discussed in more detail in the next chapter. For now, however, all we need do is decide on the driving force for momentum transport. Since all of our systems will always have constant mass, it is velocity <u>differences</u> which will determine whether or not momentum will be transported and the rate of transport will depend on the magnitude of the driving force; i.e., the magnitude of the velocity differences.

REFERENCES

[1] Bird, R.B., Stewart, W.E., and E.N. Lightfoot, *Transport Phenomena*, John Wiley & Sons, N.Y., 1960.

[2] McCabe, W.L., Smith, and J.C. Harriot, *Unit Operations of Chemical Engineering*, 3rd ed., McGraw-Hill Inc., N.Y., 1976.

[3] Coulson, J.M. and J.F. Richardson, *"Chemical Engineering, Vol. 2 - Unit Operations"*, 3rd ed., Pergammon Press, N.Y., 1977.

[4] Foust, A.S., Wenzel, L.A., Clump, C.W., Maus, L. and L.B. Andersen, *"Principles of Unit Operations"*, 2nd ed., John Wiley & Sons, N.Y., 1980.

[5] Byrd, R.B., Stewart, W.E., and E.N. Lightfoot, *"Transport Phenomena"*, John Wiley & Sons, N.Y., 1960, pp. 563-569.

CHAPTER 2

Transport Phenomena Laws

2-1 THE DEFINITION OF FLUXES

As we shall see, the differential conservation balances, as well as the phenomenological laws which describe the molecular transport of energy, mass, and momentum are usually expressed in terms of the <u>fluxes</u> of energy, mass and momentum. Consequently, let us consider the definition of a flux in more general terms. Generalizing will be useful since, as we have already pointed out, transport can occur by means of convective motion as well as by molecular mechanisms.

The basic definition of a flux can be stated as:

"A flux, F_Λ, of property Λ is the quantity of extensive property, Λ, which crosses a unit area per unit time."

Note that we can only talk about fluxes of <u>extensive</u> properties; the flux of an intensive property (such as temperature) would be meaningless. However energy, mass, and momentum all qualify as extensive properties since each depends on the total quantities involved. While the above definition may be formal, it is far more useful if it is expressed in terms of more familiar properties. For instance we can show that a flux of Λ can always be obtained by multiplying the volumetric concentration of Λ (i.e., Λ/vol)[#] by the velocity of transport of Λ across the area perpendicular to the velocity, $A_\perp$. To show this, we revert back to our formal definition. According to this definition we require the quantity of extensive property Λ per unit time per unit area, or

$$F_\Lambda = \frac{\Lambda}{\delta t \, A_\perp} \tag{2.1}$$

Since $\Lambda / \delta t$ is just the flow rate of 'Λ', $\dot{\Lambda}$, and the flow rate of 'Λ' can always be expressed in terms of its volumetric concentration (ρ_Λ) and the <u>volumetric</u> flow rate of the fluid within which it is contained, $\dot{V}$, Equation (2.1) can be written as

$$F_\Lambda = \frac{\dot{\Lambda}}{A_\perp} = \frac{\rho_\Lambda \dot{V}}{A_\perp} \tag{2.2}$$

But the volumetric flow rate is just the velocity of the fluid stream multiplied by the area across which 'Λ' is flowing ($\dot{V} = v \, A_\perp$). Consequently

$$F_\Lambda = \frac{\rho_\Lambda \, \dot{V}}{A_\perp} = \frac{\rho_\Lambda \, v \, A_\perp}{A_\perp} = \rho_\Lambda \, v \tag{2.3}$$

or, in words

$$\text{FLUX of } \Lambda = \left[\text{VOL CONC of } \Lambda\right] \circ \left[\text{TRNSPT VEL}\right]$$

$$F_\Lambda \quad = \quad\quad \rho_\Lambda \quad\quad\quad \circ \quad\quad v$$

The first thing which should be apparent is that we have arbitrarily selected some velocity without specifying its direction. Thus a flux is not completely specified until something is said about its direction. Note that since we have taken the area to be perpendicular to the velocity, an area can also be thought of as having direction (and hence, vector properties). All of the various forms of energy as well as mass and momentum meet the requirements of being extensive and of being transported from point to point. However both energy and mass are <u>scalar</u> quantities while momentum is a <u>vector</u>. The flux of both energy and mass is then simply a vector, but problems arise with the flux of momentum since here we are dealing with the directional transport of a quantity which already has directional significance. In order to handle such problems, mathematicians have invented quantities called <u>tensors</u>. It should always be kept

---

[#] For example, the volumetric concentration of momentum would be $m \, \vec{v}$ /vol; or, $\rho\vec{v}$

in mind that these are strictly inventions and if they are useful they ought to result in some simplifications. As it turns out, they do have some useful properties and, in fact, a tensor is more general than a vector (a vector is a first-order tensor). But for most purposes, its main utility is in the shorthand notation it affords in writing very long equations. Other than the ominous ring the word connotes, no problems should be encountered as long as it is always kept in mind that it is associated with a combination of directions.

The primary advantage of Equation (2.3) is in dealing with convective fluxes. While the general definition of a flux as given by Equation (2.3) is also applicable for molecular fluxes, in this case the velocities of transport take place on a molecular level. For gases, these velocities are the molecular velocities as described by the kinetic theory of gases. However, for liquids and solids, the "velocities" are really complex combinations of molecular motion and vibration. Viewing molecular transport from this point of view proves to be useful for predicting numerical values for their molecular transport properties, as will be discussed in Chapter 6. But for now let's apply the definition to the convective fluxes of energy, mass, and momentum. First of all, in view of Equation (2.3), we require the velocities of transport, and these are simply the velocities associated with the fluid motion. To complete the definition then, all we need do is to identify the volumetric concentrations of energy, mass, and momentum.

Energy

Although there are many forms of energy only kinetic energy and enthalpy are illustrated here. Kinetic energy is defined as $mv^2 / 2$ but we require its volumetric concentration, kinetic energy/volume. Thus, if the kinetic energy is divided by volume,

$$\text{Concentration of K.E.} = \frac{K.E.}{Vol} = \frac{1}{2}\frac{m}{V} v^2 = \frac{\rho v^2}{2} \tag{2.4}$$

Then the convective flux of kinetic energy in the x-direction, $f_E(K)_x$, is given by

$$\text{Convective Kinetic Energy Flux} \equiv f_E(K)_x = \frac{\rho v^2}{2} v_x \tag{2.5}$$

Similarly, the enthalpy is given by $H = m\hat{H}$; and thus the volumetric concentration of enthalpy is $\rho\hat{H}$ and the convective flux of enthalpy in the x-direction is $f_E(H)_x = (\rho\hat{H})v_x$. Thermodynamic principles allow for the expression of enthalpy in terms of temperature by

$\hat{H} = \hat{c}_p (T - T_{ref})$ and thus the volumetric concentration of enthalpy can be written as $\rho \hat{c}_p (T - T_{ref})$[#]

Mass

In this case the volumetric concentration of the mass of species A is simply ρ_A (or, C_A, in molar units) and thus the convective flux of A in the x-direction is

$$\text{Convective Mass (Molar) Flux} \equiv f_m(A)_x = \rho_A v_x \quad (f_M(A)_x = C_A v_x) \qquad (2.6)$$

Momentum

Since the x-component of momentum is given by mv_x, the volumetric concentration of x-momentum is ρv_x and the convective flux of x-momentum in the x-direction, $(f_{mom}(mv_x))_x$, is

$$\text{Convective x - Momentum Flux} \equiv (f_{mom}(mv_x))_x = (\rho v_x) v_x \qquad (2.7)$$

Similarly, the convective flux of x-momentum in the y-direction would be $(f_{mom}(mv_x))_y = (\rho v_x) v_y$ which is the same as the convective flux of y-momentum in the x-direction, $(f_{mom}(mv_y))_x = (\rho v_y) v_x$.

A summary of both the convective and molecular fluxes for all three transport phenomena is given in Table 2-1.

[#]   Keep in mind that T_{ref} is arbitrary and can be chosen equal to zero if desired (as in Table 2-1).

TABLE 2 - 1 Convective and Molecular Fluxes (z-Component)

	CONVECTIVE	MOLECULAR
Energy	$$f_E(K)_z = (\frac{\rho v^2}{2}) v_z$$ $$f_E(H)_z = (\rho \hat{H}) v_z$$ $$= (\rho \hat{c}_P T) v_z$$	$$q_z = k \frac{\partial T}{\partial z}$$
(Mass)$_A$	$$f_M(A)_z = (C_A) v_z$$ $$f_M(A)_z = (\rho_A) v_z$$	$$J_{A_z} = CD_{Am} \frac{\partial X_A}{\partial z}$$ $$j_{A_z} = \rho D_{Am} \frac{\partial \omega_A}{\partial z}$$
Momentum	$$(f_{mom}(mv_x))_z = (\rho v_x) v_z$$ $$(f_{mom}(mv_y))_z = (\rho v_y) v_z$$ $$(f_{mom}(m v_z))_z = (\rho v_z) v_z$$	$$\tau_{zx} = \mu \frac{\partial v_x}{\partial z}$$ $$\tau_{zy} = \mu \frac{\partial v_y}{\partial z}$$ $$\tau_{zz} = \mu \frac{\partial v_z}{\partial z}$$

2-2 THE PHENOMENOLOGICAL LAWS

2-2.a Phenomenological Laws in One-Dimension

The phenomenological laws are those mathematical expressions which describe how matter goes about the business of transporting energy, mass, and momentum on a molecular level. The "velocities of transport" do not explicitly appear in these equations because they are inherently contained within the phenomenological coefficients themselves. That is, the molecular velocities do the transport, and we only need to get down to this level when we seek to predict numerical values of the coefficients (see Chapter 6). In this section of the chapter, we introduce these laws in their one-dimensional form and then extend them to their more general forms.

In the previous chapter, Fourier's observations were mentioned relative to the molecular transport of energy. To reiterate, he observed that the energy transport rate, $\dot{Q}$, was found to be proportional to both the area across which the energy is being transported and the temperature difference between the hot and cold sides of the material. In addition, $\dot{Q}$ was also found to be inversely proportional to the distance between the hot and cold sides of the material. These observations can be linked together by defining a proportionality coefficient, k, which is known as the thermal conductivity and is a fundamental material property (i.e., numerical values for various materials can usually be found in a handbook) which can vary with temperature. It is this phenomenological parameter which accounts for the transport "velocities" in molecular transport. Fourier's Law can be generalized by applying it to an infinitesimally thin piece of material, δx,

$$\dot{Q} = \left| \ kA_x \frac{\delta T}{\delta x} \ \right| \tag{2.8}$$

where A_x is the area perpendicular to the direction of transport; in this case, x. Since this law is really based on observation, it is known as a "phenomenological" law; i.e., one which is arrived at by observing natural phenomena. Since transport implies that something (energy in this case) is moved from one point to another, there must be a direction associated with the transport. This necessity results in a slightly different form of Equation (2.8), one that has come to be the formal statement of Fourier's Law in one dimension and which applies at any point in the material. In this statement, Equation (2.8) is divided by A_x and δx is allowed to approach zero so that

$$\frac{\lim}{\delta x \to 0} \frac{\delta T}{\delta x} \ \to \ \frac{dT}{dx} \tag{2.9}$$

and

$$q_x = \frac{\dot{Q}}{A_x} = -k \frac{dT}{dx} \tag{2.10}$$

where q_x is the molecular *flux* of energy in the x-direction. Note that the units of the flux are energy/time-area. Two other features of Equation (2.10) should also be pointed out. First of all, the subscript on q indicates the direction of transport and the negative sign is a convention which results in positive values of q_x (transport in the positive x-direction). Thus, when the value of the temperature is higher at x than it is at $x + \delta x$ (i.e., the temperature decreases as x increases), the value of dT/dx is inherently negative and a positive q_x results.

Since Fourier's Law is based on observation and not derived from basic principles, it would not be too surprising if it did not apply to some systems. Actually it has been found to apply universally although, as we shall see, the same is not true for the phenomenological laws in molecular mass and momentum transport.

There are two analogous phenomenological laws to describe both the molecular transport of mass and momentum, but unfortunately, they are not as straightforward as Fourier's Law. To describe the molecular mass transport (or "diffusion") of a particular species, A, we use <u>Fick's Law</u>

$$J_{A_x} = -C\,D_{Am}\frac{dX_A}{dx} \tag{2.11}$$

which is a mathematical expression describing how species 'A' moves through a media by molecular mechanisms. In this equation, J_{A_x} is the molecular flux of species A in the x-direction (moles of A/area-time), X_A is the mole fraction of A at any point, D_{Am} is the diffusivity of A in the media and, again, the negative sign assures a positive flux in the direction of high concentrations to low concentrations. It should be pointed out here that the flux, J_A, is the flux of A relative to the bulk fluid motion and some of those consequences will be addressed in Chapter 4. It should also be noted that, if the total concentration, C, is constant, then Equation 2.11 becomes

$$J_{A_x} = -D_{Am}\frac{dC_A}{dx} \tag{2.12}$$

where C_A is the volumetric concentration of 'A'. The diffusivity is analogous to the thermal conductivity in Fourier's Law, and thus it is a property of the material being analyzed. However, unlike thermal conductivity, its value depends on the molecular properties of species 'A' <u>as well as</u> the concentrations and molecular properties of other species in the media. These considerations, and the fact that all of the other species may also be diffusing, can make the analysis of mass transport in multicomponent mixtures a very complex undertaking.

For certain types of fluids, the molecular transport of momentum can be expressed in a similar manner; namely, by <u>Newton's Law of Viscosity</u>

$$\tau_{xy} = -\mu\frac{dv_y}{dx} \tag{2.13}$$

As might be expected, fluids which obey this law are known as <u>newtonian</u> fluids. In fluids which do not obey this law (non-newtonian fluids), the "apparent" viscosity generally depends on the velocity gradient itself and this will be covered in Chapter 6. In Equation (2.13), τ_{xy} is the y-component of the molecular momentum flux transported in the x-direction (y-momentum/area-time), v_y is the y-component of velocity and μ is a property of the fluid, called the viscosity. Note the increased complexity of molecular momentum transport; there are two subscripts associated with τ and we must specify which velocity component we are dealing with. The reason for this is that momentum is a vector quantity and has a direction independent of the direction of molecular transport. A more thorough description of these dual directions will be given in Chapter 5. For now, you should appreciate the differences, as well as the similarities between the molecular transport of energy, mass, and momentum.

2-2.b Analogies

With respect to similarities, the equations are certainly of similar form; the fluxes are all proportional to differential changes in the individual driving forces and the proportionality constants are properties of the material. Moreover, they can all be expressed in the classical electric circuit manner as

$$\text{CURRENT} = \frac{\text{POTENTIAL}}{\text{RESISTANCE}}$$

(2.14)

$$\text{FLUX} = \frac{\text{DRIVING FORCE}}{\dfrac{1}{\text{CONDUCTANCE}}}$$

Note that while the conductance term is obvious in Fourier's law (viz., k), the diffusivity and viscosity can also be viewed as the ability of a material to "conduct" mass or momentum. Again, while all three phenomenological laws are very similar, the fact that the flux of both energy and mass are vector quantities while the flux of momentum is a <u>tensor</u>, is a major difference; necessitating the specification of two directions. This electrical analog is summarized below in Table 2-2.

Even a stronger analogy between the phenomenological laws can be made if we view the transport as <u>diffusion</u> of energy, mass, and momentum. To do this we have to draw on the form of Fick's Law and express our driving forces in terms of the gradients of the <u>volumetric concentrations</u> of energy, mass, and momentum; i.e., in the same manner as Equation (2.12). For energy transport, the volumetric concentration of energy is $\rho \hat{H}$, or $\rho \hat{c}_p T$, where $\hat{c}_p$ is the heat capacity at constant pressure. Thus if Fourier's law [Equation (2.10)] is multiplied and

TABLE 2-2 Phenomenological Laws In One-Dimension

	Energy	**Mass**	**Momentum**
Flux	Vector (q_x)	Vector (J_{A_x})	Tensor (τ_{xy})
Differential Driving Force	$\dfrac{dT}{dx}$	$\dfrac{dC_A}{dx}$	$\dfrac{dv_y}{dx}$
Conductance	k	D_{Am}	μ

divided by $\rho\, \hat{c}_p$, we obtain, for constant properties,

$$q_x = -\frac{k}{\rho\, \hat{c}_p} \frac{d}{dx}(\rho\, \hat{c}_p T\,) \tag{2.15}$$

and, in analogy with the diffusivity in Fick's Law, $\dfrac{k}{\rho\, \hat{c}_p}$ is called α, the *thermal diffusivity*.

Similarly, we can do the same with Newton's Law of viscosity. Here the concentration of y-momentum is ρv_y and Equation (2.13) can be written

$$\tau_{xy} = -\frac{\mu}{\rho} \frac{d}{dx}(\rho v_y) = -v\frac{d}{dx}(\rho v_y) \tag{2.16}$$

where v is known as the <u>kinematic viscosity</u> or, in keeping with the analogy, as the <u>momentum diffusivity</u>. With this analogy, equations (2.12), (2.15) and (2.16) express the phenomenological laws in terms of diffusion and the coefficients in front of the gradients all have units of L^2/t.

2-2.c Extension of the Phenomenological Laws to Other Coordinates and Dimensions

Up to this point, we have only talked about one-dimensional transport, specifically in the x-direction. Of course, it is possible to have transport in all directions and Cartesian coordinates, such as employed in Equations (2.10 - 2.13) may not be the most convenient for other regular geometries such as cylinders and spheres. For problems in these coordinate systems, the most common situation is radial transport. For both cylindrical and spherical geometries, the radial component of the molecular fluxes are very similar to Equations (2.10 - 2.13) with the x-derivatives replaced by derivatives with respect to the radial coordinate, r. For example

$$q_r = -k\frac{dT}{dr}$$

$$J_{A_r} = -C\,D_{Am}\frac{dX_A}{dr} \qquad (2.17)$$

$$\tau_{ry} = -\mu\frac{dv_y}{dr}$$

Although we will primarily be restricting ourselves to one-dimensional problems, it is important to be aware of the extension of the three phenomenological laws to three-dimensions. This is done by utilizing the mathematical definition of the del operator (or "DEL"), $\vec{\Delta}$. DEL is a differential operator with vector properties and, in Cartesian coordinates can be expressed as

$$\vec{\Delta} = [\frac{\partial}{\partial x}\vec{i} + \frac{\partial}{\partial y}\vec{j} + \frac{\partial}{\partial z}\vec{k}] \qquad (2.18)$$

so that

$$\vec{q} = -k\,\vec{\Delta}T$$

$$\vec{J}_A = -C\,D_{Am}\,\vec{\Delta}\,X_A \qquad (2.19)$$

$$\bar{\tau}_{ij} = -\mu\,\vec{\Delta}\,\vec{v}$$

Note that DEL is operating on scalar quantities in both energy and mass transport but operates on a vector in momentum transport. When Del operates on a scalar it results in the gradient of the scalar (called GRAD) and when it operates on a vector, it has other properties which will not be considered in this text. Table 2-3 lists the components of GRAD in each of the three common coordinate systems and is only applicable to the phenomenological laws related to the molecular transport of energy and mass.

Table 2-3 Components of the GRAD Operator In Various Coordinate Systems

COORDINATE SYSTEM	COMPONENTS OF GRAD ($\vec{\Delta}$)
Cartesian	$\dfrac{\partial}{\partial x}\vec{i} + \dfrac{\partial}{\partial y}\vec{j} + \dfrac{\partial}{\partial z}\vec{k}$
Cylindrical	$\dfrac{\partial}{\partial r}\vec{i} + \dfrac{1}{r}\dfrac{\partial}{\partial\theta}\vec{j} + \dfrac{\partial}{\partial z}\vec{k}$
Spherical	$\dfrac{\partial}{\partial r}\vec{i} + \dfrac{1}{r}\dfrac{\partial}{\partial\theta}\vec{j} + \dfrac{1}{r\sin\theta}\dfrac{\partial}{\partial\phi}\vec{k}$

2-3 DIFFERENTIAL BALANCES AND THE CONSERVATION LAWS

The principles and applications of the conservation of energy, mass, and momentum should be familiar topics. Whereas initial exposure to the conservation of momentum principle is likely to have been presented in the context of solid dynamics, applications of the principle to a fluid may be somewhat new. These very same conservation laws are used to describe transport phenomena except the balances are taken around differential volume elements. The goal here is to derive differential equations which describe the conservation of energy, mass and momentum <u>at a point</u>. Equations which apply at a point are then capable of producing solutions to a wider variety of problems; i.e., by solving the differential equations over the region of interest and incorporating the specific boundary conditions which apply. The general conservation equation can be expressed as shown in Equation (2.20)

$$[\text{INPUT}] - [\text{OUTPUT}] + [\text{SOURCES}] - [\text{SINKS}] = [\text{ACCUMULATION}] \qquad (2.20)$$

which holds for all conserved quantities (energy, mass, momentum). We will discuss what constitutes each of these terms shortly but first let us concentrate on the systems to which it will be applied.

In order to derive a differential equation from Equation (2.20) it is necessary to apply it to a differential volume element. Thus the first task is to decide what this element should look like. The construction of the element will not only depend on the geometry of the system being analyzed, but also on the anticipated direction of the transport. For example, if we expect energy transport to be only in the x-direction (because of temperature differences in that direction), we should construct an element with only one differential dimension, say δx. There is no need to have differential dimensions in the y and z-directions since there will be no transport in those directions and thus solutions (integration) will not be required in those directions.[#]

Figure 2-1 shows some typical differential volume elements for one-dimensional transport in different geometries. Deciding on the appropriate geometrical shape of the element is usually straightforward since it will be dictated by the geometry of the physical system. However, deciding on the direction of transport requires insight and experience. For example, the element in Figure 2-1b would be appropriate for the transport of energy through a solid metal rod which has its cylindrical surfaces insulated. However, Figure 2-1c would be more applicable in a situation where a very long copper wire was being electrically heated and cooled at its radial surfaces.

The next step in the differential analysis is to apply the conservation equation to the differential element over some time period, δt. The most important thing to keep in mind here is that we are conserving ENERGY, MASS and MOMENTUM. Therefore the units of each term in Equation (2.20) must have the units of energy, mass, or momentum. Now lets turn our attention to the individual terms in this equation.

[INPUT], [OUTPUT]

The first two terms in the conservation equation represent the flow of material across the boundaries of the system (differential element). Because the phenomenological laws are all written in terms of fluxes, and we will eventually need these laws to solve molecular transport problems, it is common practice to specify these flow terms as fluxes. Since we require units of energy, mass or momentum and the flux of energy, for example, has units of energy/area-time, each flux term must be multiplied by the area across which it is flowing and by the time period over which the balance is being taken, before being entered into Equation (2.20). Figure 2-2 shows the molecular energy flux, q_z, entering and leaving the differential volume element shown in Figure 2-1b and the form it will take when entered into Equation (2.20).There are a few important features in Figure 2-2 which should be pointed out. First of all, note that we have drawn the molecular energy flux vector in the positive z-direction. This direction is not arbitrary, but is dependent on how we draw the coordinate system (which is arbitrary). As you can see, the flux is drawn into and out of the element in a direction parallel to the z-axis. Once the location and direction of the coordinate system is decided upon, fluxes must be drawn in the positive direction of the coordinate system (in order to be consistent with the negative sign convention in the phenomenological laws). The fact that this direction is different from the

---

[#]   Another approach to these problems would be to derive a general differential equation and then eliminate the unnecessary terms. This approach is described in detail in Appendix A.

Figure 2-1 Examples of Differential Volume Elements :

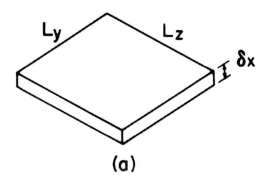

(a)

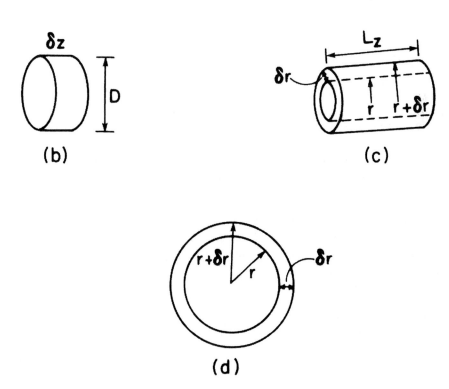

(b)   (c)

(d)

[(a) Flat Plate, (b) Cylindrical Pellet, (c) Cylinder, (d) Sphere]

direction of the actual transport should not be a concern. If consistency is maintained and there are no careless sign errors, the solution will provide the correct direction of transport. Of course if, by analyzing the physical problem, the direction of transport is obvious, then it makes sense to draw the coordinate system in that same direction. However, in complex problems the direction of transport may not be obvious, and paying attention to the rules and sign conventions will often "save the day." The second point to note is the location and the identification of the fluxes entering and leaving the element. Since we are analyzing the energy transport in the z-direction, the element can be located at <u>any</u> point within the solid cylindrical rod. It is shown in Figure 2-2 as being located at an arbitrary distance, z, from the origin of the coordinate system. To distinguish between the z-component of the flux entering and leaving the element, the former is identified as $q_{z\,|_z}$ and the latter (i.e., at the position $z + \delta z$) as $q_{z\,|_{z+\delta z}}$.

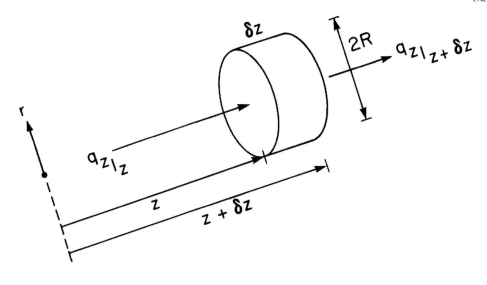

$$[\text{INPUT}] - [\text{OUTPUT}] = q_{z\,|_z}\,\pi R^2 \delta t - q_{z\,|_{z+\delta z}}\,\pi R^2 \delta t$$

Figure 2-2 Molecular Energy Flux Entering/Leaving a Differential Volume Element

[SOURCES], [SINKS]

Sources and sinks in the conservation equation refer to the <u>generation</u> and <u>consumption</u> of either energy, mass, or momentum *within* the differential element. In order to be considered as sources and sinks in the conservation equation, these phenomena must occur <u>uniformly</u> throughout the physical system being analyzed. Because of this uniformity, we can always express these terms on a per unit volume basis (since the SOURCES/SINKS are the same everywhere throughout the volume element). Furthermore, since the balance is applied over δ time, we must also express the SOURCES/SINKS as <u>rates</u>; i.e., generation/consumption per unit time. In an energy transport problem, a source might be a homogeneous (uniform) exothermic chemical reaction. Mathematically, if the rate of reaction of species A is R_{Av} (moles of A reacting/volume-time) and if the heat of reaction is $\Delta \tilde{H}_r$ (energy released per mole of A reacting), then the source of energy is simply $R_{Av}\Delta \tilde{H}_r$ (energy released/volume-time). If the reaction is endothermic, then the same mathematical term results, but it is placed into the conservation equation as a sink. Note that when we follow the sign conventions in Equation (2.20), the heat of reaction is an absolute value. Similarly, sources and sink terms in the conservation of mass are chemical reactions which either produce or consume a given species (eg., $\pm R_A$). Sources and sinks in momentum transport are somewhat more complex, and we will postpone that discussion until Chapter 5. Equation (2.21) shows the exothermic reaction term in the form it would have when entered in Equation (2.20) as a SOURCE.

$$[\text{SOURCE}] = R_{Av}\Delta \tilde{H}_r \quad \delta V \quad \delta t$$

$$(2.21)$$

$$(\text{Energy}) \quad = \frac{\text{Energy}}{\text{vol - time}} \ (\text{vol}) \ (\text{time})$$

[ACCUMULATION]

The accumulation term which appears on the right-hand side of Equation (2.20) refers <u>only</u> to accumulation with time. In other words it accounts for an increase or decrease of the conserved quantity <u>within the differential element</u> as time increases. If there are no changes in any of the variables with time, the system is said to be at "steady state." Because the ACCUMULATION term refers to what is happening within the element as time changes, and the element is infinitesimally small, we can consider the conserved quantity to be uniformly distributed within the element. Therefore, just as we did with the SOURCE and SINK terms, we can express the conserved quantity on a per unit volume basis ("volumetric concentration") and then multiply by the volume of the element in order to obtain the correct units. Consequently the ACCUMULATION term will involve $\rho \hat{c}_p (T - T_{ref})$, C_A (or ρ_A), ρv_i when conserving energy, mass, and momentum, respectively.

As an illustration, consider the conservation of species A in a spherical geometry where the differential element drawn in Figure 2-1d applies. If our problem is an unsteady state problem (there are changes with time), then the accumulation with time refers to changes occurring over an infinitesimal time increment, δt. Furthermore, accumulation implies changes as time moves forward. If the volumetric concentration of A is C_A, then the accumulation term becomes

$$[\text{ACCUMULATION}] = (C_A\, 4\pi r^2\, \delta r\,)_{t+\delta t} - (C_A\, 4\pi r^2\, \delta r\,)_t$$

The final step in deriving the applicable differential equation is to divide the entire equation by the volume of the element and the differential time increment and then let the differential quantities approach zero in the limit. It is at this point that errors can occur due to neglect of sign conventions. Recalling the "mean value theorem" from calculus, the definition of the derivative is

$$\frac{dy}{dx} = \lim_{\delta x \to 0} \frac{y_{x+\delta x} - y_x}{\delta x}$$

Note that the derivative is positive when the dependent variable at x is subtracted from the dependent variable at x + Δx. When the opposite occurs, the derivative will be negative.

The information given above is summarized in a "step-by-step" manner in Table 2-4.

TABLE 2-4 Step-By-Step Procedure For Differential Balances

1. The Physical System
   a.   Decide on what is being transported.
   b.   Decide on the direction of transport.
   c.   Choose the location and orientation of an appropriate coordinate system.

2. Construct the Differential Volume Element
   a.   Choose a regular geometry appropriate to the physical problem (sketch it).

3. Draw Fluxes Into and Out of the Differential Element
   a.   Be sure that flux directions are in the positive direction of the coordinate system chosen in Step 1c
   b.   Are the fluxes molecular or convective or both?

4. Decide Whether There are Sources and/or Sinks
   a.   Ask whether energy, mass or momentum is being <u>uniformly</u> generated or consumed *within* the boundaries of the physical system.
   b.   Once you decide on the existence of sources/sinks, express them on a per unit volume basis.

5. Decide Whether There is Accumulation (is this a steady or unsteady state problem?)

6. Apply Conservation Equation (2.20) to the Volume Element
   a.   Heed sign conventions.
   b.   Check units (energy, mass, momentum) in every term.

7. Divide the Equation in 6 by $\delta V\ \delta t$ and Simplify

8. Take the Limit as $\delta V\ \delta t \to 0$
   a.   <u>Be careful of signs</u> (Mean Value Theorem).
   b.   Express the differential equation in terms of <u>one</u> dependent variable (use the phenomenological laws if necessary).[#]

[#]   In some problems it may be more convenient to integrate before substituting the phenomenological laws

PROBLEMS

2-1   Describe in your own words how molecular and convective momentum transport takes place. This description must be *typewritten* (or printed from a word processor) and will be graded on the basis of *English* as well as technical content. It should be no longer than one typewritten page.

2-2   With respect to the phenomenological laws:

(a) Write the analogy to $\dot{Q}$ in Equation (2.8) for molecular momentum transport, then use Newton's second law to express the result in terms of a force.

(b) Temperature measurements taken in a complicated energy transport problem have been fitted with a polynomial with the result $T = 60 + .2x^2 - .03x^4$, where T is in °C and x is in cm. If the thermal conductivity of the media is 60 W/m-K, what is the energy flux, q_x, at x = 5 cm? Is the media where the measurements were taken, a gas, liquid or solid?

(c) The "fully developed", steady state turbulent velocity profile in a 2.5" diameter pipe can be fit to a "power law" model; specifically, $v_z = \bar{v}\ (\frac{R-r}{R})^{\frac{1}{7}}$, and the *turbulent viscosity*, μ_t, can be defined by $\tau_{rz} \equiv -\mu_t \frac{dv_z}{dr}$. Under these conditions it can be shown that the molecular momentum flux, τ_{rz}, varies with radius according to $\tau = \tau_w\, r\, /\, R$ where τ_w is the molecular momentum flux at the wall. If the average velocity, $\bar{v}$, is 50 ft/sec, and the molecular momentum flux at the wall, τ_w, is 23 lb$_m$/ft-s², calculate the turbulent viscosity at the r / R values of .2, .7, .95 and .999. Compare these values with the viscosity of liquid water, which is 1 cp.

2-3   Apply Equation (2.14) to the steady state molecular energy transport through three adjacent solid materials (designated as 'A', 'B' and 'C') to express the molecular energy flux in terms of the <u>overall</u> temperature driving force between the three walls, $T_1 - T_4$. HINT: use the fact that resistances in series are additive.

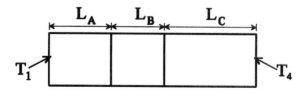

2-4   Referring to Figure 2-1 and the general conservation equation (Eq 2.20), write the appropriate terms for:

(a) The ACCUMULATION of z-momentum in the differential volume element shown in Figure 2-1c.

(b) The SOURCE of momentum in the differential volume element shown in Figure 2-1d as a result of a force, F, applied uniformly to the volume element.

(c) The ACCUMULATION of the mass of A (m_A) in the differential volume element shown in Figure 2-1a.

(d) The IN - OUT terms for the molecular transport of energy in the differential volume element shown in Figure 2-1.c

(e) The IN - OUT terms for the convective transport of the moles of 'A' in the differential volume element shown in Fig 2-1b.

(f) The SOURCE of energy due to an exothermic chemical reaction (heat of reaction = $\Delta \tilde{\tilde{H}}_r$, reaction rate = R_{AV} moles/vol-t) for the DVE in Figure 2-1c.

(g) The INPUT-OUTPUT terms for the molecular transport of r-momentum for the DVE in Figure 2-1d.

(h) The ACCUMULATION of the moles of species A for the DVE in Figure 2-1c.

(i) The INPUT-OUTPUT terms for the transport of convective z-momentum in Figure 2-1b.

BE SURE TO <u>CLEARLY</u> IDENTIFY THE NOMENCLATURE YOU USE

2-5

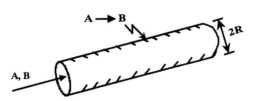

A liquid containing 'A' and 'B' is flowing in laminar flow through a cylindrical tube as shown above. The walls of the tube are catalytic and the reaction, $A \rightarrow B$, is first order with respect to 'A'

and occurs only at the wall. At one downstream location, measurements taken of the concentration of 'A' (C_A) across the cross section of the tube were fit to a parabolic equation which resulted in

$$C_A = .08\left[1-\left(\frac{r}{R}\right)^2\right]+.03$$

where C_A is in gmoles/liter. Since, under steady-state conditions, the <u>surface</u> reaction rate at the wall (R_{AS} moles / l^2- t) must equal the molecular transport rate of 'A' at the wall, derive an expression (in terms of D_{AB} and R) for the reaction rate constant, k_r, at this downstream position. [NOTE: For a first-order reaction, $R_{AS} = k_r C_{AS}$]

2-6   Given the following physical situation:

Steady-state energy transport in a solid rod of radius, R, and length, L, with a perfectly insulated cylindrical surface which is at steady state with one end at T_1 and the other at T_2, ($T_1 > T_2$).

(a) Provide a sketch of the physical problem and draw the appropriate Differential Volume Element (DVE) <u>within</u> the sketch, labeling all its dimensions and showing where it is located relative to your coordinate system.

(b) Label the DVE and show the directions of all fluxes in and out of the DVE.

(c) Write the appropriate expression(s) for each of the five terms of the general conservation equation.

(d) Starting with the results in (c), show <u>all</u> steps in deriving the applicable differential equation for this physical situation.

2-7   Repeat 2-6 for the same rod described in 2-6 except, in this case,

the rod is initially at T_1 and suddenly has one end immersed in a temperature bath at temperature, T_2.

2-8   Repeat 2-6 for the following physical situation:

Steady-state energy transport through a flat, thin metallic plate which is used as a cover on a rectangular vat of hot fluid. The bottom of the cover is at 70 °C and the top is at room temperature (30 °C).

2-9   Repeat 2-6 for the following physical situation

Energy transport in a fluid at uniform velocity, $\bar{v}$, and density, ρ, which flows at steady-state through a perfectly insulated tube of radius, R, and length, L. The fluid contains a radioactive species which gives off heat at a uniform rate, $(\dot{Q}_N)_V$.

2-10   Liquid sodium is being used as a heat transfer fluid and, in one application, the temperature distribution in the flow direction (z-direction) has been empirically fit to give $T = 100 - 10z^2$, where T is in degrees Fahrenheit and z is in feet. If the fluid is flowing at a velocity of 0.1 ft/s, compare the magnitude of the molecular energy flux in the z-direction to that of the convective energy flux in the z-direction at a position, $z = 1$ ft. Assume an enthalpy reference temperature of 0 °F. [FLUID PROPERTIES: $\hat{c}_p = 0.3$ kcal/kg-K, $\rho = .8$ g/cm^3, $k = 85$ w/m-K]

2-11   Referring to the following differential equation

$$
\underset{[1]}{\frac{\partial C_A}{\partial t}} = \underset{[2]}{-\bar{v}\,\frac{\partial C_A}{\partial z}} + \underset{[3]}{\frac{1}{r}\frac{\partial}{\partial r}\left(r\,D_{Am}\frac{\partial C_A}{\partial z}\right)} + \underset{[4]}{\frac{\partial}{\partial z}\left(D_{Am}\frac{\partial C_A}{\partial z}\right)} - \underset{[5]}{R_{AV}}
$$

answer the following questions using the numbers above each term in the differential equation.

(a) Assign each term in the equation to the individual elements in the general conservation equation, Eq. (2.20).

(b) Which are the dependent and independent variables?

(c) If D_{Am} is constant and $R_{AV} = k_r(C_A)^{1/2}$, is the equation linear or nonlinear?

(d) If R_{AV} is constant and D_{Am} varies linearly with C_A, is the equation linear or nonlinear?

CHAPTER 3

One-Dimensional Molecular Energy Transport

3-1 MODELING PHYSICAL SYSTEMS

3-1.a The Adequacy of the Model

One of the tools that should be placed in the hands of every engineer is the ability to describe physical problems in mathematical terms. Not only does this help give added insight into the problem at hand but, with the very high cost of doing industrial experimentation, modeling has become a necessary part of scale-up procedures. The key to successful modeling is to strike a balance between realism and elegance. That is, the model must be able to describe the physical system to the degree that is consistent with our knowledge of the system and the kind of information we wish to extract from the model. In years past, our ability to solve the complex mathematics which describe many physical systems was not very good and thus engineers spent much of their time trying to simplify the models so as to obtain a solution. Often the simplifications were so great that the resulting model failed to give a reasonable description of the physical system and the information obtained from the model did not have much credibility. There have been many advances in computational methods in recent years, and this problem is no longer the impediment it once was. Nevertheless, good engineering still dictates a balance between what is needed and what it takes (time, money) to obtain what is needed.

This chapter deals with the application of the transport phenomena approach to energy transport problems. In so doing, we will actually be describing physical systems in mathematical terms by using differential conservation balances. Since we will be restricting ourselves to one-dimensional molecular transport problems, certain simplifications will have to be made. As a result, questions should arise: Is this a physically realistic description? Will the simplified model provide the type of information desired? To answer these questions it is necessary to be able to appreciate the physical phenomena and then decide how it conforms to the mathematics being employed. During this process we have to distinguish between the phenomena which is occurring within the system and that which is only taking place at the boundaries of the system. The latter is every bit as important as the former since we will not have a solution to the model until we account for both.

3-1.b Boundary Conditions

Whereas the differential equation resulting from the differential conservation balance describes what is happening within the system, it is the "boundary conditions" which mathematically describe the physical phenomena occurring at the boundaries of the system. You will probably recall from your course in differential equations that the solutions to differential equations can be either "general" or "particular." A general solution means that the solution has a general form (or "shape") but is not tied down to any particular set of numerical values. Consequently, a general solution is said to be a "family" of solutions and is accompanied by arbitrary constants (or "constants of integration"). Once these arbitrary constants are assigned for a given problem, then the solution is a "particular" solution; that is, one of the members of the family. The values of the arbitrary constants are determined by the specification of the boundary conditions. The number of arbitrary constants is dictated by the order of the differential equation and therefore, the number of specified boundary conditions must be equal to the order of the differential equation.

As engineers we typically deal with specific problems and consequently boundary conditions are every bit as important as modeling the physical system to obtain the applicable differential equation. In transport phenomena, we look to the physical phenomena occurring at the boundaries of the physical system in order to specify the boundary conditions for a given problem. It is important to realize that these conditions must have an assignable numerical value in order to be viable boundary conditions. That is, either the dependent variable (temperature, concentration, or velocity) and/or their derivatives (i.e., fluxes) must be specified. It is not enough to merely specify a "statement of fact." For example, stating that a temperature will be a maximum at some location in the physical system is <u>not</u> a valid boundary condition even though the statement may be correct. However, if we have a separate knowledge of the <u>value</u> of that maximum temperature, then we do have a valid boundary condition. In other words, we must be able to assess what is happening at the physical boundaries of the system and to then specify numerical values for the dependent variables and/or their derivatives.

Sometimes the numerical values at the boundaries are given symbols so that specific numerical values can be substituted as they become available. An example of this is to state that the temperature at a particular surface of a solid material is a constant equal to T_0. In this case, the particular solution is obtained in terms of T_0, a parameter, which can be any numerical value just as long as it is maintained constant for the physical system being studied. The engineering task is to evaluate the physical phenomena and to then decide what is a realistic boundary condition. For example, we should ask ourselves how the surface is being kept at a constant temperature. Is it physically realistic? Are there situations where it would be impossible to maintain a constant temperature, and is this one of them?

Now, since we are dealing with energy transport in this chapter, lets look at some examples of specified boundary conditions. Hopefully the extensions to mass and momentum transport will be apparent. Consider a cylindrical electric heating rod which is sheathed by a concentric tube of thickness, L_R. The entire assembly is immersed in a fluid and the system is at steady-state, as shown in Figure 3-1. We wish to determine the temperature distribution <u>within</u> the sheath. First lets formulate some questions we should be asking ourselves about this system.

- Is the temperature of the heater constant in the z-direction?
- Is the temperature at the inner wall of the tube (R_1) the same as the heater temperature?
- Is the temperature of the fluid a constant with respect to time and all along the tube? If so, how is this achieved?
- Is the temperature at R_2 the same as the fluid temperature?

Assume that we have asked these questions and arrived at the following estimates:

1. The temperature of the heater is constant.
2. The temperature at R_1 is the same as the temperature of the heater, T_H.
3. The fluid temperature is constant at T_F and this is the temperature of the surrounding sheath at R_2.

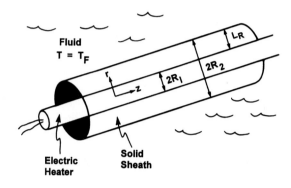

Figure 3-1 Circular Tube Being Heated from the Inside by an Electric Heater

With these conditions and, at steady-state, it can be shown (from a differential energy balance) that the differential equation applying to this problem is second order

$$\frac{d}{dr}(r\frac{dT}{dr}) = 0 \qquad (3.1)$$

and therefore two boundary conditions will be needed in order to obtain a specific solution. Since we have knowledge concerning the temperatures at two points in the system (at $r = R_1$ and R_2), we can satisfy the need by stipulating the following:

$$at\ r = R_1,\ T = T_H$$

$$at\ r = R_2,\ T = T_f$$

Notice how the boundary conditions are stipulated. We must first stipulate a specific value of the independent variable, r, and then state a specific value for the dependent variable, T, (or its derivatives), <u>at</u> that location. Another point to note is that the specifications <u>must be within the physical system being analyzed</u>. In this case the physical system lies between R_1 and R_2 (or $R_1 \le r \le R_2$) and the heater is <u>not</u> part of the system being analyzed.

This illustration is fairly straightforward since we have the required information for two values of the dependent variable (temperature). However, in many other physical problems we will only have information on the transport fluxes. For example, if we have a perfectly insulated surface in an energy transport problem, we can stipulate that the molecular energy flux is zero at that boundary. In some cases (usually rare), we may even know a particular value of the flux at the boundary. In these cases the formal statement of the boundary condition takes the form

$$at\ r = R_2,\ q_r\mid_{R_2} = -k\left(\frac{dT}{dr}\right)\mid_{R_2} = CONSTANT \tag{3.2}$$

where, of course, setting the constant to zero will also accommodate a perfectly insulated surface.

The extension of these principles to mass and momentum transport requires a little more thought but should be evident. Table 3-1 summarizes some of the typical boundary conditions which can be encountered in all three modes of transport phenomena and describes physical examples pertaining to each.

Table 3-1   Boundary Conditions In Momentum, Mass, And Energy Transport

	STIPULATIONS	
	Dependent Variable	Flux (Derivatives)
MOMENTUM		
Math Definition	v_j = Constant	$\tau_{ij} = -\mu\ dv_j\ /\ dx_i$
Physical Example	Moving or stationary surface (no slip)	Liquid-gas interfaces (complete slip; $\tau_{ij} = 0$)
MASS		
Math Definition	C_A = Constant	$J_A = -D\ dC_A\ /\ dx_i$ = Constant
Physical Example	Equilibrium at an Interface	Impermeable boundary ($J_A = 0$); Known reaction rate at a surface
ENERGY		
Math Definition	T = Constant	$q_i = -k\ dT\ /\ dx_i$ = Constant
Physical Example	Isothermal surfaces	Perfectly insulated surface ($q_i = 0$)

One final point relative to boundary conditions should be made. Since the particular solution which is obtained applies everywhere within the physical system, *including* the boundaries, the final solution should reduce to the boundary conditions when it is applied at the boundaries. This is a good way to check that algebraic mistakes haven't been made.

3-2 STEADY-STATE MOLECULAR ENERGY TRANSPORT

In this section, we will illustrate the approach to modeling physical systems by applying the principles of Chapter 2 to three separate molecular energy transport problems. The objective in each will be to obtain mathematical descriptions for the temperature distributions and/or the energy fluxes. The first problem, "Axial Energy Transport in a Rod," is a simple one-dimensional transport problem in rectangular coordinates while the second problem, "Energy Transport Through an Insulated Furnace Wall," extends this approach to a composite material. The third problem, "Radial Temperature Distribution in an Electric Wire," introduces a problem in cylindrical coordinates and also illustrates how to include an energy SOURCE into an energy transport problem.

3-2.a Axial Energy Transport in a Rod

Figure 3-2 shows a cylindrical rod of radius, R, and with an insulated cylindrical surface. One end of the rod is kept at a temperature higher than the other ($T_H > T_C$) and the system is at steady-state. We wish to solve for the temperature distribution in the rod.

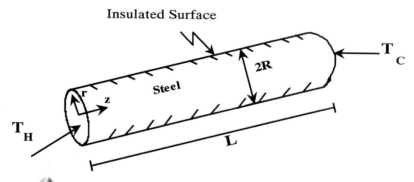

Figure 3-2 Axial Conduction in a Steel Rod

The first step in the solution is to decide on the direction of transport. Since the cylindrical surface is insulated, there is no energy transport in the radial direction, and thus we can conclude that the transport is unidirectional and along the axial coordinate of the rod. The next decision is where to place the coordinate system. As shown in Figure 3-2, we have (arbitrarily) placed the coordinate system at the hot end of the rod. The appropriate DVE is chosen at this point and, referring to Figure 2-1, it is the cylindrical disk (Fig. 2-1c) which is the correct choice. Figure 3-3 shows a sketch of the DVE, which itself is located at some distance, z, from the origin. Since the rod is a solid, there will be no convective transport (recall that convective transport requires a velocity) and so the only energy flux will be molecular, which is shown entering and leaving the DVE in the direction of the positive coordinate system. The notation indicates that the flux, q_z, entering the DVE is the value at position z = z and, allowing for the flux to change, the flux leaving the DVE has the designated value at z = z + δz. Note that once the coordinate system is chosen, the fluxes <u>must be</u> in the direction of the positive coordinate system. In this case, we chose the same direction that we expect the flux to be. If, instead, the coordinate system had been placed at the cold face of the rod, then the fluxes in Figure 3-3 would have to be drawn in the opposite direction; even though the energy would actually move in the opposite direction. In that case, the solution would give negative values of q_z, indicating the correct direction of energy transport.

At this point we are ready to apply the general conservation equation, Equation 2.20. The individual terms for this physical application are:

$$INPUT = q_{z\,|\,z}\,\pi R^2 \delta t$$

$$OUTPUT = q_{z\,|\,z+\delta z}\,\pi R^2 \delta t$$

$$SOURCES = 0$$

$$SINKS = 0$$

$$ACCUMULATION = 0$$

Note that every term has units of energy. The last three terms are zero due to the absence of SOURCES or SINKS within the DVE and to the steady-state conditions which makes the ACCUMULATION term zero.

When these terms are inserted into Equation (2.20), we get

$$(q_{z\,|\,z} - q_{z\,|\,z+\delta z})\pi R^2 \delta t = 0$$

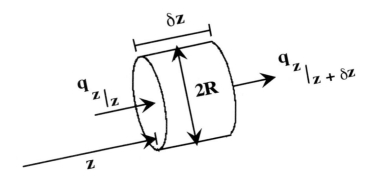

Figure 3-3 DVE for Conduction in a Rod

Dividing through by the volume of the DVE and $\delta t\,(\pi\,R^2\,\delta z\delta t)$ and taking the limit as δz approaches zero, the following differential equation results

$$-\frac{dq_z}{dz}=0 \tag{3.3}$$

Equation (3.3) is a correct description of energy transport in the cylindrical fin, and if we integrate it, we find that the flux is a constant; i.e.,

$$q_z=C_1 \tag{3.4}$$

where C_1 is a constant of integration. However we do not have a specific value for this constant and so we proceed to the next step which is to express Equation (3.4) in terms of temperature. This can be done by employing Fourier's Law, which from Chapter 2-2 is simply, $q_z=-k\dfrac{dT}{dz}$, where k is the thermal conductivity of the steel rod. Substituting this into Equation (3.4) and integrating once more to solve for temperature

$$-k\frac{dT}{dz}=C_1$$

$$\frac{dT}{dz}=-\frac{C_1}{k}$$

$$T = -\frac{C_1}{k}z + C_2 \tag{3.5}$$

where C_2 is another constant of integration. Boundary conditions are needed in order to solve for C_1 and C_2. Remember that these must be <u>values</u>[*]. In this problem we know the values of the temperature at each end of the rod; namely, T_H and T_C. Mathematically, this can be stated by

$$at \ z = 0, \ T = T_H$$
$$at \ z = L, \ T = T_C$$

Substituting the boundary conditions into Equation (3.5), C_1 and C_2 are

$$C_1 = \frac{k}{L}(T_C - T_H)$$

$$C_2 = T_H$$

so that the solution becomes

$$T = T_H - (T_H - T_C)\frac{z}{L}$$

Note that the temperature distribution is linear. This could also have been recognized from Equation (4.4), since q_z is a constant and therefore, by Fourier's Law, dT/dz must also be constant.

Now let's turn our attention to the heat flux. We know it is a constant but we do not yet have a value for it. Equation (3.4) tells us that the flux is equal to the constant, C_1, and therefor

$$q_z = \frac{k}{L}(T_H - T_C)$$

Note that since we picked the coordinate system in the same direction as the energy flux, q_z is a positive value. Note also that we could have arrived at this result from the fact that the temperature gradient is a constant and thus

$$q_z = -k\frac{dT}{dz} = -k\frac{\Delta T}{\Delta z} = -k\frac{(T_C - T_H)}{L}$$

---

[*] Sometimes, the appropriate boundary condition is an equation which relates values to one another.

3-2.b Composite Materials: Energy Loss through a Furnace Wall

Figure 3-4 shows a sketch of the cross section of the insulated wall of a high temperature furnace. In order to measure the heat losses from the furnace, thermocouples are placed at the outer furnace wall and at the outer surface of the insulation as shown. Since the measurement of TC_1 is not very reliable, it is desired to calculate the heat loss in terms of the temperature measurements of the furnace gases, T_F and the ambient temperature, T_a. As a first approximation, it can be assumed that the inner furnace wall is at T_F and the temperature at the outer insulation is equal to T_a.

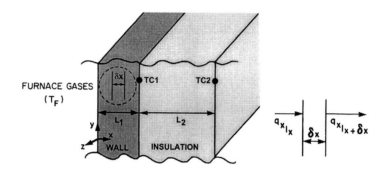

Figure 3-4 Furnace Wall

Since the entire furnace wall is exposed to the same temperature, the only direction of energy transport is through the wall thickness. Consequently a rectangular differential volume element is chosen, having a thickness δx and arbitrary dimensions (L_y, L_z) in the other two directions. Selecting the wall-furnace interface as the origin of the coordinate system, each of the terms in Equation (2.20) are:

ACCUMULATION $= 0$ (steady-state)

SOURCES/SINKS $= 0$ (no generation/absorption of energy within the wall)

INPUT $=$ $q_{x\,|_x} L_y L_z \delta t$

OUTPUT $=$ $q_{x\,|_{x+\delta x}} L_y L_z \delta t$

Applying a differential energy balance to the element located in the wall, we have

$$[q_x \mid_x - q_x \mid_{x+\delta x}] L_y L_z \delta t = 0$$

so, dividing by $L_y L_z \delta x \delta t$ and taking the limit,

$$\lim_{\delta x \to 0} \frac{(q_x \mid_x - q_x \mid_{x+\delta x})}{\delta x} = -\frac{dq_x^w}{dx} = 0$$

where the superscript, w, denotes that this equation is valid <u>only</u> for the furnace wall. The solution to this differential equation is quite simple since q_x must be a constant.

$$q_x^w = C_1$$

However, the boundary conditions are not so simple since we are really dealing with three separate materials: the furnace gas, the furnace wall and the insulation. First of all, we do not have a boundary condition for q_x and so we use Fourier's Law and integrate once more

$$q_x^w = -k^w \frac{dT^w}{dx} = C_1$$

$$T^w = -\frac{C_1}{k^w} x + C_2 \tag{3.6}$$

where C_2 is a second constant of integration. In order to solve for C_1 and C_2, two boundary conditions are required. Although we do have two boundary conditions, one at $x = 0$ and the other at $x = L_1 + L_2$, the second is not applicable to Equation (3.6) because it applies to the outer surface of the insulation and the temperature distribution described by Equation (3.6) is only valid at the wall; i.e., $0 \le x \le L_1$. In order to obtain the desired solution we must link the energy transport in the wall to that occurring in the insulation. To do this we need to take a second differential energy balance; this time within the insulation. Again, at steady-state and with no SOURCES/SINKS, we obtain an identical differential equation

$$-\frac{d}{dx}(-k^I \frac{dT^I}{dx}) = 0$$

which can be integrated twice to give

$$T^I = -\frac{C_3}{k^I} x + C_4 \tag{3.7}$$

Applying the boundary conditions:

$$At \ x = 0, \ T^w = T_F$$

$$At \ x = L_1 + L_2, \ T^I = T_a$$

to Equations (3.6) and (3.7), we find that

$$C_2 = T_F$$

(3.8)

$$T_a = -\frac{C_3}{k^I}(L_1 + L_2) + C_4$$

At this point, we have only been able to solve for one of the constants of integration and have used two boundary conditions. Since there are four constants of integration, four boundary conditions are needed and thus two more are needed. The two additional boundary conditions can be obtained by making use of the fact that nature is more or less continuous, even at interfaces.[#] Consequently, both temperature and the molecular energy flux must be continuous at $x = L_1$, i.e.,

$$q_x^w \Big|_{L_1} = q_x^I \Big|_{L_1}$$

$$T^w \Big|_{L_1} = T^I \Big|_{L_1}$$

Using these continuity boundary conditions at $x = L_1$, together with Equations (3.6-3.9), we can solve for C_1, C_3 and C_4 to obtain the temperature distribution in the wall

$$T^w = -\frac{(T_F - T_a)}{L_1 + \frac{k^w}{k^I}L_2} x + T_F$$

[#]   There is some controversy here, but on the scale we are operating on, this is a reasonable assumption. Also we shall see that there are apparent discontinuities when dealing with mass transport at interfaces between gas-liquid phases. But this is due to the use of concentration driving forces instead of chemical potential.

The prediction of the temperature at $x = L_1$ is therefore

$$T^w{}_{|L_1} = -\frac{(T_F - T_a)}{1 + \dfrac{k^w}{k^I}\dfrac{L_2}{L_1}} + T_F$$

and the heat flux is just C_1, or

$$q_w = \frac{(T_F - T_a)}{\dfrac{L_1}{k^w} + \dfrac{L_2}{k^I}}$$

$$q^w_x = -\,k^w\frac{dT^w}{dx} = -\,k^I\frac{dT^I}{dx} \tag{3.9}$$

There is a simpler way to arrive at the heat flux in terms of the temperature difference, $T_F - T_a$, and that is to make use of the concept of "resistances in series" which was introduced in Chapter 2 (see 2-2.b). Since at steady-state, the heat flux is a constant through the two materials, the energy flux is constant and consequently, the temperature gradients are constant, or

$$\frac{dT^w}{dx} = \frac{\Delta T}{\Delta x} = \frac{T^w{}_{|L_1} - T_F}{L_1}$$

Thus Equation (3.9) is in the form of:

$$Current\ (flux) = \frac{Potential\ (\Delta T)}{Resistance\ (\dfrac{L}{k})}$$

This means we can make use of the concept of "resistances in series" for an electrical circuit; i.e.,

$$I = \frac{\Delta E}{\displaystyle\sum_i R_i}$$

where ΔE is the <u>overall</u> potential and R_i are the individual electrical resistances. In terms of the energy transport problem, this can be written

$$q_x^w = \frac{(T_F - T_a)}{\sum\limits_i R_i} \qquad (3.10)$$

where

$$\sum\limits_i R_i = \frac{L_1}{k^w} + \frac{L_2}{k^I}$$

Note that the <u>overall</u> temperature driving force, T_F - T_a , is analogous to the overall potential.

3-2.c Radial Temperature Distribution in a Wire

The earlier example of molecular energy transport (conduction) in a rod was an occasion where radial temperature gradients were not present because the radial surface was perfectly insulated. Now let's analyze a situation where radial gradients must be considered. The physical problem consists of a long wire of radius, R, (where $L/R \ll 1$) which is carrying an electric current. The flow of electric current results in "joule heating", which is a uniform source of energy, $(\dot{Q}_e)_V$, and the outer surface of the wire is maintained at a constant temperature, T_a.

Obviously there is transport in the radial direction since the outer wire surface is maintained at ambient temperature and we know that the joule heating will release heat within the wire. Furthermore, because the length of the wire is very long relative to the radius, molecular transport in the longitudinal direction can be neglected. In order to derive a differential equation which describes the steady-state temperature distribution in the wire, the DVE given in Figure 2-1c must be employed and the sketch in Figure 3-5 shows its dimensions and the applicable fluxes entering and leaving. Under steady-state conditions, then, the individual terms in Equation (2-20) are:

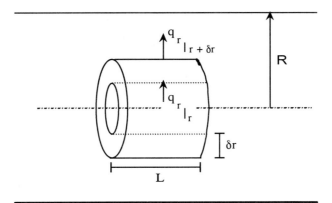

Figure 3-5 DVE for Radial Energy Transport

ACCUMULATION =   0   (steady-state)

SOURCES   = $(\dot{Q}_e)_V\ 2\pi r\delta r\delta z\delta t$

INPUT   = $q_{r|_r}\ 2\pi r\delta z\delta t$

OUTPUT   = $q_{r|_{r+\delta r}}\ 2\pi(r + \delta r)\delta z\delta t$

and the differential energy conservation equation is

$$q_{r|_r}\ 2\pi r\delta z\delta t - q_{r|_{r+\delta r}}\ 2\pi(r+\delta r)\delta z\delta t + (\dot{Q}_e)_V\ 2\pi r\delta r\delta z\delta t = 0$$

Upon dividing through by the volume of the DVE and taking the limit as $\delta r \to 0$ we have to account for the fact that the inner and outer radii of the DVE are not the same and so

$$\frac{r\,q_{r|_r} - r\,q_{r|_{r+\delta r}}}{r\delta r} - \frac{q_{r|_{r+\delta r}}\ \delta r}{r\delta r} + (\dot{Q}_e)_V = 0$$

(3.11)

$$-\left[\frac{dq_r}{dr} + \frac{1}{r}q_r\right] + (\dot{Q}_e)_V = 0$$

Recognizing that [*]

$$\frac{1}{r} q_r + \frac{dq_r}{dr} = \frac{1}{r} \frac{d}{dr} (rq_r)$$

Equation (3-11) is separable and integration yields

$$q_r = \frac{(\dot{Q}_e)_V \ r}{2} + \frac{C_1}{r}$$

where, $C_1 = 0$ by virtue of the symmetry boundary condition, $q_{r_{|r=0}} = 0$. Substitution of Fourier's Law and a second integration subject to the boundary condition, $T = T_a$, at $r = R$, results in

$$T = T_a + \frac{(\dot{Q}_e)_V \ R^2}{4k} \left[1 - \left(\frac{r}{R} \right)^2 \right]$$

where k is the thermal conductivity of the wire.

3-3 STEADY-STATE ENERGY TRANSPORT IN A COOLING FIN

3-3.a General Considerations

In Chapter 2, we saw that the energy transport rate, $\dot{Q}$, is directly proportional to the surface area over which the energy is being transported. Occasionally situations arise when it is desired to cool a fluid or solid, but design constraints dictate a compact unit with limited surface area. A good example of this is the standard automobile radiator. To compensate for the limited surface area, a series of thin metallic "fins" are welded to the radiator. The objective here is to effect molecular energy transport ("conduction") in the solid fins so that the energy can be transported to the surrounding air via the "extended" surface area of the fins.

---

[*]   This is not always easy to recognize. See Example 3.1 for an alternate method of deriving this differential equation.

EXAMPLE 3-1: Alternate Derivation of Equation (3.11)

Use the concept of the "derivative of a product" to obtain an alternative derivation of Equation (3.11)

Solution

Rather than write the OUTPUT term as was done in the derivation of (3.11), it can also be written as follows

$$\text{OUTPUT} = q_{r\,|\,(r+\delta r)}\, 2\pi\, r_{\,|\,(r+\delta r)}\, L\delta t \;=\; 2\pi (r\, q_r)_{\,|\,(r+\delta r)}\, L\delta t$$

If we write the INPUT term in a similar way; i.e.,

$$\text{INPUT} \;=\; (r\, q_r)_{\,|\,r}\, L\delta t,$$

then, in the process of dividing through by the elemental volume, we obtain

$$\lim_{\delta r \to 0}\; \frac{(r\, q_r)_{|\,r} - (r\, q_r)_{|\,(r+\delta r)}}{r\delta r} \;=\; -\frac{1}{r}\frac{d}{dr}(r\, q_r)$$

which is can be directly integrated and avoids the necessity of recognizing that the two bracketed terms in Equation (3.17) can be regrouped into one.

Commercially, these types of heat exchangers are called "extended area" heat exchangers. However, if energy cannot be successfully transported to the extended surface areas, then the fins will not be effective and we might as well save our money.

Figure 3-6a shows a sketch of a cylindrical cooling fin which consists of an aluminum rod, 90 cm long and 2.54 cm in diameter. One end of the rod is immersed in a steam chest where it comes into contact with saturated steam at 231 kPa (125 °C). Before looking at the specifics, let us examine, in a qualitative manner, the design parameters which will affect the efficiency of the rod as a cooling fin. From the previous discussions of the phenomenological laws in Chapter 2, we know that the heat flux will depend on the magnitude of the temperature difference between the rod surface and the ambient air. Thus we would like to keep the surface temperature of the rod as high as possible. If the rod is constructed of a material with a low thermal conductivity, it will keep the temperatures high in the region close to the steam chest, but they will be lower over most of the fin. In addition, this can also result in lower surface temperatures due to significant radial temperature gradients. This is not desirable because the

Figure 3-6

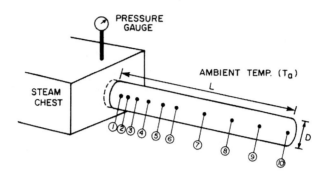

(a) Cooling Fin Experimental Apparatus

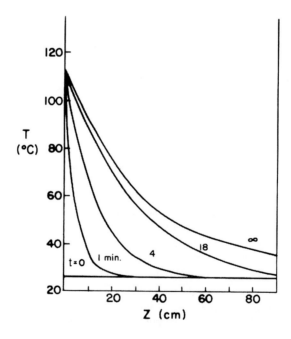

(b) Unsteady State Temperatures

heat rate is proportional to the magnitude of the temperature differences between the surface of the rod and the bulk air. In other words, we would not be making "effective" use of the available surface area, thus defeating the purpose of the fin.

Clearly then, we want to use a high thermal conductivity material. Certainly the larger the fin, the higher the surface area. However, there are problems here, too. Fourier's observations point out that the conductive "resistance" is directly proportional to the length of the conduction path. Thus larger values of R and L increase the conductive resistances in the radial and axial directions and induce larger temperature differences along the conductive paths. In the radial direction this causes a lowering of the temperature driving force between the rod surface and the air. In the z-direction, this produces "steep" temperature gradients and could possibly result in a large portion of the rod surface area being under-utilized. That is, the temperatures will be lower near the end of the fin and less heat will be transferred to the ambient air in those locations. As can be seen, there is evidently a trade-off between the temperature driving force and surface area. Thus we wish to develop an analysis which would be useful in deciding on an optimum design; that is, determine the dependency of the effective surface area utilization on the pertinent design variables. This is what we mean by efficiency.

Returning to the specific situation depicted in Figure 3-6, the rod is instrumented with thermocouples to measure the temperatures at the centerline of the rod as a function of distance from the immersion point. An experiment was conducted with such a rod, utilizing the laboratory apparatus described by Crosby [1]. Figure 3-6b shows a plot of the temperature profiles measured as a function of time from the start of the experiment (i.e., at the point when steam was admitted to the steam chest) and Table 3-2 gives numerical values for the steady-state temperature data.

Table 3-2 Steady-State Temperature Profile in a Cylindrical Cooling Fin

TC POSITION NO.	DISTANCE (fm Z = 0, cm)	TEMPERATURE (°C)
TC1	0.5	112
TC2	3.5	105
TC3	9.5	92
TC4	15.0	81
TC5	23.0	69
TC6	30.0	60
TC8	60.0	41
TC10	90.0	36

Before entering into a quantitative analysis of this problem, let's first speculate on what kind of steady-state temperature profiles might be expected from such an experiment. Figure 3-7 shows some possible profiles, for a few different physical situations. Profile '1' is of course what we would expect at steady-state if energy could not leave the rod at all, while profile '2' is the expected result if energy could only leave the rod at its end. Of course, in this experiment, energy will continually leave the radial surfaces and the rate at which it leaves will be proportional to the temperature differences between the rod and the ambient air ($\Delta T = T - T_a$). Thus we should expect profiles more like '3' and '4'. The deviation from a straight-line profile is due to the fact that a disproportionate quantity of energy will be lost at areas closer to $z = 0$, where ΔT is larger. Whether the actual profile will look more like '3' or '4' will depend on the dimensions and properties of the rod as well as the rate at which energy is able to be transported to the ambient air. Note, that if profile '4' is the result of our design, then we are not effectively utilizating the total fin area. One other point should be made relative to the profiles shown in Figure 3-7; viz., in profile '3', the temperature at $z = L$ is <u>not</u> equal to the ambient temperature. This is because energy will also be leaving the rod at this point (although it will be much less than at the radial surfaces), and consequently the temperature at this location must be higher than the ambient temperature.

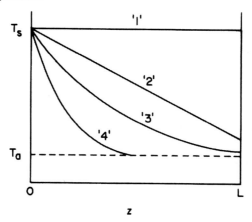

Figure 3-7 Steady-State Temperature Profiles in a Cooling Fin

3-3.b Cooling Fin Temperature Profile

In order to derive the differential equation which describes the transport, the general conservation law, Equation (2.20), is applied to an energy balance over the appropriate infinitesimal differential volume element. Following the steps in Table 2-4, question 1.a is easy to answer; we are analyzing <u>energy</u> transport. However, to answer question 1.b we need to look at which assumptions might be appropriate. The initial assumptions we will make are that the system is at steady-state, the material properties are both uniform and independent of temperature and that the steam chest is capable of imposing a steady, uniform temperature across the fin at $z = 0$.[#]

In reality, the direction of transport is in both the radial and axial directions. However, this leads to a partial differential equation (two independent variables, z and r) and we wish to simplify matters. The problem can be greatly simplified by realizing that the fin should have a high thermal conductivity and that in order to maximize the surface area/mass of the fin, D should be small. Thus, in view of Fourier's Law, there should be minimal temperature gradients in the radial direction. Actually, the assumption that we are really making is that the radial temperature gradients are confined to a region very close to the outer surface of the fin, as shown in Figure 3.8. While this sounds like a reasonable assumption, it leaves us with no mechanism to get the energy to the radial surfaces.

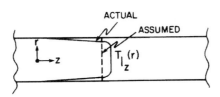

Figure 3-8 Radial Temperature Profiles in a Cooling Fin

If radial temperature gradients are ignored, the only temperature gradients are in the axial direction and thus transport occurs only in this direction. Proceeding with step 2a in Table 2-4, the shape of the differential volume element should be a disk (Figure 2-1b) and the origin of the coordinate system will be placed at the point where the fin meets the steam. The DVE in this case is exactly as that shown in Figure 3-3 and in order to account for the energy that is being lost at the cylindrical surface as we proceed down the rod in the z-direction, the energy loss can be viewed as a SINK (Step 4 in Table 2-4). According to the rules for SOURCES and SINKS (STEP 4b in Table 2-4), the sink must apply uniformly within the DVE and, must be expressed on a per unit volume basis before being inserted into the energy balance over the DVE. Uniformity is assured by virtue of the assumption that there are no temperature gradients in the radial direction. For now, we will use the symbol, $\dot{Q}_v$, (units of energy/vol-time) for the sink term.

The individual terms in the general conservation equation are, for this problem:

[#]   This temperature will equal the steam temperature, provided that there is a negligible thermal resistance between the steam and the rod

$$INPUT = q_{z\,|\,z}\,\pi\,R^2\,\delta t$$

$$OUTPUT = q_{z\,|\,z+\delta z}\,\pi\,R^2\,\delta t$$

$$SOURCES = 0$$

$$SINKS = \dot{Q}_v\,\pi\,R^2\,\delta z \delta t$$

$$ACCUMULATION = 0$$

So that, insertion into Equation (2.20) gives

$$(q_{z\,|\,z} - q_{z\,|\,z+\delta z})\pi R^2 \delta t - \dot{Q}_v \pi R^2\,\delta z \delta t = 0$$

Dividing through by the volume of the element and $\delta t (\pi R^2 \delta z \delta t)$ and taking the limit as δz approaches zero, we obtain

$$-\frac{dq_z}{dz} - \dot{Q}_v = 0 \tag{3.12}$$

While Equation (3.12) is a correct description of energy transport in the cylindrical fin, it cannot be solved until we get it in terms of <u>one dependent</u> variable and/or its derivatives (Step 8.b in Table 2-4). Although both q_z and $\dot{Q}_v$ in Equation (3.12) are functions of temperature[#], Fourier's Law allows us to express q_z in terms of the temperature gradient which in this case (rectangular coordinates) is

$$q_z = -k\frac{dT}{dz} \tag{3.13}$$

To express $\dot{Q}_v$ in terms of temperature, we need to recognize that the energy is really being lost at the surface of the DVE , and the rate of energy loss there is primarily a function of the degree of convective energy transport between the surface and the ambient air. An empirical relationship which describes this phenomena is <u>Newton's Law of Cooling</u>

$$\dot{Q}_s = h\,\Delta T\,A_s \tag{3.14}$$

[#] Remember that $\dot{Q}_v$ represents the radial heat loss, and it will vary with the fin temperature

where A_s is the surface area (in this case, of the DVE) and ΔT is the temperature difference between the surface of the fin and the ambient air. The proportionality parameter, h, is called the <u>heat transfer coefficient</u>, and is dependent on the degree of convective energy transport. This concept will be discussed in more detail in Chapter 10. However, $\dot{Q}_V$ must be expressed on a per unit volume basis and so

$$\dot{Q}_V = \frac{\dot{Q}_s}{\dfrac{\pi D^2}{4} \delta z}$$

with $A_s = \pi D \delta z$, and $\Delta T = T - T_a$, the expression for $\dot{Q}_V$ is then

$$\dot{Q}_V = \frac{h(\pi D \delta z)(T - T_a)}{(\pi D^2 \delta z / 4)} = \frac{4h}{D}(T - T_a) \tag{3.15}$$

Using Equations (3.13) and (3.15), Equation (3.12) becomes, for constant k,

$$\frac{d^2 T}{dz^2} - \frac{4h}{kD}(T - T_a) = 0 \tag{3.16}$$

To solve this equation, two boundary conditions are needed for T and/or its derivatives at two locations of z. Assuming that the rod temperature at z = 0 is equal to the steam chest temperature gives us one boundary condition, viz.

$$\text{at } z = 0; \ T = T_S \tag{3.17}$$

If the diameter of the rod is small relative to its length, then the surface area at the end of the rod will be much less than the remaining surface area. Under these conditions it is a good assumption that the rod behaves as though it were insulated at z = L. Thus the second boundary condition is

$$\text{at } z = L, \ \frac{dT}{dz} = 0 \tag{3.18}$$

It should be noted that Equation (3.18) is a less restrictive boundary condition than the assumption that the rod temperature equals the ambient temperature at z = L, since the latter will only occur if the situation is such that profile "4" in Fig. 3-7 occurs[#].

[#] Note that if this <u>is</u> the case, boundary condition 3.18 still applies.

Equation (3.16) is a second-order, homogeneous, ordinary differential equation with a general solution [2][*]

$$T - T_a = C_1 \sinh(\Gamma_h\ z) + C_2 \cosh(\Gamma_h\ z) \tag{3.19}$$

where

$$\Gamma_h = \sqrt{\frac{4h}{kD}} \tag{3.20}$$

and C_1 and C_2 are constants of integration. Applying boundary conditions (3.17) and (3.18) gives

$$C_1 = -(T_s - T_a)\tanh(\Gamma_h\ L)$$

$$C_2 = T_s - T_a$$

so that the particular solution we desire from (3.19) is

$$\frac{T - T_a}{T_s - T_a} = \cosh(\Gamma_h\ z) - \tanh(\Gamma_h\ L)\sinh(\Gamma_h\ z) \tag{3.21}$$

Now let's compare the results given by Equation (3.21) with the experimental data given in Table 3-2. For the 2.54 cm aluminum rod, k = 206 J/m-sec-K, and, if the energy transport mechanism surface of the rod to the ambient air is governed by "natural convection," a good value of h would be about 17 J/m^2-sec-K [3]. Thus

$$\Gamma_h = 2\sqrt{\frac{17}{(206)(.0254)}} = 3.6\ m^{-1}$$

Figure 3-9 compares the temperature profile predicted by Equation (3.21) (solid line) with the experimental data. As can be seen, the value of the heat transfer coefficient used in Equation (3.21) does a very reasonable job of matching the experimental data. Notice however that the temperatures are over predicted near the steam chest but are under predicted near the end of the rod. This would lead one to suspect that the heat transfer coefficient may not be a constant value over the length of the rod as was assumed in the derivation of Equation (3.21) This is dealt with in a more detailed manner in Section 3-5.

---

[*]   Transform the variable, T, into a new variable θ by defining θ = T - T$_a$

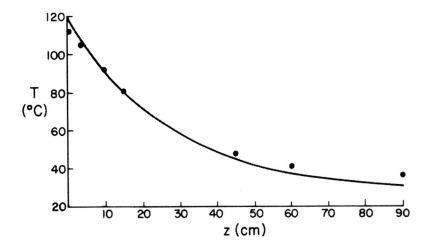

Figure 3-9 Predicted Versus Experimental Temperature Profiles

3-3.c Fin Effectiveness

With Equation (3.21), it is a simple matter to determine the heat dissipated by the fin. This should be a natural outcome of our analysis since, after all, we are trying to see how effectively the fin will dissipate energy. The heat loss at any point along the fin is simply the heat leaving the cylindrical surface of the DVE, or

$$Rate\ of\ heat\ loss\ at\ a\ point,\ z,\ = \delta\dot{Q} = h\pi D\delta z\,(T_{|z} - T_a)$$

This overall loss rate is simply the sum of this expression over $n\ \delta z$, i.e.,

$$\dot{Q} = \sum_n \delta\dot{Q} = \pi Dh \sum_n (T_{|z} - T_a)\ \delta z$$

which, as n approaches infinity, becomes

$$\dot{Q} = h\pi D \int_0^L (T - T_a)\,dz \tag{3.22}$$

Another method to calculate the heat dissipation can be used if it is recognized that, since all the heat actually dissipated by the rod had to enter the rod at z = 0,

$$\dot{Q} = \frac{\pi D^2}{4} q_z \mid_{z=0} = \frac{\pi D^2}{4} (-k \frac{dT}{dz}) \mid_{z=0}$$ (3.23)

We can now use the solution given by Equation (3.21) to calculate $\dot{Q}$, either by integrating according to Equation (3.22), or differentiating it at z = 0 according to Equation (3.23). In either case, we obtain

$$\dot{Q} = \frac{\pi D^2}{4} k(T_s - T_a) \Gamma_h \tanh(\Gamma_h L)$$ (3.24)

Note that for a large dissipation of heat we would want both k and Γ_h to be large.

This brings us back to the word "efficient," If we were asked whether a particular fin design is efficient or not, one reply might be based on the quantity of heat which the fin is capable of dissipating. But what if the fin was extremely long so that it dissipated a large quantity of heat but most of the dissipation occurred over the initial 10% of the fin length? Clearly, from a material utilization point of view, this would not be efficient at all. Consequently, let us define a quantity called the "fin effectiveness," η, which is given by

$$\eta = \frac{[heat\,actually\,dissipated]}{[heat\,dissipated\,if\,entire\,fin\,were\,at\,T_s]}$$

$$\eta = \frac{\dot{Q}}{\pi Dh(T_s - T_a)L}$$ (3.25)

The fin effectiveness can thus be calculated from our solution; i.e., by substituting Equation (3.24) into (3.25), we obtain

$$\eta = \frac{\tanh(\Gamma_h L)}{(\Gamma_h L)}$$ (3.26)

Figure 3-10 shows a plot of η versus $\Gamma_h L$ and, according to this figure, the maximum effectiveness, $\eta \sim 1.0$, occurs when $\Gamma_h L$ is small; i.e., when h and L are both low, and k and D are both high. The desirability of small values of h would at first seem puzzling since all our intuition would lead us to conclude that a high heat transfer coefficient (for example, by blowing air past the fins with a high-speed blower) is better. The reason for this seeming anomaly is in the definition of the "fin effectiveness." By this definition, the fin is being evaluated on the basis of its ability to maintain the temperature of the rod at the steam temperature. Thus if the fin were insulated perfectly, h would be zero and the fin would be

100% effective. In the same vein, if the <u>length</u> of the fin were zero it would also be 100% effective. In other words, the most "efficient" operation of the fin is simply not to operate at all.

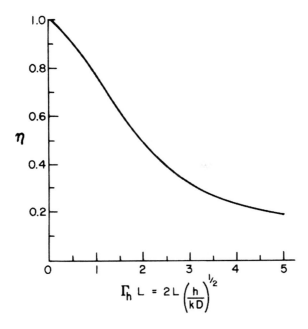

FIGURE 3-10 Cooling Fin Effectiveness Factor

By now, it should be appreciated that this paradox has come about because, in the definition of the fin effectiveness, no consideration has been given to the <u>quantity</u> of heat dissipated by the fin. Since this was the point of designing the fin in the first place, we have lost sight of our goals. The optimal design is a combination of a high η (to better utilize fin material) and a high dissipation of heat. This illustrates the caution which should always accompany any discussion of "efficiency." It is important that there is a clear understanding of <u>what kind</u> of efficiency is being dealt with before coming to any conclusions about a particular design.

3-4 UNSTEADY-STATE MOLECULAR ENERGY TRANSPORT

Unsteady-state molecular transport problems generally result in partial differential equations unless we are able to make simplifying assumptions. That is, there must be some justification for neglecting spatial temperature gradients or some other, less accurate way to account for energy being transported across the surfaces of the system. Methods of solving problems with such simplifying assumptions will be illustrated here, and the more rigorous solution to these types of problems are introduced in Chapter 7.

3-4.a Unsteady-State Heating of an Al-Sphere

Consider an experiment where an aluminum sphere of diameter 5.0 cm, initially at temperature T_0, is suddenly immersed in a hot temperature bath (at temperature, T_H). Figure 3-11 shows the experimental configuration as well as the tabulated data collected from a thermocouple located in the <u>center</u> of the sphere. If the bath is well-mixed and has a much larger mass than that of the sphere, then it is valid to assume that the bath remains at T_H for the duration of the experiment. Furthermore if the rate at which energy is delivered to the sphere can be described by Newton's Law of Cooling, then, $\dot{Q} = 4\pi R^2 h (T_H - T_{|R})$. One way of determining a value for h, is to construct a mathematical model which will allow for the calculation ("prediction") of the temperature distribution in the sphere and to then decide on the value of h which provides a "best" match of the calculations to the experimental data listed in Figure 3-11.

In this case, the appropriate DVE is given by Figure 2-1d and the INPUT - OUTPUT terms in Equation 2-20 become

$$INPUT : q_{r_{|r}} 4\pi r^2 \delta t = (r^2 q_r)_{|r} 4\pi \delta t$$

(3.27)

$$OUTPUT : q_{r_{|r + \delta r}} 4\pi (r + \delta r)^2 \delta t = (r^2 q_r)_{|r + \delta r} 4\pi \delta t$$

$$T_H = 57 \ C$$

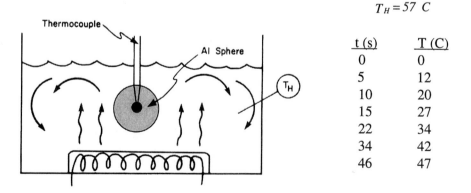

t (s)	T (C)
0	0
5	12
10	20
15	27
22	34
34	42
46	47

Figure 3-11 Unsteady Heating of an Aluminum Sphere

While there are no SOURCES or SINKS internal to the sphere, the energy transported to the sphere does serve to increase its internal energy, U, which then accumulates with time. Expressing U on a per unit mass basis and the accumulation term on a per unit volume basis, the general conservation equation, becomes,

$$(r^2 q_r)|_r - (r^2 q_r)|_{r+\delta r}\, 4\pi \delta t = \rho\left[\hat{U}|_{t+\delta t} - \hat{U}|_t\right] 4\pi r^2 \delta r$$

Dividing through by the volume of the sphere and δt and taking the limit as $\delta t \to 0$, we obtain

$$-\frac{1}{r^2}\frac{\partial}{\partial r}(r^2 q_r) = \rho \frac{\partial \hat{U}}{\partial t} \qquad (3.28)$$

Once again we need to write this differential equation in terms of one dependent variable and to do this we turn to Fourier's Law to express q_r in terms of the temperature gradient and to thermodynamics to express $\hat{U}$ in terms of T. For $\hat{U}$,

$$\hat{U} = \hat{c}_v(T - T_{\text{ref}}) \qquad (3.29)$$

and since $\hat{c}_v \sim \hat{c}_p$ for an incompressible system, Equation (3.28) becomes[#]

$$\frac{\partial T}{\partial t} = \frac{k}{\rho \hat{c}_p}\frac{1}{r^2}\frac{\partial}{\partial r}\left(r^2 \frac{\partial T}{\partial r}\right) \qquad (3.30)$$

The initial and boundary conditions are then

$$\text{at } t = 0,\ T = T_0;$$

$$\text{at } r = 0,\ \frac{\partial T}{\partial r} = 0;\ \text{at } r = R,\ -k\frac{\partial T}{\partial r} = h(T_H - T|_R)$$

Note that since internal temperature gradients are not neglected, we have generated a partial differential equation. Although there are methods for obtaining analytical solutions for these type of equations (usually in terms of converging, infinite series[##]), we shall leave these solutions until the discussion of "similarity techniques" in Chapter 7. The solution would give us T(r, t), which we would evaluate at r = 0 (where the experimental data were taken) and then adjust the parameter, h, until a good match to the experimental data are obtained.

[#]   Note that all the material properties are assumed to be invariant with r, t, and T.

[##]   See Problem 3-5 for an example of one such solution.

EXAMPLE 3-2: Neglect of Internal temperature Gradients in a Heated Sphere

Assume that internal temperature gradients of the aluminum sphere are negligible and then use the methods illustrated for the cooling fin (Section 3-3) to determine an approximate temperature-time relationship.

<u>Solution</u>

Since there are no internal gradients in this case, the system is simply the entire sphere. Thus there are no INPUT or OUTPUT terms, and there is no need for a differential volume element or a coordinate system. Just as was done with the cooling fin in Section 3-2, we can treat the energy transported to the sphere as if it were a uniform source, i.e.,

$$\frac{\dot{Q}_s}{Vol} = \frac{h \ (Surf.Area) \ (\Delta T)}{Volume}$$

(3.31)

$$(\dot{Q}_s)_V = \frac{h\pi D^2 (T_H - T)}{\dfrac{\pi D^3}{6}} = \frac{energy}{time\text{-}vol}$$

The accumulation is, again, expressed in terms of the internal energy, $\hat{U}$ and, in turn, by the heat capacity at constant pressure so that the differential conservation balance becomes

$$\frac{6\,h(T_H - T)}{D} \frac{\pi D^3}{6} \delta t = \rho(\hat{U}_{(t+\delta t)} - \hat{U}_t) \frac{\pi D^3}{6}$$

(3.32)

$$[SOURCE] \quad = \quad [ACCUMULATION]$$

or, as the differential equation

$$\rho\,\hat{c}_p \frac{dT}{dt} = \frac{6\,h(T_H - T)}{D}$$

(3.33)

This is a first order, separable differential equation with a general solution

$$-\ln(T_H - T) = \Phi t + C_1$$

(3.34)

where

$$\Phi = \frac{6\,h}{\rho\,\hat{c}_p\,D}$$

The constant of integration, C_1, can be determined from the initial condition

$$\text{at } t = 0, \quad T = T_o$$

so that

$$C_1 = -\ln(T_H - T_o)$$

With the result that

$$\ln\frac{(T_H - T)}{(T_H - T_o)} = -\Phi t \qquad (3.35)$$

We can now examine the validity of the assumptions leading to this solution by comparing the tabulated data in Figure 3-11 with Equation 3.35. This is most easily done by determining whether the experimental data plot in a logarithmic fashion, as predicted by Equation (3.35). If it does, then we should obtain a straight line when we plot $\ln[(T_H - T)/(T_H - T_o)]$ versus time and the slope of the line should equal Φ. Figure 3-12 shows such a plot and, as can be seen, a reasonable straight line can be fit to the data where the slope has a numerical value of 0.039. Note that this now gives us the opportunity of determining the proportionality constant, h. That is

$$\Phi = 0.039 = \frac{6\,h}{\rho\,\hat{C}_p\,D}$$

which, with the physical properties of aluminum ($\rho = 2.65$ g/cm^3, $\hat{C}_p = .92$ J/g-K), can be solved for h to give

$$h = \frac{(.039)(2.65)(.92)(5.0)}{6} = .079 \, J / cm^2 \text{-} s \text{-} K$$

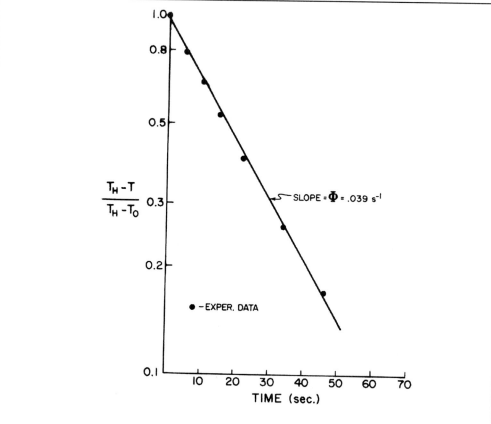

Figure 3-12 Comparison of Model with Experimental Data

EXAMPLE 3-3: Energy Transport with a Uniform Sink

A porous $CaCO_3$ cylinder (diameter D, length L) at room temperature is suddenly thrust into a 600 °C oven. At this temperature $CaCO_3$ endothermically decomposes (heat of reaction = ΔH_r) into CaO and CO_2. Assume that the porosity is such that there are no internal gradients of temperature or of CO_2 and that the decomposition reaction rate is uniform and constant at R_A. As with the problem discussed in Example 3.2, assume that the energy is transported from the oven to the cylinder at a rate proportional to the product of the surface area and the temperature difference between the oven and the cylinder, where the proportionality constant is h. Solve for the temperature of the cylinder as a function of time.

Solution

As with the heated sphere in Example 3.2, there is no need for a differential volume element or a coordinate system since we are neglecting internal gradients. Thus, the individual terms in Equation (2.20) are:

INPUT = OUTPUT = 0

SOURCE $\quad = [\ \dfrac{h\pi DL(T_{ov}-T)}{\dfrac{\pi D^2 L}{4}}\ ]\ \dfrac{\pi D^2 L}{4}\ \delta t$

SINK $\quad = R_A\ \Delta\tilde{H}_r\ \dfrac{\pi D^2 L}{4}\ \delta t$

ACCUMULATION $= \rho\hat{c}_p\ (T\mid_{t+\delta t} - T\mid_t)\dfrac{\pi D^2 L}{4}$

When these terms are substituted into the general conservation equation and δt is allowed to approach zero, the applicable differential equation is

$$\frac{dT}{dt} = -\frac{R_A\Delta\tilde{H}_r}{\rho\hat{c}_p} + \frac{4h}{\rho\hat{c}_p D}\ (T_{ov} - T)$$

This is a first-order, nonhomogeneous differential equation of the form

$$\frac{d\theta}{dt} + \Phi_h\theta = \Phi_r$$

where

$$\theta = (T_{ov}-T)$$

$$\Phi_h = \frac{4h}{\rho\hat{c}_p D}$$

$$\Phi_r = \frac{R_A\Delta\tilde{H}_r}{\rho\hat{c}_p}$$

the solution is

$$\theta = C_I\exp[-\Phi_h t] + \frac{\Phi_r}{\Phi_h}$$

which, when combined with the initial condition, $\theta \mid_{(t\,=\,0)}=(T_{ov}-T_o)=\theta_o$, yields

$$\theta = (\,\theta_o - \frac{\Phi_r}{\Phi_h}\,)\exp[\,-\Phi_h\,t\,] + \frac{\Phi_r}{\Phi_h} \tag{3.36}$$

3-5 NON-LINEAR ENERGY TRANSPORT PROBLEMS

3-5.a Cooling Fin with Nonlinear Cooling

Referring back to our analysis of the cooling fin problem in Section 3-3.a, recall that energy was transported from the rod to the surrounding air at a rate that was proportional to $\Delta T = T - T_a$. As we shall see when we take up convective transport in more detail, this empirical law (Newton's Law of Cooling) accounts for the convective transport of energy <u>within</u> the surrounding air by virtue of the dependence of the proportionality constant, h, on the properties and dynamic state of the fluid. However, if there is no forced movement of the air (by a blower, for example), then the only movement of the air is due to natural convection. In this case, it is density differences which lead to the fluid motion and, in a gas, this is caused by temperature gradients since density is proportional to absolute temperature. Thus, the value of h will be a function of the fluid density gradients, or ΔT. For the cooling fin situation, this will mean that h is not really a constant. That is, the local temperature gradients in the air near the steam chest will be higher than at the end of the rod where the rod temperature is closer to the ambient temperature. Correlations that relate h to ΔT^n are given in many texts (for example, see Foust et. al., [4]), and $n \sim 1/2$ is a typical dependency.

The cooling fin can be reanalyzed, accounting for the variable h. First of all, the differential equation obtained in Section 3-3, i.e., Equation (3.16) is still valid since the energy balance is taken around the differential volume element where all properties are constant.

$$\frac{d^2T}{dz^2} - \frac{4h}{kD}\,(\,T - T_a\,) = 0 \tag{3.16}$$

Although this equation is perfectly valid at any <u>point</u> along the rod, we need to account for its variation with temperature when we solve the differential equation. From the discussion above, $h \propto (\Delta T)^{1/2}$ and, using K_l as the proportionality constant, Equation (3.16) becomes

$$\frac{d^2T}{dz^2} - \Phi_F \, (\, T - T_a \,)^{3/2} = 0 \tag{3.37}$$

where

$$\Phi_F = \frac{4\,K_1}{k\,D}$$

This differential equation is now nonlinear and numerical techniques are usually required to solve it. As mentioned early in the text, powerful simulation software packages are now available and are capable of solving many non-linear differential equations. The ODE Solver in MATLAB is used in this text although there are a number of other similar packages available; for example, the ACSL differential equation solver [5]. A description of how to set up a set of differential equations for solution with MATLAB is given in Appendix B. With that as a reference, a MATLAB computer listing to solve Equation (3.37) and the boundary conditions, [(3.17) and (3.18)], is shown below for the 2.54 cm aluminum rod. In this listing, all dimensions are in cm and $K_1 = 7.6 \times 10^5$ J/cm^2-s-K$^{2.5}$.

It should be appreciated that these differential equation solvers are <u>initial value</u> numerical techniques; i.e., they start at one end of the geometry and "march" to the other. As a result, it is required that all boundary conditions be specified at z = 0. However, in this problem we actually have a boundary condition at either end of the physical system. Consequently, in order to use MATLAB, we have to guess at the temperature derivative at z = 0, and then check to see if the guess matches the boundary condition at z = L (i.e., $\frac{dT}{dz}\big|_L = 0$) and then iterate until it does. The MATLAB listing shows the "guess" of the initial temperature gradient (DTDZ0 = - 4.04 K/sec) which gave a reasonable match to the zero derivative at L.

Figure 3-13 compares the results of the MATLAB solution to both the mathematical solution, Equation 3.16, and the experimental data. As can be seen, accounting for the variation of h with temperature does a better job of fitting the experimental data than did the solution for a constant h.

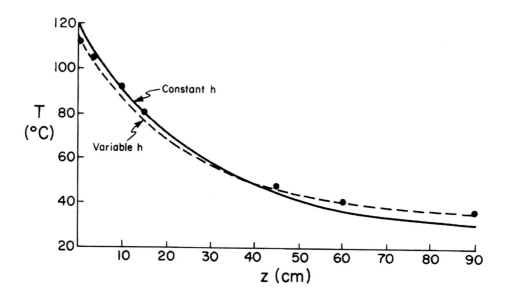

Figure 3-13 Temperature Profiles In A Cooling Fin

3-5.b Radiant Heating

Another illustration of a nonlinear energy transport problem is a radiant heating problem. Although we will not delve into radiation transport in this text,[*] suffice it to say that when temperatures are high, as in a furnace for example, radiation often overwhelms convective or molecular transport as the dominant mechanism. Simplistically, the radiant energy flux between a source at T_s and a target at T_T, can be calculated from

$$q = \varepsilon\sigma \left(T_s^4 - T_T^4 \right) \tag{3.38}$$

---

[*]   See References [6] or [7], for more detailed descriptions of radiant energy transport.

```
MATLAB Ode Program For Non-Linear Cooling Fin Problem

M - File

function dT=COOLFIN(z,T)
dT=zeros(2,1);
K1=7.6e-5; k=.206; D=2.54; TA=26;
PHIF=(4*K1)/(k*D);
dT(1)=PHIF*(T(2)-TA)^1.5;
dT(2)=T(1);

Command File

[z,T]=ode45('COOL',[0 90],[-6.31 120])
res[z,T]
ploty(z,T)
```

where ε is the <u>emissivity</u> of the target and σ is the <u>Stephan-Boltzmann Constant</u>. As can be seen from the equation, the radiant energy flux is highly non-linear with respect to temperature. Typically, radiation is a "surface-to-surface" phenomenon[*], and thus it is more properly a <u>boundary condition</u> (see Example 3.4 to see how to set up a problem in this manner).

To illustrate the incorporation of both a SOURCE and SINK into a differential energy balance, let's apply radiation energy transport to the problem described in Example 3.3, the heating of a $CaCO_3$ pellet. In this case, assume that the pellet is a 1 cm diameter sphere, that it is placed in a furnace where the temperature of the electrical heating elements is at 975 K, and that we wish to predict its temperature-time behavior.

The differential conservation balance is very similar to that presented in the context of Example 3.2, except that the SOURCE term is now due to radiation rather than convection. In order to include the radiant heating as a source term, it is necessary to assume that the radiation heat flux is delivered to the entire sphere as opposed to the outer surface of the sphere. In this case the SOURCE term can be expressed as

$$SOURCE = \varepsilon\sigma \frac{(T_F^4 - T^4)\pi D^2}{\frac{\pi D^3}{6}} \delta V \delta t$$

so that the differential equation becomes

---

[*]   In actuality, most materials do have some <u>transmissivity</u>, but this is usually neglected in engineering energy transport applications.

$$\frac{dT}{dt} = -\frac{R_A \Delta \tilde{H}_r}{\rho \hat{c}_p} + \frac{6\varepsilon\sigma}{\rho \hat{c}_p D} (T_F{}^4 - T^4) \tag{3.39}$$

The data to solve this problem is given below along with the MATLAB listing and the temperature-time results are shown in Figure 3-14. Note that the sphere heats up very rapidly due to the fourth-order dependency on temperature driving force and then slows down abruptly under the influence of the reduced driving force and the endothermic heat of reaction. With respect to the latter, it should be pointed out that the assumption of a constant reaction rate is unrealistic since it does not account for the depletion of $CaCO_3$. That is, there should be some point in the reaction where the rate would depend on the concentration of the remaining $CaCO_3$. Thus we would need to do a differential mass balance in order to keep track of it. Furthermore, the assumption of a constant rate is also unrealistic in that it results in a finite reaction rate even at unrealistically low temperatures. Actually, $CaCO_3$ decomposition does not proceed very rapidly until about 825K. Note that this is taken into account in the MATLAB program through the use of the <u>logical</u>, *"IF -"ELSE"* statement. The proper way to deal with this problem is to solve the energy and mass balance equations <u>simultaneously</u>. This is left as a problem exercise in Chapter 4 (Problem 4-21).

DATA FOR CaCO$_3$ SPHERE

$$\rho = 1.9 \ g/cm^3 \ , \ \ \hat{c}_p = 0.267 \ cal/g\text{-}K \ , \ \ D = 1 \ cm, \ T_0 = 298 \ K$$

$$\varepsilon = .7 \ , \ \ \Delta \tilde{H}_r = 40 \ Kcal/mole \ , \ \ R_{Av} = 1.86 \times 10^{-6} \ moles/cm^3\text{-}s$$

MATLAB ODE FOR THE RADIANT HEATING of a CaCO$_3$ PELLET

<u>M - FILE</u>

```
function dT=RADHEAT(t,T)
dT=zeros(1,1);
rho=1.9; cp=0.267; d=1.0; eps=.7;dhr=4e4; TF=975;
PHIRAD=6*1.355e-12*eps/(rho*cp*d);

if T(1)<825
  ra=0;
else
  ra=1.8e-6;
end

PHIR=ra*dhr/(rho*cp);
dT(1)=PHIRAD*(TF^4-T(1)^4)-phir;
```

COMMAND FILE

```
[t,T]=ode45('RADHEAT',[0 200],[273])
res[t,T]
ploty(t,T)
```

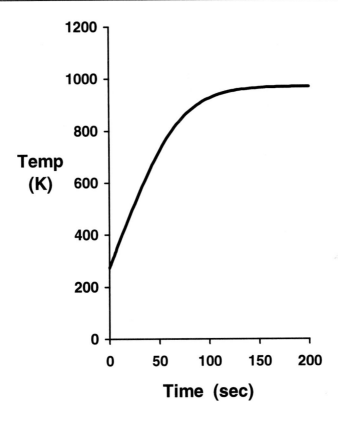

Figure 3-14 Radiant Heating Of A $CaCO_3$ Sphere

EXAMPLE 3-4: Radiant heating of a Sphere with Internal Temperature Gradients

Derive the differential equation and state the accompanying boundary conditions to account for internal temperature gradients during the radiant heating of the $CaCO_3$ sphere of Section 3-5.b

Solution

Since we are accounting for internal gradients, the appropriate DVE is that in Figure 2-1d. The INPUT - OUTPUT terms are identical to those used in Section 3-4a for the aluminum sphere and the SINK due to the uniform, endothermic chemical reaction is applied to the DVE, so that

$$SINK = \tilde{R}_A \Delta \tilde{H} \; 4\pi \, r^2 \, \delta r \delta t$$

The unsteady-state differential balance is then

$$\rho \, \hat{c}_P \left[T_{|_{t+\delta t}} - T_{|_t} \right] 4\pi \, r^2 \, \delta r = \left[(r^2 q_r)_{|_r} - (r^2 q_r)_{|_{r+\delta r}} \right] 4\pi \delta t \; \tilde{R}_A \; \Delta \tilde{H} \; 4\pi \, r^2 \delta r \delta t$$

which, after employing Fourier's Law, yields the partial differential equation

$$\frac{\partial T}{\partial t} = \frac{\alpha}{r^2} \frac{\partial}{\partial r} (r^2 \frac{\partial T}{\partial r}) \; \tilde{R}_A \; \Delta \tilde{H}$$

with initial and boundary conditions given by

$$at \; t=0 \; , \; T=T_0$$

$$at \; r=0 \; , \; \left(\frac{\partial T}{\partial r} \right)_{|_{r=0}} = 0$$

$$at \; r=\frac{D}{2} \; , \; k \left(\frac{\partial T}{\partial r} \right)_{|_{r=\frac{D}{2}}} = \sigma \varepsilon \left(T s^4 - T^4_{|_{r=\frac{D}{2}}} \right)$$

When the problem is posed in this manner, use is made of the energy <u>flux</u> stipulation as a boundary condition (see Table 3-1) and it is here where the nonlinearity occurs. While analytical solutions are available for partial differential equations of this type with linear boundary conditions, the non-linear boundary condition here necessitates the use of numerical methods in order to obtain solutions.

REFERENCES

[1] Crosby, E.J. *Experiments in Transport Phenomena*, John Wiley & Sons, N.Y., 1961.

[2] Kreyszig, E., *Advanced Engineering Mathematics*, 5th ed., p. 56, John Wiley & Sons, N.Y., 1983.

[3] Perry, R.H. and Chilton, C.H., *Chemical Engineer's Handbook*, 5th ed., McGraw-Hill Book Co., N.Y., 1973, p. 10-11.

[4] Foust, A.S., Wenzel, L.A., Clump, C.W., Maus, L. and L.B. Andersen, *Principles of Unit Operations*, 2nd ed., John Wiley & Sons, N.Y., 1980, p. 290.

[5] "DACSL, A New Method for Numerical Parameter Estimation", Dow Chemical Co., 1986

[6] Bird, R.B., Stewart, W.E. and E.N. Lightfoot, *Transport Phenomena*, John Wiley & Sons, N.Y., 1960, p. 426.

[7] Perry, R.H. and Chilton, C.H., *Chemical Engineer's Handbook*, 5th ed., McGraw-Hill Book Co., 1973, p. 10-48.

PROBLEMS*

3-1

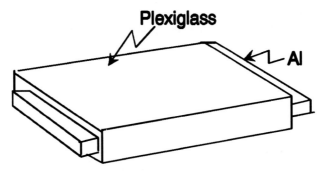

Describe how you would mathematically model the following physical problem; clearly stating (and <u>justifying</u>) all of the assumptions you make:

A thin aluminum strip, 3mm thick by 2 cm wide by 25 cm long, and encased in plexiglass (as shown in the sketch) has one end immersed in boiling water and the other end immersed in ice water. We wish to know the temperature distribution within the aluminum at steady-state.

---

* C* indicates that the problem requires a computer solution

3-2 Derive the differential equation and state the boundary conditions necessary to solve the transient problem described in 3-1 if the material is initially at ambient temperature, T_a. DO NOT SOLVE THE EQUATION.

3-3 With respect to the cooling fin problem in Section 3-3 of the text:

(a) use a differential energy balance to derive the differential equation which describes the *unsteady-state* temperature distribution in the fin.

(b) state the boundary conditions necessary to solve the differential equation in (a)

3-4 In the discussion of the cooling fin problem, it was assumed that there were no radial temperature gradients. Re-derive the differential equation and the applicable boundary conditions for the steady-state problem <u>without</u> making that assumption.

3-5 Operators in a chemical plant are issued gloves made of polybenzimidazole (PBI) which are designed to withstand relatively high temperatures. We wish to assess the effectiveness of these gloves at protecting the operator for a "worse-case" scenario. The maximum temperature the operator's hand can be exposed to is 48 °C which will be taken to be the inside glove temperature. The properties of this material are: k = .39 W/m-K, ρ = 360 kg/m³, $\hat{c}_p$ = 1300 J/kg-K. A proposed solution to the transient temperature distribution for the problem is

$$\frac{T - T_0}{T_1 - T_0} = 1 - \frac{2}{L} \sum_{n=0}^{n=\infty} \text{Exp}(-\lambda_n t) \frac{2L(-1)^n}{(2n+1)\pi} \cos\left[\frac{(2n+1)\pi(x-L)}{2L}\right]$$

where $\lambda_n = \dfrac{\alpha(2n+1)^2\pi^2}{4L^2}$, and α is the thermal diffusivity of the glove ($\dfrac{k}{\rho\hat{c}_p}$). This corresponds to initial and boundary conditions: T = T_0 at t = 0, all x; T = T_1 at x=0, all t; dT/dx = 0 at x = L, all t.

(a) Describe (in words) what is being assumed for the physical situation.

(b) Check the solution to insure that it satisfies the boundary conditions at x = 0 and x = L.

(c) If the glove is initially at 28 °C, how long can the operator hold an object which remains at 300 °C before the inside glove temperature reaches 48 °C? [The thickness of the glove is 5-mm, and, for estimation purposes, all terms in the series with $n > 0$ can be neglected.]

3-6   As discussed in Section 3-3 of the text, there are conditions where the temperature in a cooling fin will reach the ambient temperature at a distance less than the length of the fin. For such cases a reasonable boundary condition is at $z = \infty$; $T = T_a$.

(a) Use this boundary condition in place of Equation (3.18) and re-solve for the temperature distribution in the cylindrical cooling fin.

(b) In view of the logarithmic solution obtained in (a), plot the logarithm of temperature data in Table 3-2 versus z to obtain a value of the heat transfer coefficient, h. The ambient temperature during this experiment was 26 °C. Compare your answer to the heat transfer coefficient assumed in Section 3-3 of the text.

3-7   Calculate the cooling fin effectiveness for the experimental data presented in TABLE 3-2

(a) By using the results of the mathematical model [Equation (3.26)].

(b) By applying Equation (3.22) to the experimental data.

(c) By applying Equation (3.23) to the experimental data.

(d) Compare your results from these three methods and recommend the preferred method. BE SURE TO JUSTIFY YOUR RECOMMENDATION.

3-8   In Section 3-5.b of the text it was assumed that there were no radial temperature gradients in the spherical pellet. In order to check the validity of this assumption, we wish to solve for the "psuedo" steady-state temperature distribution in the sphere, _after_ the reaction begins to take place. For this estimate we will also neglect the source terms; accounting for the heat input in terms of a constant surface temperature.

(a) Apply a differential energy balance to the sphere and show that the following type of differential equation results:

$$\frac{1}{r^2}\frac{d}{dr}(r^2\frac{dT}{dr}) - P_1 = 0$$

where P_1 is a parameter which is constant.

(b) Assuming that the outer surface of the sphere is maintained at 975 K, and the values of R_A and ΔH_r are the same as shown in the MATLAB program listing (consistent units in cal-sec-cm) and that the effective thermal conductivity of the pellet is 0.8 W/m-K, solve for the radial temperature distribution. Is the neglect of radial temperature gradients a good assumption for a 5cm diameter sphere?

3-9 (**C***) For the $CaCO_3$ spherical pellet described in Section 3-5.b of the text

(a) Modify the differential equation, Equation 3.39, to account for an additional SOURCE of energy resulting from "convective" heating of the pellet. Use "Newton's Law of Cooling" to describe the convective heating. Assume that the heat transfer coefficient is constant at a value of 170 J/m^2-K

(b) Determine the temperature-time history of the pellet, using MATLAB.

3-10 (**C***) Determine the temperature-time history in the cylindrical $CaCO_3$ pellet described in Example 3-3 for the same radiation conditions used in Section 3-5.b of the text for the spherical pellet. Assume that the diameter of the cylinder is also 1" and that its length is 4".

3-11 Consider the same electrical wire discussed in Section 3-2.c. Solve for the difference between the maximum and surface temperature in the wire if it is 10 Gauge copper (resistance = .001 ohms/ft) and the current is 15 amps.

3-12 A spherical acorn squash of radius, R, and initially at temperature T_o, is placed in a microwave oven which generates a uniform source of energy within the squash of $(\dot{Q}_{mw})_V$. Neglecting radial temperature gradients, derive an equation for the temperature of the squash as a function of time if energy is also lost at the surface of the squash at a rate $h(T - T_a)4\pi R^2$, [E/t] where T_a is the temperature in the microwave cavity. Assume all properties are constant.

3-13

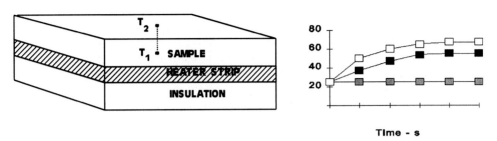

The above sketch shows an apparatus being used to measure a value of the thermal conductivity of a solid material and the accompanying plot shows the measured values of T_1, T_2 and ambient temperature, T_a as a function of time. The sample is configured as a large, thin slab with thickness, L_x, and it is placed firmly on top of a heater strip and there is a thermocouple at the very bottom of the strip (T_1) as well as at the top surface (T_2). If the heater power is **W** watts

(a) Derive an equation for the steady-state temperature distribution in the sample slab and then derive an equation which can be used to calculate the thermal conductivity of the sample from the measured temperatures.

(b) Using the thermocouple data (it should be obvious which temperatures are plotted), calculate the thermal conductivity of the sample if its thickness is 0.3 cm and the heater strip power is 30 watts.

(c) Re-derive the equation to calculate k if the top thermocouple is placed in the surrounding air, just above the sample and the energy flux lost at the top surface is proportional to the temperature difference between the top of the sample and the surrounding air (T_a). Assume that the proportionality coefficient, h, is known.

3-14

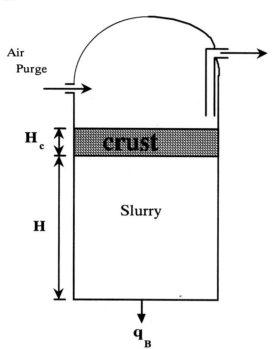

The sketch shows Tank SY-101, a cylindrical radioactive waste storage tank at the Hanford Nuclear site. Periodically this tank "burps" a gas consisting of approximately equal volumes of H_2, N_2O and N_2 and so there is much interest in the prevailing transport phenomena associated with it. Due to radioactive decay, there is a uniform source of energy within the slurry, $(\dot{Q}_N)_V$, and the tank is sufficiently large in diameter (75 ft) that it can be assumed there is no heat leakage through the side walls. However there is a known heat leakage through the bottom of the tank, q_B, and the temperature at the bottom of the crust is also known (T_c). The solids concentration in the slurry is sufficiently high so that the energy transport is essentially by molecular mechanisms with an effective thermal conductivity, k_{eff}. For steady-state conditions:

(a) Derive the differential equation and state the boundary conditions which describe the energy transport in the slurry.

(b) Solve for the temperature distribution in the slurry in terms of $(\dot{Q})_V$, k_{eff}, T_c and q_B

(c) In SY-101, the total rate of energy released due to radioactive decay is 50,000 BTU/hr. and 15% of this energy leaves at the bottom of the tank. If H = 40 ft, k_{eff} = 3 BTU/hr-ft-F and T_c = 120 ^{0}F; what is the location and value of the maximum temperature in the tank?

3-15

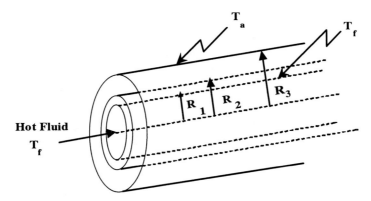

The above diagram shows a copper pipe surrounded by insulation. The inner surface of the pipe can be assumed to be at the temperature of the fluid, T_f, and the outer surface of the insulation can be assumed to be at the temperature of the ambient air, T_a.

(a) Derive the differential equation which describes steady-state energy transport in the <u>insulation</u> and then solve for temperature distribution within the insulation.

(b) If: $T_f = 350$ K, $T_a = 290$, $R_1 = 25$ mm, $R_2 = 27$ mm, $R_3 = 50$ m, and k' = 0.025 w/m-K

(i) Calculate the rate of heat loss per meter of pipe length (give units)

(ii) Decide whether it would be reasonable to assume that the temperature at the inner surface of the insulation is equal to the fluid temperature.

3-16 In order to facilitate the pumping of a viscous crude oil through a 2", schedule 80 stainless steel pipe, the pipe wall is heated by passing an electric current through it to generate 4.14×10^4 J/m^3-s.

(a) If the inside wall temperature is known (T_1, a design parameter) and the energy flux to the surroundings has been estimated for the worst case scenario, solve for the steady-state temperature distribution within the pipe wall in terms of the inner and outer radii, the energy source within the pipe [$(\dot{Q}_e)_V$] and the energy flux at the outer wall (q_a).

(b) If the inner wall temperature is 335 °K and the energy flux at the outer wall is equal to 2.5×10^4 J/m^2-s, what is the outer wall temperature under steady-state conditions?

(c) How much energy savings per length of pipe can be realized if 4 cm of insulation (k_I = .034 w/m-K) is added. In this case assume that the outer surface of the insulation is at ambient temperature (T_a = 270 K) and that the temperature in the pipe wall is uniform at 335 K.

3-17

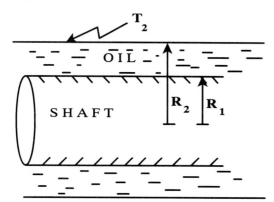

A lubricating oil with thermal conductivity, k, fills the annulus between a long, cylindrical rotating shaft and an outer sleeve as shown above. The friction due to the rotation generates a uniform energy source within the oil, $(\dot{Q}_\mu)_V = \mu \left(\dfrac{\Omega R_I}{R_2 - R_I} \right)^2$ $[E/l^3 \text{-} t]$, where Ω is the shaft rpm and μ is the viscosity of the oil. If the sleeve is cooled and kept at T_2 and the surface of the shaft is perfectly insulated:

(a) Solve for the steady-state temperature distribution within the oil.

(b) Solve for the energy flux at R_2 .

(c) Show that your answer in (b) is consistent with an overall energy balance.

3-18

FLUID (T_H)

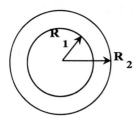

The thin hollow metallic sphere shown in the sketch, initially at temperature, T_0 , is suddenly immersed in a well mixed hot temperature bath which is at T_H. The inner surface of the shell (at R_I) can be assumed to be perfectly insulating and the heat transfer coefficient between the hot fluid and the sphere is h. If temperature gradients within the shell can be neglected, solve for the temperature of the shell as a function of time.

3-19

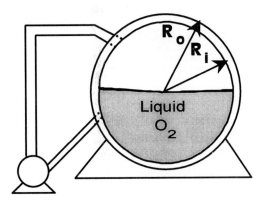

A pressurized spherical storage tank contains liquid oxygen and the ambient air temperature is at T_a. Provisions exist so that the vapor that boils off (heat of vaporization = $\Delta \hat{H}_v$) is continuously vented, compressed, refrigerated, and returned to the tank as a liquid. Assuming that heat is conducted into the vessel uniformly in all directions,

(a) Derive an expression for the steady-state temperature distribution within the tank wall as a function of r and in terms of the boil-off rate, $\dot{m}$.

(b) Determine the boil-off rate if the tank is constructed of stainless steel and if the temperature of the inner shell wall is constant at an oxygen temperature of -60 °C. In addition, the following data apply:

T_a = 25 C, R_o = 50 ft., R_i = 49.5 ft, $\Delta \hat{H}_V$ = 2.14 x 10^5 J/kg

3-20 Referring to problem 3-19:

(a) Solve for the temperature distribution within the stainless steel wall if a thickness of insulation, L_1, is added to the outer surface of the tank.

(b) By what percent can the boil-off rate be reduced if L_1 = 6 cm and the thermal conductivity of the insulation is .02 w/m-K?

3-21

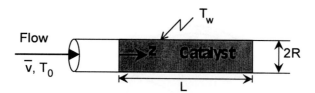

The above sketch shows a tubular flow chemical reactor in which an exothermic reaction takes place (heat of reaction = $\Delta \tilde{H}_r$) at a constant rate, $(R_A)_V$ moles/l³-t. The energy flux leaving at the outer wall of the reactor is proportional to the difference between the reactor temperature and the wall temperature which is maintained at a constant temperature, T_w. If the velocity in the reactor, $\bar{v}$, is constant everywhere and if radial temperature gradients are ignored and molecular transport in the flow direction can be neglected, solve for the steady-state reactor temperature as a function of distance along the reactor. The inlet fluid temperature is at T_0.

3-22 In catalytic reactors such as described in 3-21, the catalyst packing can cause <u>dispersion</u> in the flow direction which can be modeled in the same fashion as molecular transport; i.e, $q_{dz} = - k_d \dfrac{dT}{dz}$

where q_{dz} is the dispersion flux of energy and k_d is the thermal dispersion coefficient. If, in this case, the reactor is adiabatic and all other conditions are as in 3-21.

(a) Solve for the steady-state temperature distribution along the reactor in terms of the arbitrary constants of integration.

(b) Propose the two boundary conditions needed to solve this problem and state the PHYSICAL BASIS for your choice.

3-23

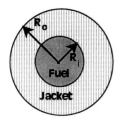

A spherical nuclear fuel pellet produces heat as a result of radioactive decay. The pellet is covered by a spherical jacket of thermal conductivity, k, with the dimensions shown in the sketch. Under steady-state conditions, the inner and outer radii of the jacket are maintained at T_i and T_o respectively. Solve for the temperature distribution within the jacket.

3-24 For the same pellet as in 3-23, assume that the temperature at R_I is not known but that the radioactive energy source is known and equal to $(\dot{Q}_N)_V$. If again, the temperature at R_o is T_o, solve for the steady-state temperature distribution within the jacket.

3-25 In the "Dynamic Radial Heat Flow" method for determining the thermal diffusivity of solids [Sheffield, G.S. and J.R. Schorr, Ceramic Bulletin, $\underline{70}$, No. 1, p. 102, 1991], a long cylindrical sample is heated radially and the heating rate, $\dfrac{dT}{dt}$, is maintained constant. Derive the differential equation which describes this experiment and show that the authors' Equation is valid.

$$\alpha = \frac{R^2 \dfrac{dT}{dt}}{4\Delta T}$$

In this equation, α is the thermal diffusivity and ΔT is the measured temperature gradient between the center of the sample and its surface (i.e., at $r = R$).

3-26 A brass cylinder with $R = 0.5$ cm, L = 15.2 cm, is subjected to an unsteady-state heating such as described in Example 3.2 of the text. Given the brass temperature-time data listed below, determine the heat transfer coefficient between the cylinder and the surrounding temperature bath. [brass properties: $k = 104$ w/m-k, $\hat{c}_p = .38$ J/g-K, $\rho = 8.47$ g/cm^3]

TIME(s)	TEMPERATURE(C)
0	0
25	10
50	27
75	35
100	40
150	47
250	50
350	51

3-27

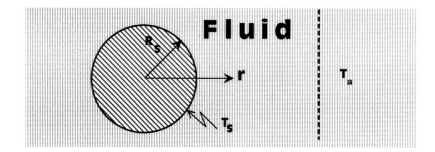

The sphere shown in the above sketch is placed in a stagnant fluid (thermal conductivity = k) under steady-state conditions and its surface temperature is kept constant at T_s. If the temperature of the fluid very far from the sphere is also constant at , T_a:

(a) Solve for the temperature distribution in the fluid.

(b) Solve for the energy flux at the surface of the sphere.

(c) Solve for the heat transfer coefficient, h, between the sphere and the fluid if it is defined as

$$h = \frac{q_{r|_{RS}}}{(T_S - T_\infty)}$$, and then solve for the dimensionless heat transfer coefficient, $\dfrac{h D_s}{k}$, which

is commonly known as the <u>Nusselt Number</u>.

3-28

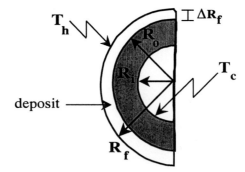

The sketch shows a fouling deposit on the surface of a heat exchanger tube. According to Sanatgar and Somerscales (Chem. Engr. Progr., p.53, December 1991), one viable model for describing the thickness of the deposit with time is the <u>Linear Fouling Model</u> where the thickness varies linearly with time according to $\Delta R_f = at$. If the deposit grows very slowly, the system can be said to be at quasi-steady-state conditions; that is, accumulation terms in the conservation equation can be neglected. Assuming that the temperatures at the edge of the fouling deposit and at the inside wall of the tube are maintained constant at T_h and T_c, respectively,

(a) Solve for the energy flux at R_i as a function of R_i, R_o, and R_f.

(b) Assuming that $\Delta R_f \gg R_o$ and that $\Delta R_f = at$, simplify the answer in (a) to obtain an expression for the energy flux at R_i as a function of time.

CHAPTER 4

Molecular Mass Transport

Without a doubt, mass transport is the most difficult transport process to analyze, if for no other reason than it always involves <u>mixtures</u>. Consequently, we can no longer count on the uniformity of physical properties; rather we must account for the variation of properties which may exist in a given system. When a system contains three or more components, as many industrial fluid streams do, the problem becomes unwieldy very quickly. As you might expect, the engineering approach to multicomponent problems is to attempt to reduce them to representative binary (i.e., two component) problems. As a result, much of the discussion will be confined to binary systems.

Just as momentum and energy transport have two mechanisms for transport- molecular and convective, so does mass transport. However, as we shall see, there are convective fluxes in mass transport, even on a molecular level. The reason for this is that in mass transport, whenever there is a driving force, there is always a <u>net</u> movement of the mass of a particular species which results in a bulk motion of molecules. The flux resulting from this bulk motion is called the <u>bulk diffusion flux</u>. Of course, there can also be convective mass transport due to macroscopic fluid motion. So, in mass transport we can have two types of convective fluxes; one due to molecular motion and one due to fluid motion. We can consider bulk diffusion as either a component of the convective mass transport (since it <u>is</u> a form of convection) or, since it is due to molecular transport, as a component of the molecular mass transport. Since the focus of this chapter is on molecular transport, bulk diffusion will be analyzed in terms of molecular transport.

In an attempt to ask <u>why</u> molecular mass transport takes place, it helps to first consider the conditions under which it will <u>not</u> take place. The answer, of course, is when the system is at equilibrium, which in this case is when the chemical potential of each species is the same everywhere. In this respect, the equality of chemical potential can be thought of as a quantitative measurement of complete randomness of species. Note that it is much the same as many of the other thermodynamic properties; that is, purely an invention to quantify what we have already observed qualitatively.

4-1 COMPONENT FLUXES

In Chapter 2, fluxes were defined in terms of the movement of "aggregate" mixtures. Molecular fluxes corresponded to aggregate mixtures of molecules, which move on a molecular scale, and convective fluxes corresponded to macroscopic aggregate movement. In order to deal with the movement of a particular "species" of molecules, it is necessary to define still another type of flux, the <u>diffusion flux</u>, which is described in Section 4-1.a. As might be expected, the diffusion flux can be related to macroscopic fluxes and their interrelationship is developed in Section 4-1.b.

4-1.a Diffusion Fluxes

If the chemical potential of a given molecular species differs from point to point in a system, then that molecule will tend to move in a direction that would ultimately result in a uniform chemical potential everywhere. Note that this is over and above any macroscopic fluid motion that the system as a whole might have. That is, the motion due to chemical potential gradients is a motion <u>relative</u> to the system motion. The flux of mass of a particular species which moves due to chemical potential gradients is molecular mass transport and is called a <u>diffusion</u> flux. It is associated with a velocity relative to the system velocity.

To quantify this discussion it is necessary to define some terms. First of all, we know from Chapter 2 that a flux can always be expressed in terms of a volumetric concentration multiplied by the velocity of transport. Diffusional fluxes present two problems we have not encountered before. One is that we can consider either mass concentrations, ρ_i, or molar concentrations, C_i (among others) and the other is that the velocity of transport is a <u>relative</u> velocity. The following equations define these fluxes and the nomenclature to be used

<u>Diffusional Mass Flux</u>

$$\vec{j}_i = \rho_i \circ [relative\,velocity] \tag{4.1}$$

Diffusional Molar Flux

$$\vec{J}_i = C_i \circ [relative\,velocity] \tag{4.2}$$

Note also that, unlike energy and momentum transport, it is necessary to identify the particular species, i, which is being transported. To complete these expressions, we must decide what we mean by relative velocity. Since the velocity is relative to the system velocity, a means must be established for calculating the system velocity. If $\vec{v}_i$ is defined as the velocity of i with respect to stationary fixed coordinates, and if the system velocity is some average velocity representative of all the species in the system, then this average should be calculated by "weighting" it according to the concentrations of all the species. Since there are two ways of expressing concentration (mass or molar), there are two ways of calculating average velocities.

Mass Average Velocity

$$\rho\,\vec{v} = \sum_{i=1}^{n}(\rho_i\,\vec{v}_i) \tag{4.3}$$

Molar Average Velocity

$$C\,\vec{v}^* = \sum_{i=1}^{n}(C_i\,\vec{v}_i) \tag{4.4}$$

In these two equations, n is the total number of components in the system and ρ and C represent the total density and the total molar concentration, respectively. For most engineering problems, there will be little difference in $\vec{v}^*$ or $\vec{v}$ and so the mass average velocity, $\vec{v}$, will be used in all further discussions. Thus we may also write

$$\vec{v} = \sum_{i=1}^{n}(\frac{C_i\,\vec{v}_i}{C}) \tag{4.5}$$

With these definitions, equations (4.1) and (4.2) become

$$\vec{j}_i = \rho_i\,(\vec{v}_i - \vec{v}) \tag{4.6}$$

$$\vec{J}_i = C_i(\vec{v}_i - \vec{v}) \tag{4.7}$$

4-1.b System Fluxes

Practically speaking, we do not have a technique for measuring diffusion fluxes directly; the best we can do is back-calculate them. To do this, we must relate them to a flux which can be measured directly; namely, a flux with respect to stationary fixed coordinates. If we define $\vec{N}_i$ as the molar flux with respect to stationary fixed coordinates then, by the definition of a flux

$$\vec{N}_i = C_i \vec{v}_i \tag{4.8}$$

To obtain a relationship between $\vec{N}_i$ and $\vec{J}_i$ let us return to Equation (4.7). If we expand (4.7) and substitute for $\vec{v}$ from (4.5), then

$$\vec{J}_i = C_i \vec{v}_i - C_i \left[\sum_{i=1}^{n} \frac{C_i \vec{v}_i}{C} \right] \tag{4.9}$$

Since $C_i / C = X_i$ and $C_i \vec{v}_i = \vec{N}_i$, (4.9) becomes

$$\vec{N}_i = \vec{J}_i + X_i \sum_{i=1}^{n} \vec{N}_i \tag{4.10}$$

Substituting Fick's Law for $\vec{J}_i$ (Equation 2.19), we obtain for a binary system

$$\vec{N}_A = -C D_{AB} \Delta X_A + X_A (\vec{N}_A + \vec{N}_B) \tag{4.11}$$

The appearance of the second term on the right hand side of (4.11) is a mathematical representation of the <u>bulk diffusion flux</u>, arising from the molecular motion of all of the species which are present. Written in this form, Equation 4.11 is useful for analyzing purely molecular mass transport. However, if the system as a whole is moving due to convective motion, then the mass average velocity will be due to external factors, a pump or gravity, for example. In this case, it makes more sense to use Equation 4.8, and then, 4.5, so that 4.10 can be written as

$$\vec{N}_i = \vec{J}_i + X_i \sum C_i \vec{v}_i$$

$$= \vec{J}_i + X_i C \vec{v}$$

Since $X_i C$ is simply C_i, then the appropriate equation for a convecting system is

$$\vec{N}_i = \vec{J}_i + C_i \vec{v} \tag{4.12}$$

and it is easily seen that the flux of i relative to fixed coordinates consists of two components, a diffusion flux, J_i, and a convective flux, $C_i \vec{v}$. A summary of the diffusion and convective fluxes are shown in Table 4-1.

Table 4-1 Convective and Molecular Mass Transport

<u>Fluxes</u>

	Molar	Mass
Convective	$\vec{f}_{M_i} = C_i \vec{v}$	$\vec{f}_{m_i} = \rho_i \vec{v}$
Molecular	$\vec{J}_i = C_i [\vec{v}_i \ \vec{v}]$	$\vec{j}_i = \rho_i [\vec{v}_i \ \vec{v}]$
	$\vec{N}_i = C_i \vec{v}_i$	$\vec{n}_i = \rho_i \vec{v}_i$

<u>Bulk Velocities</u>

<u>Convection</u>

[Macroscopic Fluid Motion]

$\vec{v}$ - Determined from momentum transport analysis of fluid motion

<u>Diffusion</u>

[Molecular Motion]

$$\vec{v} \equiv \frac{1}{\rho} \sum_i \rho_i \vec{v}_i - \frac{1}{C} \sum_i C_i \vec{v}_i$$

4-2 STAGNANT FILM DIFFUSION

Now let's return to a purely diffusive system and explain the concept of <u>bulk diffusion</u> in a more physical manner. Picture a physical situation as shown in Figure 4-1. Here we have a mixture of A and B to the right of the semipermeable membrane, I - I', which is permeable to species A but impermeable to B. We imagine a "film" of thickness, δ, across which, A can be molecularly transported, from the bulk ($z = \delta$) to the interface. The mole fractions of A and B at the bulk and interface are $X_{A\delta}$ and X_{AI}, respectively. In other words, all of the concentration gradients are localized within the film. If the system is at steady-state and there are no chemical reactions, then N_{Az} and N_{Bz} will be constant[#] and since $N_{Bz} = 0$ at the interface, I - I', it is zero everywhere. Thus, B is stagnant and this situation is often referred to as <u>stagnant film diffusion</u>. If we write (4.11) for both A and B, then for this one-dimensional problem we have

---

[#]   This can be shown by taking a differential mass balance on both A and B, using the methods of Chapter 2.

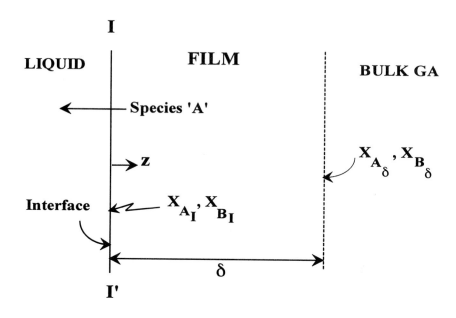

Figure 4-1 Molecular Mass Transport through a Film

$$N_{Az} = J_{Az} + X_A N_{Az} \qquad (4.13)$$

$$0 = J_{Bz} + X_B N_{Az} \qquad (4.14)$$

Now, in order for mass transport of A to take place, $X_{Al\,\delta} > X_{Al\,I}$. Furthermore, since in a binary system, $X_A + X_B = 1$, X_{BI} must be greater than $X_{B\delta}$. If this is true, there must be a diffusion flux of B in the positive z-direction. But we know that there can be no net flow of B, since under steady-state conditions there is no way for B to be replenished at the interface and thus, B is stagnant (i.e.; $N_{Bz} = 0$). Equation (4.14) provides the solution to this dilemma and shows that there is indeed a diffusion flux of B but that it is exactly counterbalanced by $X_B N_{Az}$ so that there is no _net_ motion of B. In other words, the diffusion of B in the positive z-direction is counteracted by the bulk diffusion of B in the negative z-direction. The physical explanation can be found in the fact that as A moves in the negative z-direction, it "carries" some species B with it (bulk diffusion) so that this bulk diffusion continually replaces the B which is tending to move in the positive z-direction by diffusion. This phenomenon also affects the steady-state mass transport of A, for if we solve (4.13) for N_{Az} we find that

$$N_{A_z} = \frac{J_{A_z}}{(1 - X_A)}$$

Thus the fact that there is bulk diffusion not only counterbalances the diffusion of B in the positive z-direction but it also <u>increases</u> the flux of A in the negative z-direction over and above the flux due solely to molecular diffusion (since $1 - X_A < 1$). This phenomena might be viewed as a "snowballing" effect. That is, a number of A molecules want to move in the negative z-direction since the mole fractions are lower there. In doing so, they drag the molecules around them, some of which are also other A molecules. Since the dragged A molecules are also trying to diffuse, the entire motion of A in the negative z-direction is enhanced.

We can more easily see the quantitative effect of bulk diffusion in a stagnant film if we solve for the concentration profile of A within the film. A steady-state differential mass balance for species A within the film yields

$$\frac{dN_{A_z}}{dz} = 0$$

which implies that N_{A_z} is not a function of z and is thus a constant in this steady-state system. Integrating once and substituting for N_{A_z} from Equation (4.14),

$$N_{A_z} = A_1 = \frac{J_{A_z}}{1 - X_A}$$

or, using Fick's Law,

$$N_{A_z} = A_1 = \frac{C D_{AB}}{1 - X_A} \frac{dX_A}{dz} \tag{4.15}$$

A second integration gives

$$\ln(1 - X_A) = \frac{A_1}{C D_{AB}} z + A_2$$

The two boundary conditions pertaining to this system; i.e., at $z = 0$, $X_A = X_{A_1}$ and at $z = \delta$, $X_A = X_{A\delta}$, can be used to determine the constants of integration A_1 and A_2 to obtain an explicit relationship for X_A as a function of z

$$X_A = 1 - (1 - X_{A_I}) \left[\frac{1 - X_{A\delta}}{1 - X_{A_I}} \right]^{\frac{z}{\delta}} \qquad (4.16)$$

We can also obtain a quantitative expression for the flux of A, by recognizing that the constant of integration, A_I, is equal to N_{A_z}. To obtain an expression for N_{A_z}, we can either evaluate A_1 at some specific location (z = 0, for example), or we can integrate Equation (4.15) between limits, as follows

$$\int_0^\delta N_{A_z} = -CD_{AB} \int_{X_{A_I}}^{X_{A\delta}} \frac{dX_A}{1 - X_A}$$

$$N_{A_z} = \frac{CD_{AB}}{\delta} \ln \left[\frac{1 - X_{A\delta}}{1 - X_{A_I}} \right]$$

The other extreme which can occur in a situation such as shown in Figure 4-1 is where the interface (I - I') is also permeable to the transport of species B which is supplied to the left-hand side of I - I' <u>at the same rate</u> at which A diffuses in from the right. As a result, we have <u>Equimolar Counter Diffusion</u> i.e.,

$$N_{A_z} = - N_{B_z}$$

In this case, the bulk diffusion of A in Equation (4.11) drops out and the flux of A across the interface is due only to molecular diffusion. For the same values of δ, $X_{A\delta}$ and X_{A_I}, the mass transport of A in this case will be less than that which would occur if B were stagnant. Physically we can view this as a net motion of B in the positive z-direction which opposes the motion of A in the negative z-direction and offers additional resistance. The net effect of the counterdiffusion is to produce a <u>linear</u> concentration profile for species A (see Example 4-1), whereas the concentration profile pertaining to stagnant film diffusion is <u>nonlinear</u>, as can be seen from Equation 4.16

While we have examined in some detail the two extremes of stagnant and equimolar counter diffusion there are an infinite number of possibilities. In general, Equation (4.10) must be written for each species and solved simultaneously. Of course there must also be some additional information on the relationship between N_A and N_B. If A and B were participants in a steady-state, chemically reacting system, for example, then the stoichiometry might provide such a relationship.[*]

To handle the situation where the relationship between N_A and N_B is completely general (including equimolar counter diffusion or stagnant gas diffusion), let us define

$$\Phi = \frac{\vec{N}_A + \vec{N}_B}{\vec{N}_A} \tag{4.17}$$

In one dimension then, Equation (4.11) becomes

$$N_{A_z} = -\frac{c D_{AB}}{1 - \Phi X_A} \frac{dX_A}{dz} \tag{4.18}$$

EXAMPLE 4-1: Concentration Profiles in Equimolar Counterdiffusion

Determine the concentration distribution for species A within the film shown in Figure 4-1 if species B diffuses opposite and equal to A.

Solution

A differential mass balance for A within the film again gives $\dfrac{d N_{A_z}}{dz} = 0$, or N_{A_z} is a constant (call it A_1) and, by virtue of the fact that $N_{A_z} = N_{B_z}$, Equation (4.11) becomes

$$N_{A_z} = C D_{AB} \frac{dX_A}{dz} + X_A[N_{A_z} + (-N_{A_z})]$$

$$A_1 = C D_{AB} \frac{dX_A}{dz}$$

which can be integrated directly, using the same two boundary conditions given above, to give

$$X_A = (X_{A\delta} - X_{A1}) \frac{z}{\delta} + X_{A1}$$

---

[*]   For example, if $2N_A \rightarrow N_B$, then, since there can be no accumulation in a steady-state system, $\vec{N}_A = -2\vec{N}_B$

In this case, the flux, $N_{A_z} = A_1$, is simply

$$N_{A_z} = C D_{AB} (X_{A_\delta} - X_{A_1})$$

Equation (4.18) is thus the starting point for the general case of diffusional mass transport in one dimension. Note that

$\Phi = 0$: EQUIMOLAR COUNTER DIFFUSION

$\Phi = 1$: STAGNANT DIFFUSION

4-3 DIFFUSION IN A CYLINDRICAL PORE

In this section, we analyze diffusion of a gaseous reactant in a cylindrical catalyst pore which is the mass transport equivalent to the molecular energy transport situation discussed in Chapter 3.[#] The physical description of the phenomena is presented in Figure 4-2 which shows a spherical, porous solid catalyst and an enlargement of a typical pore in the catalyst. Heterogeneous catalysts (in this case a solid) are used to promote the rates of fluid phase (usually a gas) chemical reactions which would be undetectably slow under the existent operating conditions of temperature and pressure. One of the keys to a good catalyst is to be able to disperse small quantities of the "active" species (often, a noble metal) over a large surface area. In order to accomplish this, the active species are anchored to a porous support material. The support materials usually have excellent thermal stability and the surface areas of some of the supports can be in excess of 1000 m^2/g. This large surface area is possible as a result of the very fine pores which can exist in catalyst supports (diameters on the order of microns). Of course, once the active species has been successfully dispersed over the support, it is still necessary that the gaseous reactants be able to reach these sites. Because of the small diameters and excessive lengths of the pores, the molecular transport (diffusion) of the reactants to the active sites can be the limiting rate step in many catalytic systems.

As should be realized, the straight cylindrical pore shown in Figure 4-2 is an idealization. But nevertheless, describing the mass transport in such a system can lend valuable insight into the parameters which affect the overall catalytic reaction rates and even suggest methods by which to improve the process.

[#]   It is a worthwhile exercise to attempt to compare each aspect of the analogy as we proceed through the analysis.

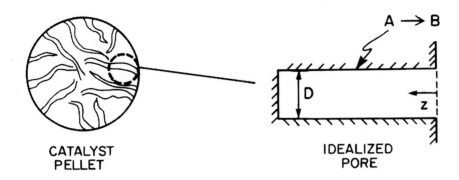

Figure 4-2 Porous Catalyst Pellet

Let us assume we are dealing with a reaction, A $\rightarrow$ B, where the rate of the reaction, $(R_A)_s$ is proportional to the concentration of A[*] and to the surface area of the catalyst. Because the pores have such small diameters, it is not unreasonable to assume that there are no radial concentration gradients. However, the actual catalytic reaction takes place at the walls of the pores and thus, as A diffuses down the pore, it will be consumed by the reaction at the walls.

Just as in the case of the cooling fin problem in Chapter 3, we must account for the consumption of A in some other way since we are not allowing for radial concentration gradients. Again, we can do this by treating the reaction as if it were occurring <u>uniformly</u> within the pore volume. Thus we have a uniform "sink" for A

$$(R_A)_V = (R_A)_s \; \frac{\pi D L}{\dfrac{\pi D^2 L}{4}} = \frac{4 k_r}{D} C_A \qquad (4.19)$$

where k_r is the proportionality constant (the *rate constant*) and $(R_A)_V$ has the units of moles "A" reacting per unit time per unit volume of pore.

With these assumptions and simplifications, the mathematical description is reduced to a one-dimensional system and thus the differential volume element is identical to that shown in Figure 3-2 with the molecular energy fluxes replaced by molar fluxes with respect to fixed coordinates (see Figure 4-3).

---

[*] This is called a <u>first order reaction</u>; i.e., a reaction where the rate is proportional to the first power of concentration [1].

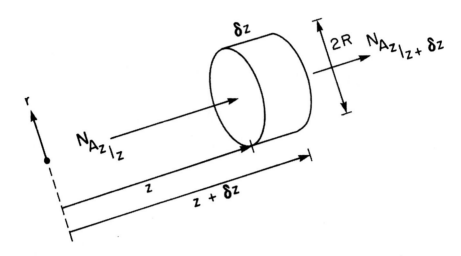

Figure 4-3 Cylindrical Pore Volume Element

The individual terms in the general conservation Equation (2.20) are, for steady-state conditions:

ACCUMULATION $= 0$ (steady-state)

INPUT $= N_{Az \mid z} \dfrac{\pi D^2}{4} \delta t$

OUTPUT $= N_{Az \mid (z + \delta z)} \dfrac{\pi D^2}{4} \delta t$

SINK $= (R_A)_V \dfrac{\pi D^2 \delta z}{4} \delta t$

Substituting the above relations into Equation (2.20), dividing through by the volume of the differential volume element and δt, and then taking the limit as $\delta z \rightarrow 0$ gives us the applicable differential equation

$$-\frac{d}{dz} N_{Az} - (R_A)_V = 0 \qquad (4.20)$$

The flux can be related to the one-dimensional concentration gradient by employing Equation (4.18) and recognizing that, for the one-to-one stoichiometry of this reaction, N_{Az}

must equal - N_{B_z} under steady-state conditions. Since, in this case, $\Phi = 0$, direct substitution of (4.18) into (4.20) gives

$$\frac{d}{dz}(C D_{AB}\frac{dX_A}{dz}) - (R_A)_V = 0 \tag{4.21}$$

Since the reaction rate is first order and thus dependent on C_A, we can express Equation (4.21) in terms of one dependent variable, C_A, by using Equation (4.19)

$$D_{AB}\frac{d^2 C_A}{dz^2} - \frac{4 k_r}{D}C_A = 0 \tag{4.22}$$

where we have also assumed that both the total concentration, C, and the diffusivity are constant. At this point, it is necessary to discuss the diffusivity in more detail. Although we are dealing with a binary system, the use of the molecular diffusivity, D_{AB}, may not be appropriate for two reasons. For one, if the pore diameters are very small, molecules of A may collide with the wall more frequently than they collide with other molecules. In that case we have <u>Knudsen</u> diffusion and the diffusivity depends on the pore diameter.[*] A second complication is due to our idealization of the pores as straight and cylindrical (they are usually irregular and tortuous). To compensate for this, an "<u>effective</u>" diffusivity, D_{eff}, is normally utilized and can be related to the molecular diffusivities as well as to the physical parameters of the pores [2]. For our purposes we will simply use D_{eff} with no further comment.

Since we have a second-order differential equation, we will need two boundary conditions. Physically, we have a constant concentration of A at the entrance to the pore and a zero flux at the end of the pore; the latter as a result of the relatively small reaction rate there. Mathematically, these can be expressed as

$$at\ z=0;\ \ C_A = C_{A0}$$

$$at\ z=L;\ \ \frac{dC_A}{dz} = 0$$

Equation (4.22) and its associated boundary conditions can be placed in a form which is identical to Equations (3.7) - (3.9) in Chapter 3. Thus, for the catalyst pore we have

$$\frac{d^2 C_A}{dz^2} - (\Gamma_c)^2 C_A = 0 \tag{4.23}$$

---

[*]   See Chapter 6 for additional discussion and the criterion to decide whether or not Knudsen diffusion prevails.

where $\Gamma_c = \sqrt{\dfrac{4\,k_r}{D\,D_{e\!f\!f}}}$

While in the Cooling Fin, the equation is

$$\frac{d^2\,\theta}{dz^2} - \Gamma_h{}^2\,\theta = 0 \tag{4.24}$$

where $\Gamma_h = \sqrt{\dfrac{4\,h}{k\,D}}$

As a result, the solution obtained there is absolutely applicable here once we identify the differences in the variables and parameters. Table 4-2 shows this comparison, term-by-term, so that the solution is

$$\frac{C_A}{C_{A0}} = \cosh(\Gamma_c\ z) - \tanh(\Gamma_c\ L)\ \sinh(\Gamma_c\ z) \tag{4.25}$$

Table 4-2 Cooling Fin - Catalyst Pore Analogy

Cooling Fin	Catalyst Pore
$\theta = T - T_a$	C_A
k	$D_{e\!f\!f}$
h	k_r
$\Gamma_h = \sqrt{\dfrac{4h}{kD}}$	$\Gamma_c = \sqrt{\dfrac{4\,k_r}{D_{e\!f\!f}\,D}}$
T_s	C_{A0}

Not only is the solution for the concentration profile the same, but so is the expression for the "fin efficiency." However, in catalysis jargon it is called the *effectiveness factor.* Nevertheless its meaning is the same; viz., how effectively are we using the surface area, in this case, the internal pore area? Again, if all of the reaction takes place near the pore opening, then the catalytic species within the pores is largely underutilized. If the active species happens to be a noble metal, having a high effectiveness factor can mean a large savings in catalyst cost. As was done during the discussion of the cooling fin, lets look at the dependence of the effectiveness factor (η) on the parameters and properties of the catalyst. The variation of η with Γ_c is as shown in Figure 3-7, with Γ_h replaced by Γ_c. Note then, in order to have a high

η, we require a low value of Γ_c (low k_r, high D_{eff}, low L). If Γ_c is interpreted as a measure of the relative rates of reaction versus diffusion, then high effectiveness factors will result whenever the diffusion rates are higher than the reaction rates. There is not much one can do to influence k_r or D_{eff} in a catalyst design; the former is a consequence of the active species being employed and the latter is due to the high surface area of the support. Reducing either will give lower intrinsic catalytic rates and is usually not the goal of catalyst design. However, we do have some control over the length of the pores, since they will be roughly proportional to the size of the spherical catalyst pellet. In other words, smaller pellets mean a shorter diffusion path and a higher effectiveness factor.

4-4 MASS TRANSPORT ACROSS A CYLINDRICAL MEMBRANE

A relatively new method of gaseous separation is membrane separation [3]; so called because gases are separated from one another by virtue of their relative ability to diffuse across a porous membrane. The membranes are usually polymeric but, because of their low temperature limitations, newer ceramic membranes are also being developed [4]. Figure 4-2 shows an illustration of such a separation scheme where a mixture of H_2 and N_2 is fed through a cylindrical tube with a porous membrane wall. A concentric outer tube contains a flowing mixture of H_2 and CO which will pick up the H_2 as it diffuses across the wall from the inner tube. Of course all of the gaseous molecules have a tendency to diffuse across the wall (as long as there is a driving force), it's just that smaller molecules will do so at a faster rate. So, while in general it is necessary to analyze for all of the diffusing species, we will constrain ourselves to the analysis of H_2; specifically, to derive an expression for the molar flow rate of H_2 per unit length of tube.

Let us assume that, over the tube length of interest, the concentrations of all of the species in the two gas phases are constant at C_{As}, C_{Cs} in the outer tube and C_{A0}, C_{B0} in the inner tube; where A, B and 'C' are H_2, N_2, and CO, respectively. Since there are no concentration gradients in the flow directions, we have one-dimensional radial mass transport across the wall and the differential volume element is as shown in Figure 4-2.

At steady-state, and since there are no chemical reactions within the wall

ACCUMULATION = SOURCES = SINKS = 0

INPUT = $N_{Ar \mid r} \, 2\pi r L \, \delta t$

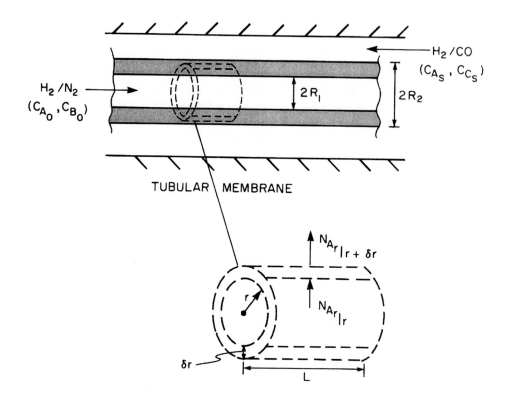

Figure 4-4 Membrane Separator

$$\text{OUTPUT} = N_{Ar \mid (r+\delta r)} \, 2\pi(r + \delta r)L \, \delta t$$

Note that, in this case, the area perpendicular to the direction of transport changes with r, and thus we must account for this change in the differential conservation balance. Placing the above terms into Equation (2.16), dividing through by the volume of the differential element, $2\pi r \delta r L$, and making use of the approach illustrated in Example 3.2,

$$\frac{(r N_{Ar}) \mid_r - (r N_{Ar}) \mid_{(r+\delta r)}}{r \delta r} = 0$$

so that, when we take the limit as $\delta r \to 0$, we obtain

$$-\frac{1}{r}\left[\frac{d}{dr}(r N_{Ar})\right] = 0 \qquad (4.26)$$

Equation (4.26) can be integrated directly to get

$$(r\, N_{Ar}) = CONSTANT = A_1 \tag{4.27}$$

Since our objective is to calculate the molar flow rate of A per unit length entering the outer tube and since

$$\frac{\dot{N}_A}{L} = N_{A\,|\,R_2}\, 2\pi\, R_2 \tag{4.28}$$

all we need to do is determine the integration constant, A_1 (since from Equation (4.27), $N_{A\,|\,R_2} = \dfrac{A_1}{R_2}$). However our only boundary conditions deal with the concentration of A at either side of the wall; viz.,

$$\text{at } r = R_1, \quad C_A = C_{A0}$$
$$\tag{4.29}$$
$$\text{at } r = R_2, \quad C_A = C_{As}$$

Thus we need to relate N_{Ar} to the concentration of A, which we can do, using Equation (4.10); which in this case, takes the form

$$N_{Ar} = J_{Ar} + X_A(N_{Ar} + N_{Br} + N_{Cr})$$

The mechanism of diffusion through the porous wall is fairly complex but we can expect that the mole fraction of A will be small there; so that

$$N_{Ar} \sim J_{Ar}$$

Furthermore, the diffusivity of A through the wall will depend on the material properties of the wall and thus there will be a particular diffusivity of A through any particular membrane; call it D_A, and using Fick's Law for constant C and D_A, Equation (4.27) becomes

$$\frac{dC_A}{dr} = -\frac{A_1}{D_A}\frac{1}{r} \tag{4.30}$$

which, upon integration, yields

$$C_A = -\frac{A_1}{D_A} \ln r + A_2$$

Now, applying the boundary conditions, Equation (4.29)

$$C_{A0} = -\frac{A_1}{D_A} \ln R_1 + A_2$$

$$C_{As} = -\frac{A_1}{D_A} \ln R_2 + A_2$$

If we eliminate A_2 between these two equations, solve for A_1 and substitute it into (4.27)

$$N_{Ar \mid R_2} = \frac{D_A}{R_2} \frac{(C_{A0} - C_{As})}{\ln \dfrac{R_2}{R_1}}$$

which, in view of (4.28), gives the desired molar flow rate per unit length of tube

$$\frac{\dot{N}_A}{L} = 2\pi D_A \frac{(C_{A0} - C_{As})}{\ln \dfrac{R_2}{R_1}} \tag{4.31}$$

Notice that it was not necessary to determine A_2 and the complete concentration profile in order to obtain the desired result.

One further point can be made here, and that concerns the radial geometry in Figure 4-4. Quite often, the unit operations calculational approach to problems such as these is to attempt to "force" all rate processes into the form

$$\dot{N}_A = (AREA)\, D_A \frac{\Delta C_A}{\Delta R} \tag{4.32}$$

In comparing Equation (4.32) with (4.31), we can see that accounting for the radial geometry results in a logarithmic term in the denominator. In order to get Equation (4.31) into the form of Equation (4.32), we can multiply and divide Equation (4.31) by ΔR (i.e., $R_2 - R_1$) so that it becomes

$$\dot{N}_A = D_A\, A_{lm} \left[\frac{(C_{A0} - C_{As})}{R_2 - R_1} \right] \tag{4.33}$$

where

$$A_{lm} \equiv 2\pi L \frac{(R_2 - R_1)}{\ln \dfrac{R_2}{R_1}} = \frac{A_2 - A_1}{\ln \dfrac{A_2}{A_1}}$$

which is known as the <u>log-mean area</u>.

4-5 DISSOLUTION OF A SPHERE IN A QUIESCENT FLUID

Consider a solid sphere, of radius R_s, which is placed in a large body of unstirred, aqueous solution as shown in Figure 4-5. The dissolution of the solid A, is slow enough that quasi-steady-state conditions can be assumed. Given that the solubility of A in the solution is C_A^*, we wish to derive an expression for the concentration distribution of A in the solution and to develop a mathematical relationship for the molar flow rate of A, entering the solution at the surface of the sphere.

First of all, we have spherical geometry and so the applicable differential volume element is as shown in Figure 4-5. Note that the analysis is being applied to the aqueous <u>solution</u> and, under steady-state conditions and with no chemical reactions producing or consuming A,

ACCUMULATION = SOURCES = SINKS = 0

INPUT = $N_{Ar}|_r \, 4\pi \, r^2 |_r \, \delta t$

OUTPUT = $N_{Ar}|_{(r + \delta r)} \, 4\pi \, r^2 |_{(r + \delta r)} \, \delta t$

Placing these terms into Equation (2-20), dividing by the differential volume and δt, and taking the limit as $\delta r \to 0$

$$\lim_{\delta r \to 0} \frac{[\,(r^2 N_{Ar})|_r - (r^2 N_{Ar})|_{(r + deltar)}\,]}{r^2 \delta r} = -\frac{1}{r^2} \frac{d}{dr}(r^2 N_{Ar}) = 0 \qquad (4.34)$$

Integrating once, we obtain

$$r^2 N_{Ar} = A_1$$

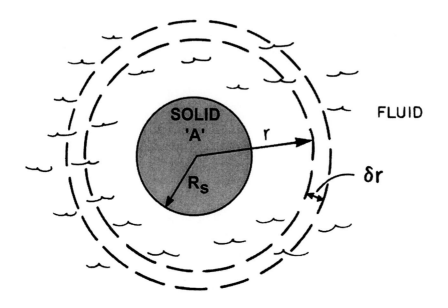

Figure 4-5 Mass Transport From A Sphere

Since we do not have any boundary conditions in terms of the molar flux, we need to use Equation (4.10) to relate N_{Ar} to concentration and, assuming that the solubility of A in the aqueous solution is low (i.e., $X_A \sim 0$),

$$N_{Ar} \approx J_{Ar} = -D_{AB} \frac{dC_A}{dr} = \frac{A_1}{r^2} \tag{4.35}$$

Integrating once more,

$$C_A = \frac{A_1}{D_{AB}} \frac{1}{r} + A_2 \tag{4.36}$$

In order to evaluate the two constants of integration, let's now look at the boundary conditions which apply to this problem. First of all, right at the interface between the solid and the solution, it is reasonable to assume that saturation (i.e., <u>equilibrium</u>) conditions prevail and

$$\text{at } r = R_s, \ C_A = C^*_A$$

where C_A^* is the saturation concentration of A (solubility) in the solution.

Since the sphere is placed in a very large body of solution, it stands to reason that at some distance removed from the sphere, $C_A \sim 0$. In fact, as far as the sphere is concerned, this distance might as well be at infinity, so we can write

$$\text{at } r = \infty, \quad C_A = 0$$

Applying the second boundary condition to (4.36), we find that

$$0 = -\frac{A_1}{D_{AB}} \frac{1}{\infty} + A_2$$

or, $A_2 = 0$. Now A_1 can be obtained from the first boundary condition; viz.,

$$C_A^* = \frac{A_1}{D_{AB}} \frac{1}{R_s}$$

$$A_1 = R_s D_{AB} C_A^*$$

The mathematical expression for the concentration distribution of A in the solution, Equation (4.36), is then

$$C_A = R_s C_A^* \frac{1}{r}$$

The flux of A at the surface of the sphere is, from Equation (4.35),

$$N_{Ar} \mid_{R_s} = \frac{A_1}{R_s^2} = \frac{D_{AB} C_A^*}{R_s} \tag{4.37}$$

Recall that in Chapter 3 we employed boundary conditions which utilized empirical observations that the energy flux was proportional to the temperature driving force and the proportionality coefficient was called the <u>heat transfer coefficient</u>. We can formulate a similar approach in this problem and define a <u>mass transfer coefficient</u>, k_{mc}, so that the molar flux at R_s is proportional to the concentration driving force between the surface of the sphere and the bulk fluid; viz.

$$N_{Ar} \mid_{R_s} = k_{mc} (C_A^* - 0) \tag{4.38}$$

Equating (4.37) with (4.38) and solving for k_{mc}

$$k_{mc} = \frac{D_{AB}}{R_S} \qquad (4.39)$$

If we now multiply Equation (4.39) through by $\dfrac{D}{D_{AB}}$, we can obtain a <u>dimensionless</u> mass transfer coefficient, Sh, which is known as the *Sherwood Number*[#]

$$Sh = \frac{k_{mc} D}{D_{AB}} \qquad (4.40)$$

We will find this to be very useful when we address convective mass transfer topics in general. Note that, for this particular problem, it is a constant equal to 2.

4-6 NonLinear Mass Transport - Ternary Film Diffusion

As was done in Chapter 3, we can solve more complex (and usually, more realistic) mass transport problems by utilizing the powerful software packages that are available to us. As an example, consider the molecular transport situation illustrated in Figure 4-1. Rather than the binary system pictured there, assume that we have a three-component system where two of the species in the bulk gas phase are being transported to the liquid phase and the third gaseous component is stagnant. A physical example of this would be the scrubbing of an "acid" gas with an aqueous caustic solution; for example, a gaseous mixture of CO_2, SO_2 in N_2. As the CO_2 and SO_2 pass across the interface into the liquid phase, they react instantaneously with the caustic so that the mole fraction of each at the interface is essentially equal to zero.

A differential mass conservation balance taken for each species <u>within the film</u>, yields

$$-\frac{dN_{A_z}}{dz} = -\frac{dN_{B_z}}{dz} = -\frac{dN_{C_z}}{dz} = 0 \qquad (4.41)$$

where A, B and 'C' represent CO_2, SO_2, and N_2, respectively. Note that, since the chemical reactions take place in the <u>liquid</u> phase, there are no sources or sinks within the gaseous film. In view of Equation (4.41) we can see (by integration) that all of the molar fluxes are constant in this problem. However, the specific constant for N_{C_z} is equal to zero since N_2 is not soluble

---

[#]   The Sherwood Number is sometimes referred to as th*e* <u>Nusselt number for mass transfer</u>

in the liquid phase and is therefore stagnant. That is, since N_{c_z} is zero at the interface, it is zero everywhere. Applying Equation (4.11) to each of the three species, we obtain

$$N_A = -CD_{Am}\frac{dX_A}{dz} + X_A(N_A + N_B)$$

$$N_B = -CD_{Bm}\frac{dX_B}{dz} + X_B(N_A + N_B) \tag{4.42}$$

$$0 = -CD_{Cm}\frac{dX_C}{dz} + X_C(N_A + N_B)$$

where the z-subscript has been dropped for clarity and the D_{im} symbols indicate multicomponent diffusivities which, in general, are functions of the concentrations of all of the species. As will be discussed in Chapter 6 [see Equation (6.40)], one method of calculating these values is[*]

$$\frac{1}{D_{im}} = \sum_{\substack{j=1 \\ j \ne i}}^{n} \frac{X_j}{D_{ij}} \tag{6.40}$$

where the D_{ij} are the individual binary diffusivities.

In order to determine the concentration profiles through the film for each of the species, Equations (4.42) must be solved simultaneously. Furthermore, the differential equations are non-linear by virtue of the dependency of D_{im} on concentration. In order to utilize MATLAB for the solution we need to rearrange Equations (4.42) so that

$$\frac{dX_A}{d\eta} = \frac{P_1}{D_{Am}}[(X_A - 1) + P_2 X_A]$$

$$\frac{dX_B}{d\eta} = \frac{P_1}{D_{Bm}}[P_2(X_B - 1) + X_B] \tag{4.43}$$

$$\frac{dX_C}{d\eta} = \frac{P_1}{D_{Cm}}[X_C (1 + P_2)]$$

---

[*] Strictly speaking, this method is only applicable to dilute concentrations of i.

where $P_1 = \dfrac{\delta\, N_A}{C}$, $P_2 = \dfrac{N_B}{N_A}$ and $\eta = \dfrac{z}{\delta}$.

The boundary conditions necessary for solving Equations (4.42) are

$$X_A = 0$$

at $\eta = 0$: $X_B = 0$

$$X_C = 1$$

It should be appreciated that it is not necessary to solve all three of the differential equations since we also have the independent condition that $\sum_i X_i = 0$. In any case, the solution will depend on the two parameters, P_1 and P_2. The parameter P_1, is a function of the mass transfer rate and, as will be seen in later chapters, can be related to the prevailing mass transfer coefficient. The parameter, P_2, on the other hand, can be found by iterating until the concentrations at the boundary of the film ($z = \delta$, or $\eta = 0$) equal the known values at that point. For our purposes here, let's solve for the relative mass transfer rate, P_2, which will yield equal concentrations of A and B at the edge of the film, at a given value of P_1. The insert, below, shows the MATLAB program to produce these solutions and Figure 4-6 shows the concentration profiles for P values of -.05 and -.20.[#] The binary diffusivity values listed in the MATLAB program (in units of cm²/sec) were either obtained from [5] (D_{AC}) or estimated from the methods presented in Chapter 6.

MATLAB Program For Ternary Diffusion in A Film

M - File

```
function dX=ternary(n,X)
X=zeros(3,1);
DAB=0.24; DAC=00.14; DBC=.34;
P2=1.8;
P1=-.05;
dX(1) = (P1/(1/(X(2)/DAB+X(3)/DAC)))*(X(1)*(P2+1)-1);
dX(2) = (P1/(1/(X(1)/DAB+X(3)/DBC)))*(X(2)+P2*(X(2)-1));
dX(3) = -(dX(1)+dX(2));
```

Command File

[#]   Note that the values of P are negative, since N_A is a vector in the negative z-direction.

```
[n,X]=ode45('ternary',[0 1],[0 0 1])
ploty(n,X)
```

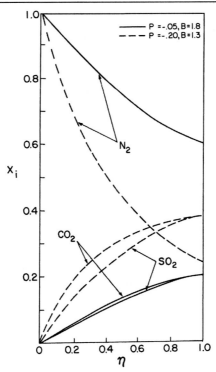

Figure 4-6 Concentration Profiles - Ternary Diffusion

References

[1]   Fogler, H.S., *Elements of Chemical Reaction Engineering*, Prentice-Hall, Englewood Cliffs, NJ, 1986, p.65

[2]   Satterfield, C.N., *Heterogeneous Catalysis in Practice*, McGraw-Hill, N.Y., 1980, pp. 334-344

[3]   Lonsdale,H.K., J. Membrane Sci., 10, 1982, pp 81-181

[4]   Rousseau, R.W., *Membrane Separations*, McGraw-Hill, N.Y., 1990

[5]   Perry, R.H. and C.H. Chilton, "Chemical Engineer's Handbook", 5th Ed., McGraw-Hill, N.Y., 1973, p. 3 - 222

PROBLEMS[*]

4-1   Using Equation (4.10) and the phenomonelogical law for molecular transport, show that $D_{AB} = D_{BA}$ for a binary system.

4-2

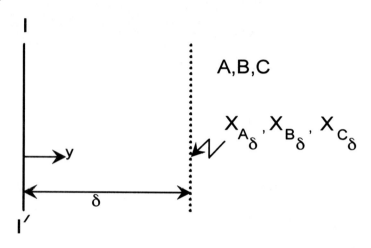

The above sketch shows steady-state diffusion of three components through a film. If there are <u>no</u> chemical reactions within the film and, at the interface,

$$N_{Ay} = -N_{By}, \quad N_{Cy} = 0$$

answer (<u>with</u> explanations) the following questions:

(a) Does $N_{Ay} = J_{Ay}$?

(b) Does $N_{By} = J_{By}$?

(c) Is X_A vs. y linear in the film?

(d) Is X_B vs. y linear in the film?

(e) How does X_C vary with y in the film

---

[*] C* indicates that the problem requires a computer solution

4-3

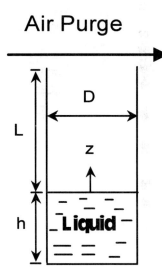

Air Purge

One method of measuring binary diffusivities of vapors in another gas is to place a volatile liquid (species `A`) at a controlled temperature at the bottom of a long graduated capillary cylinder ($L \ll h$) and allow it to vaporize. The vapor then diffuses through the gas (species `B`) above it. The vapor pressure of `A` is P_A^{*} and a purge of the other gas is gently blown across the top of the tube to ensure that the concentration of `A` at that point is zero. Assuming that the liquid level changes very slowly with time, the accumulation term in the differential mass balance can be neglected (*quasi-steady-state* assumption).

(a) Solve the resulting differential equation for $X_A(z)$ and then relate the solution to the measurement of the liquid level as a function of time so that the binary diffusivity of 'A' in 'B' can be calculated from the measurement.

(b) The following data were obtained for h as a function of time with ethanol as the liquid and nitrogen as the gas and at a constant temperature of 40 °C. Using these data and the results of (a), calculate the diffusivity of ethanol-in-nitrogen at 40 °C when the atmospheric pressure is 690 mm Hg.

L vs. Time (Ethanol at 40 C)

Time(min)	L(cm)	Time(min)	L(cm)
0	1.604	570	2.508
90	1.744	1290	3.363
450	2.372	1470	3.544

4-4   For approximately 35 years, quantities of PCBs have been entering Lake Michigan. Analysis of a core sample of the bottom mud yielded the data shown in the table.

(a) Normalize the data with respect to the PCB concentration at the lake floor and the depth at L = 9.25 cm and compare this concentration profile with that predicted by a steady-state diffusion model in the absence of SOURCES/SINKS. Are the data consistent with such a model ?

(b) Polar organic molecules can often be adsorbed by soils. Re-solve the problem using a first-order SINK; i.e., where $R_{Av} = k C_A$ (C_A = PCB concentration) to show that the normalized profile will depend on $\dfrac{z}{L}$ as well as on $\Gamma_D = \sqrt{\dfrac{k L^2}{D_A}}$.

(c) Compare the experimental value of the molar flux at the lake floor with that predicted by the model in (b) and obtain an estimated value for Γ_D .

DEPTH (fm lake floor) (cm)	PCB Concentration μ g/g
0	.36
.5	.29
.75	.20
1.25	.11
2.25	.14
3.25	.08
4.25	.09
9.25	.04

4-5

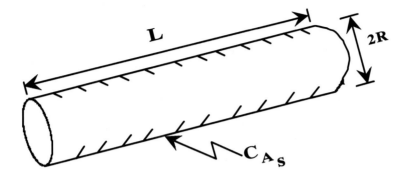

The above sketch shows a porous *extrudate* catalyst; i.e., a cylindrical catalyst where $L/2R \ll 1$. The catalyst is being used to catalyze a zero order chemical reaction, $A \rightarrow B$; that is, $R_{Av} = constant = k$. Rather than analyze the molecular transport in a single pore (as was done in Section 4-3) it can also be modelled as simultaneous molecular transport (with an effective diffusivity, D_{eff}) and chemical reaction in a porous cylinder. If the concentration of `A` at the catalyst surface is C_{AS}

(a) Solve for the steady-state concentration of `A` within the catalyst.

(b) Solve for the average concentration of `A` within the catalyst.

4-6

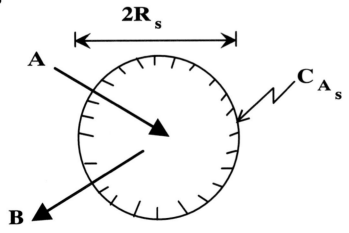

The spherical catalyst shown in the sketch is being used to catalyze the first-order reaction, $A \rightarrow B$, with a reaction rate, $R_{Av} = k \, C_A$. Analyze the problem as simultaneous molecular transport and chemical reaction in a porous sphere and

(a) Solve for the concentration of `A` within the catalyst.

[HINT: The differential equation can be solved by making the substitution $C_A = \dfrac{\Psi}{r}$ and then use the total derivative for $\Psi = f(C_s, r)$]

(b) Obtain an expression for the effectiveness factor of the catalyst.

4-7

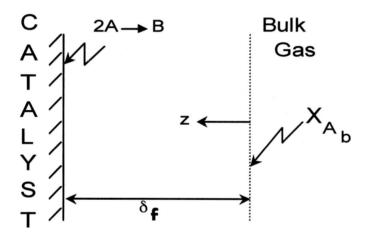

The above sketch shows the region very close to a non-porous flat catalyst surface where the reaction $2A \rightarrow B$ takes place. The species, `A` and `B` move between the bulk gas (at the edge of the film) and the catalyst surface by means of molecular transport. Assuming that the molar flux of 'A' is known (from reaction rate measurements), solve for the steady-state mole fraction of `A` at the catalyst surface in terms of C, D_{AB}, X_{Ab}, N_{A_z} and δ_f, where X_{Ab} is the mole fraction of A at the edge of the film.

4-8

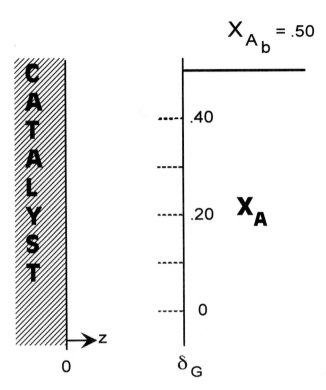

The figure shows a region very close to the surface of a solid catalyst where a reaction, $A \rightarrow 2B$, occurs at the catalyst surface; i.e., at $z = 0$. If the bulk gas contains a 50 - 50 molar mix of `A` and `B` ($X_{Ab} = .5$) , answer the following questions for steady-state conditions

(a) Express N_{A_z} in terms of the concentration gradient of `A`

(b) Using the superimposed grid shown in the figure, sketch the concentration profile for `A` if the reaction is instantaneous.

(c) Repeat (b) if the reaction occurs at a finite rate.

(d) Repeat (b) if the surface reaction is instantaneous but some reaction also occurs in the gas film and is first order with respect to `A`.

NOTE: For (b) - (d), pay attention to the <u>shape</u> of the profiles.

4-9

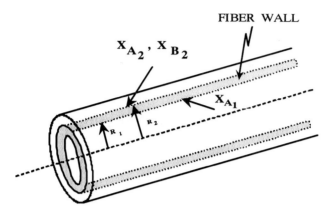

The sketch shows a cylindrical hollow fiber with membrane walls. Species `A` is undergoing steady-state difffusion through the membrane wall where its concentrations at the inner and outer membrane walls are C_{A_1} and C_{A_2}, respectively. It can be assumed that the membrane wall eliminates any bulk diffusion effects and that the diffusion flux follows Fick's Law with a membranc diffusion coefficient, D_{AM}.

(a) Solve for the concentration profile of `A` in the membrane wall in terms of the dimensions of the fiber and the known concentrations at R_1 and R_2.

(b) Solve for the molar flux of `A` leaving the outer wall of the membrane.

4-10 Referring to the hollow fiber unit in 4-9, assume that the membrane is *reactive*; that is, as `A` diffuses through the membrane it catalyzes a reaction, $A \rightarrow 2B$, with a zero order reaction rate ($R_{Av} = constant$). If all other conditions are the same, obtain an expression for the molar flux of `A` leaving the wall and compare it to the solution in 4-9 to obtain the <u>enhancement factor</u> of the reactive membrane; i.e., the ratio of fluxes with and without reaction.

4-11 Re-solve the spherical dissolution problem discussed in Section 4-5, <u>without</u> assuming that $X_A \approx 0$ (i.e., assume that there *is* bulk diffusion).

4-12

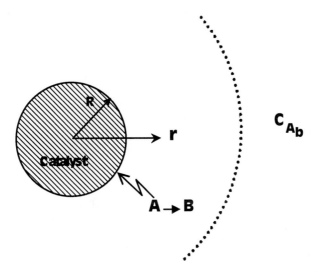

A catalytic reaction takes place at the surface of the nonporous spherical catalyst pellet shown in the sketch. The surface reaction rate, R_{A_s} (moles/area-t) is first order with respect to `A` with a reaction rate constant, k_s (l/t) and the concentration of `A` far from the surface is equal to the bulk concentration, C_{Ab}.

(a) Solve for the concentration of `A` within the fluid.

(b) If the reaction rate is very fast relative to the diffusion rate ($k_s >> D_{AB}$), sketch the concentration profile, $\dfrac{C_A}{C_{Ab}}$.

4-13

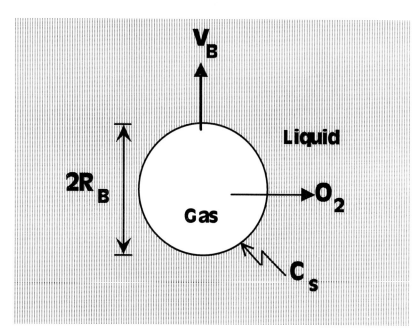

Bischoff et. al. (Chem.. Engr. Sci., <u>45</u>, p. 3115, 1991) have devised a clever means of injecting single gas bubbles into the bottom of a column of liquid in order to measure mass transfer coefficients as a function of distance from the bottom of the column. Such data are important for various applications such as the aeration of waste streams and fermentor technology. Derive a differential equation which describes the mass transfer between the bubble (see sketch) and the surrounding liquid as it rises up the column (bubble velocity = v_B) and then solve for the concentration of oxygen within the bubble, C_B, as a function of its time in the column. It can be assumed that the concentration of oxygen in the bulk liquid is constant at C_L, and that the mass transport rate between the bubble and the bulk liquid is proportional to the bubble surface area (constant) and the oxygen driving force between the concentration at the liquid-side of the bubble interface (C_s) and C_L, where the proportionality constant is the mass transfer coefficient, k_{mc}.

C_s is related to the concentration in the bubble by *Henry's Law*; i.e., $C_s = \dfrac{P_B}{H}$, where P_B is the partial pressure of oxygen in the bubble.

4-14

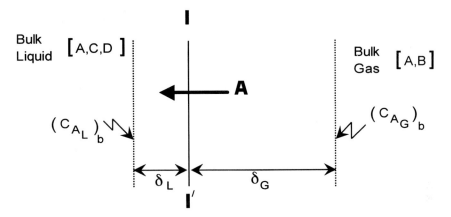

In a gas-liquid mass transfer apparatus, `A` is removed from the gas (containing `A` and `B`) through the gas-liquid interface which is impermeable to `B`. When `A` enters the liquid, it reacts with species `C` to form `D` ($A + C \rightarrow D$) at a constant (zero order) reaction rate, R_{Av}. Assume that the mole fraction in the liquid is very low and that the equilibrium conditions at the interface are

$$C_{AG}^{*} = m \, C_{AL}^{*}$$

If the system is modeled in terms of the *two film theory* as shown in the sketch,

(a) Solve for the molar flux of `A` at the gas-side of the interface in terms of $(C_{AG})_b$ and C_{AG}^{*}.

(b) Solve for the molar flux of `A` at the liquid-side of the interface in terms of $(C_{AL})_b$ and C_{AL}^{*}.

4-15   Referring to the ternary diffusion problem discussed in Section 4-6

(a) Obtain a mathematical solution for the concentration profiles of A,B and C in the by assuming constant diffusivities for all three species.

(b) Compare the numerical results with those shown in Figure 4-6 for P = -.05 cm²/s, B = 1.8 and assuming that N₂ (species C) is in large excess; i.e., $D_{Am} \approx D_{AC}, D_{Bm} \approx D_{BC}$ (see Section 4-6 for numerical values)

4-16

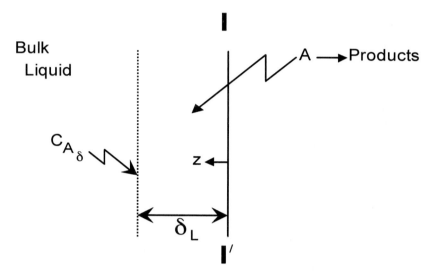

A gaseous species, A, is dissolving in a liquid where it reacts with species B in a first-order reaction ($R_{Av} = k\,C_A$), as shown in the sketch. The concentration of A at the liquid side of the interface is known ($C_A^{\,*}$) as is the concentration in the bulk liquid (i.e., at the edge of the film, $C_{A\delta}$). Assuming that the concentration of A in the liquid is quite low, obtain an expression for C_A as a function of z and for the diffusion flux at the interface, $(J_{Az})_{|_{z=0}}$.

4-17 Sensitive oxygen sensors can be assembled from stabilized zirconia by taking advantage of the fact that zirconia is a good conductor of oxygen ions at temperatures above 1000 C. In one such application this device is intended for use in a furnace in order to control the fuel-to-oxygen ratio (Wirtz, G.P. and F.M.B. Marques, J. Am. Ceram. Soc., <u>75</u>, p. 375, 1992). The current detected by the sensor is due to the flux of oxygen at the surface of the detector. If, close to the detector, we can assume that oxygen transport occurs only via molecular mechanisms, then we can consider transport through a film of thickness, δ , as was done in Section 4-2 in the text. If the furnace gas consists of a mixture of O_2, CO, and CO_2, and it may be assumed that there is always equilibrium between CO and CO_2; i.e.,

$$CO + \frac{1}{2} O_2 \rightarrow CO_2$$

$$K = \frac{X_{CO_2}}{X_{CO} \sqrt{X_{O_2}}}$$

Derive an equation for the flux of <u>atomic</u> oxygen arriving at the sensor in terms of the atomic oxygen concentrations at either edge of the film (f_{O_δ} and f_{O_0}) and the diffusivities of O_2 and CO. [HINT: use the equilibrium relationship together with the facts that $N_O = 2 N_{O_2} + N_{CO} + 2 N_{CO_2}$ and $\sum_i X_i = 0$ to express all concentrations in terms of X_{O_2}]

4-18

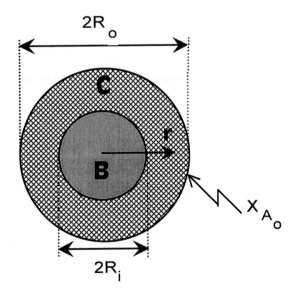

As shown in the above sketch, gaseous species, A reacts with spherical solid, B, after diffusing through the porous product layer, C, according to

$$A_{(g)} + B_{(s)} \rightarrow C_{(s)}$$

The reaction rate at the surface of `B` is instantaneous and it can be assumed that *quasi-steady-state* conditions prevail; i.e., accumulation terms in the conservation equations can be neglected. If the molar flux of `A` with respect to fixed coordinates is essentially equal to the molecular diffusion flux with an effective diffusivity, D_{eff},

(a) solve for the concentration profile of A in the product layer if its concentration at $R_o = C_{A_o}$.

(b) In reality, R_i varies with time as B is consumed and the molar flow rate of A at R_i ($\dot{N}_{A|_{R_i}}$) is equal to the molar rate of disappearance of B. If the molar density of B, C_B, is constant, take an unsteady-state molar balance on B and show that

$$C_B \frac{dR_i}{dt} = N_{A|_{R_i}}$$

(c) Integrate the expression in (b) to obtain an *implicit* expression for the fractional conversion of B, X_B, as a function of time. [**NOTE:** $X_B = 1 - \left(\dfrac{R_i}{R_o}\right)^3$].

4-19 One explanation for the red surface of Mars is that its surface is covered by a thin (2 mm) layer of iron oxide, Fe_2O_3. The Fe_2O_3 is produced by the diffusion of oxygen through the iron surface layer where the reaction

$$2Fe + \frac{3}{2}O_2 \rightarrow Fe_2O_3$$

occurs simultaneously. The thickness corresponds to the cessation of the reaction which kinetic studies have shown ocurs at $P_{O_2} \leq 0.1 \ kPa$. These same studies also show that the reaction rate is given by $r = k \ P_{O_2}$, where $k = 5 \times 10^4$ kg-moles O_2 / kPa-m^3-s

The planet surface can be modeled as a flat surface covered with crushed iron and recurring windstorms etc produce a quasi-steady-state so that the surface can always be considered to consist of pure iron. Obtain a value for the effective diffusivity of oxygen in the Martian soil if the oxygen concentration far from the surface is essentially zero and assuming that the ground level atmosphere of Mars is at 170 K and 1 kPa of oxygen.

4-20 An organic substrate is being consumed due to an enzyme which is immobilized in a spherical gel particle. The substrate diffuses through the spherical particle with a diffusivity, D_s, and reacts with a rate (moles/vol-time) given by the Mechaelis-Menten equation, $r_s = \dfrac{r_{max} C_s}{K_M + C_s}$. The concentration of substrate at the outer radius of the sphere (r = R) is C_{S_0}

(a) Solve for the concentration profile of S assuming that:

    i. $K_M \lll C_s$

    ii. $K_M \ggg C_s$ (**HINT:** let $\Psi = \dfrac{C_s}{r}$ and then use the total derivative for $\Psi = f(C_s, r)$)

(b) Define the immobilized enzyme effectiveness factor by

$$\xi = \frac{actual\ substrate\,rate}{rate\ if\ C_s\ =\ C_{S_0}\ everywhere}$$

and derive expressions for it, using the results of i. and ii. above.

4-21 (**C#**) Use MATLAB to re-solve the radiant heating of the $CaCO_3$ sphere described in 3-5.b by taking the variable reaction rate into account. The reaction rate is given by the expression: $R_{Av} = k_r C_A$ where C_A is the molar concentration of $CaCO_3$ and k_r is the reaction rate constant which can be assumed to remain constant at a value of $3.2 \times 10^4\ s^{-1}$. Note that it will be necessary to simultaneously solve for the concentration of $CaCO_3$ as a function of time.

4-22 (**C#**) Reaction rate constants are usually strong functions of temperature. Re-solve 4-21 based on the rate expression given by Wang and Thomson (Chem. Engr. Sci., 50, pp 1373-1382, 1995); that is, $k_r = 1.47 \times 10^9 \exp\left[-\dfrac{24,060}{T}\right]$, s^{-1}.

CHAPTER 5

Molecular Momentum Transport

The discussions of molecular momentum transport have been put off until last because of the difficulty in dealing with the tensor properties of the subject. Another complication in momentum transport is the distinction between convective and molecular momentum transport. That is, since convective transport requires a velocity of transport, how does <u>molecular</u> momentum transport take place? This will not pose a problem if it can be remembered that convective transport only occurs when the velocity is in the direction of transport. For these reasons, some of the direct analogies that were developed in the previous two chapters are absent here and we have to introduce the subject in a fresh manner. The chapter begins with a discussion of how the molecular momentum flux may also be viewed as a force/area and then proceeds to apply differential momentum balances to experimental situations which are in <u>fully developed, laminar flow</u>.

To begin, let us define what is meant by laminar and fully developed flow. A flow is <u>laminar</u> when the fluid travels in straight lines, or "laminae." A flow system is said to be <u>fully developed</u> when the fluid velocities no longer change <u>in the direction of the fluid motion</u>. Convective momentum transport does not take place in fully developed laminar flow even though there are velocities present. For example, in fully developed, laminar pipe flow, the only velocity component is in the direction of flow and there are no velocity gradients in the direction of flow. The absence of a velocity gradient in the flow direction precludes momentum transport in this direction. On the other hand, since the fluid at the pipe wall is stationary, the velocity is zero there. This means there is a velocity gradient from the center of the pipe to the wall and, consequently, there is momentum transport in the radial direction. However, since there are no radial velocity components, there is no convective transport in that direction. Of course turbulent flow is a much different situation since there are velocity components in all directions. The criteria for determining whether laminar or turbulent flow should exist is discussed in detail in Chapter 10.

5-1 MOMENTUM FLUX AS A SURFACE FORCE

The analogies between energy, mass and momentum transport are made stronger when we consider the transport of each in terms of fluxes. However, while the rates of energy and mass crossing an area are easy to visualize, the same is not true for momentum. This becomes even more difficult when dealing with a momentum in one direction being transported in another direction. Because of Newton's Second Law, the dilemma can be resolved by relating the time rate of change of momentum to force; i.e.,

$$\vec{F} = \frac{d}{dt}(m\vec{v}) \tag{5.1}$$

If both sides of Newton's Law are divided by area,

(FORCE/AREA) = (MOMENTUM RATE/AREA)

then the right-hand side of this equation is just the momentum flux and we can see that momentum flux is entirely equivalent to a force per unit area, or a "surface" force. This also provides some physical insight to the difficulty of visualizing momentum flux as it crosses a surface; it simply manifests itself as a force acting on the surface.

Viewing momentum flux as a surface force also provides us with a means of visualizing momentum in one direction being transported in another. Take, for example, a mail train moving in the z-direction as shown in Figure 5-1 below. As the train roars through the station, a mail bag is tossed on to a moveable cart which is chocked so that it will not move. In the diagram, the mail bag which has z-momentum by virtue of being on the moving train, is being transported in the positive y-direction. Our instincts tell us that, if the cart were not chocked, it would move in the positive z-direction as soon as the mail bag landed on its surface. Obviously we have transported z-momentum in the y-direction. Furthermore, if the wheels of the cart are chocked, then as soon as the mail bag lands, the force exerted on the cart is counteracted by the force exerted by the chocks on the wheels of the cart. In that case, the momentum transported is being manifested as a static force.

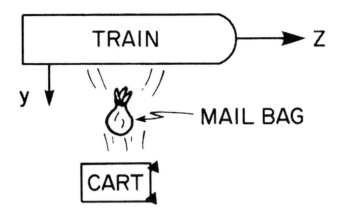

Figure 5-1 z-Momentum Transported in y-Direction

Now let's apply this same reasoning to a more conventional chemical engineering problem; that of fluid flow in a pipe. With a fluid flowing in the z-direction there can be transport of the z-momentum in the z-direction <u>as well as</u> in the radial direction. This stands to reason since there is friction at the walls of the pipe which must have resulted from momentum transport in that direction. This can be seen more clearly by referring to the diagram in Figure 5-2 which shows a radial velocity profile superimposed on the flow. The velocity of the fluid is a maximum in the center of the pipe and if there is no slip at the wall, the velocity will be zero at the wall. When the momentum flux encounters the surface of the pipe, it will be manifested as a surface force. If the pipe wall is anchored, this force will be transmitted to the anchor point(s). Of course if the pipe were not anchored, then the pipe wall would move. A good example of the latter is water flowing in a garden hose which will "snake about" if it is not held firmly.

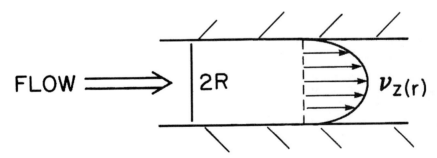

Figure 5-2 Velocity Profile in a Pipe

Since momentum flux is equivalent to a force/area, the surface force at pipe walls is called a <u>shear stress</u>. It is a stress because it is a force per unit area; it is a shear force because the adjacent fluid is moving parallel to the wall. That is, the fluid is rubbing against adjacent layers of molecules, similar to the friction which is encountered when a block is moved along a surface. So, momentum transport can be viewed in either of two equivalent ways: as a momentum flux, analogous to the other two transport phenomena, or as a shear stress. When viewed as a flux, momentum is transported in a direction perpendicular to an area. When considered as a force/area, the force acts parallel to the area and in a direction consistent with "free body" principles. Thus, in Figure 5-3, the surface force is shown to be acting in the negative direction of the momentum which was transported since this is the force exerted by the wall <u>on the fluid</u>.

In Chapter 2, it was pointed out that the symbol for the molecular momentum flux has a dual subscript in order to distinguish between the component of momentum which is being transported and the direction in which it is being transported. When viewed as a flux, τ_{ij} indicates that j-momentum is transported in the i-direction. When viewed as a force, τ_{ij} is a shear force acting in the j-direction on a surface perpendicular to the i-direction.

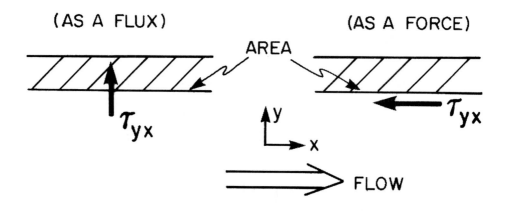

Figure 5-3 Momentum Transport as a Flux or a Force

5-2 MOMENTUM TRANSPORT IN COUETTE FLOW

One of the simplest flow systems to analyze is the case of Couette flow which, in its most elementary form, is steady-state flow between two parallel planes as shown in Figure 5-4. In this particular example, the upper plane is moving in the positive x-direction at a known velocity, V, and the bottom plane is stationary. Physically, the top surface might represent a moving belt or even a fluid which has a constant velocity which does not change for y-values greater than B. The latter is often used to describe <u>boundary layer</u> flows [1].

If the flow is at steady-state; i.e., not changing with time, and we analyze the flow at a distance, x, sufficiently removed from the entrance and exit; i.e, the velocity at any point, y, does not change with x[#], then the only direction of momentum transport is in the y-direction. Taking a momentum balance around the differential volume element shown in Figure 5-4,

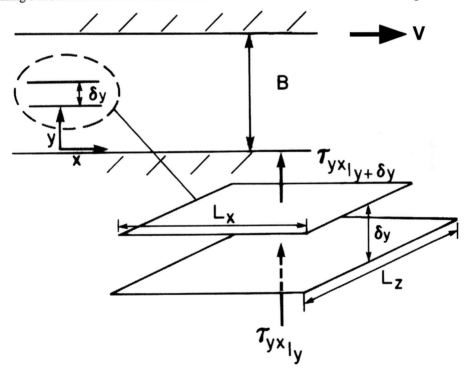

Figure 5-4 Couette Flow between Parallel Planes

[#] This is the <u>fully developed flow</u> assumption discussed at the beginning of the chapter.

$$\tau_{yx \,|\, y} \, L_x \, L_z \, \delta t \; - \; \tau_{yx \,|\, y+\delta y} \, L_x \, L_z \, \delta t$$

The differential equation is obtained by dividing by the volume of the differential element $(\delta y \, L_z \, L_x)$ and δt and then taking the limit as $\delta y \rightarrow 0$

$$-\frac{d\tau_{yx}}{dy} = 0 \tag{5.2}$$

Equation 5.2 can be integrated once to show that τ_{yx} is a constant, however we need to apply boundary conditions in order to determine the magnitude of the constant. In this problem our only boundary conditions are in terms of the fluid velocity at the two walls. That is, by virtue of the no-slip assumption

$$\text{at } y = 0; \; v_x = 0$$

$$\tag{5.3}$$

$$\text{at } y = B; \; v_x = V$$

Thus we have to express Equation 5.2 in terms of velocity, which can be done by using Newton's Law of viscosity to write τ_{yx} in terms of the velocity gradient, so that

$$\frac{d}{dy}\left(-\mu \frac{dv_x}{dy}\right) = 0$$

which, when integrated twice, gives

$$v_x = C_1 y + C_2$$

The constants of integration, C_1 and C_2, can be determined by applying the boundary conditions, Equations (5.3),

$$0 = C_1(0) + C_2$$

$$V = C_1 B + C_2$$

to give the linear velocity profile

$$v_x = V \, \frac{y}{B} \tag{5.4}$$

Once the velocity profile is known, both the volumetric flow rate, $\dot{V}$, and the average velocity, $\bar{v}$, can be calculated from

$$\dot{V} = \int_A v_x \, dA$$

(5.5)

$$\bar{v} = \frac{\int_A v_x \, dA}{\int_A dA}$$

Using $dA = L_z \, dy$ and substituting Equation (5.4) for v_x,

$$\dot{V} = \frac{V L_z B}{2}$$

$$\bar{v} = \frac{V}{2}$$

In order to generalize this flow situation somewhat, we now consider <u>Couette flow with a pressure gradient</u>. The differential volume element is identical to that shown in Figure 5-4 except for the addition of pressure forces as shown in Figure 5-5. Consistent with "free body" principles, the compressive pressure forces act in opposite directions on the differential volume element. As discussed in Chapter 2, forces can be sources or sinks of momentum, depending on whether they act as accelerating or retarding forces. The convention, consistent with the negative sign in the phenomenological laws, is that forces acting in the direction of the positive coordinate system are accelerating forces (SOURCES), and those acting in the negative direction are retarding forces (SINKS). The differential momentum balance is then

$$(\tau_{yx}|_y - \tau_{yx}|_{y + \delta y}) \, L_x L_z \delta t + P|_x L_z \delta y \delta t - P|_{x + L_x} L_z \delta y \delta t = 0$$

which yields the differential equation

$$-\frac{d\,\tau_{yx}}{dy} - \frac{\Delta P}{L_x} = 0$$

(5.6)

Notice that, in deriving Equation (5.6), the DVE had a finite dimension in the x-direction. This is consistent with the assumption of fully developed flow. That is, there are

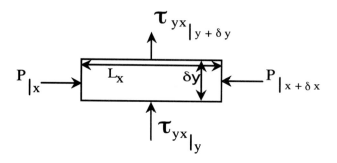

Figure 5-5 DVE for Couette Flow with a Pressure Gradient

no changes of velocities (or the forces that produce them) in the flow direction. While it is true that $P_{|x}$ must be different than $P_{|x+L_x}$ (or else there could be no flow), this difference must be a constant or else the flow would not be fully developed (see Example 5-1) Again, the two boundary conditions for this case are in terms of velocities (identical to Equations (5.3)) and Newton's Law of Viscosity must be substituted into (5.6) in order to obtain the solution in terms of velocities. Making this substitution and integrating twice,

$$v_x = \left(\frac{\Delta P}{\mu L_x}\right)\frac{y^2}{2} + C_1 y + C_2 \tag{5.7}$$

The constants of integration, C_1 and C_2 can be determined by applying the boundary conditions, Equations (5.3), to produce the velocity profile

$$v_x = \frac{B^2}{2}\left(\frac{\Delta P}{\mu L_x}\right)\left[\left(\frac{y}{B}\right)^2 - \frac{y}{B}\right] + V\frac{y}{B} \tag{5.8}$$

Using Equations (5.5), we can again solve for the volumetric flow rate and/or the average velocity for this situation by merely substituting the velocity distribution, Equation (5.8), and integrating. The result for the average velocity is

$$\bar{v} = -\frac{B^2 \Delta P}{12 \mu L_x} + \frac{V}{2} \tag{5.9}$$

Note that, in this case, the velocity profile, as well as the average velocity, depend on the pressure gradient as well as the magnitude of the velocity at the upper wall, V. Of course, when the pressure gradient is zero, as it was for simple Couette flow, Equations (5.8) and (5.9) reduce to the same answers found for simple Couette flow.

EXAMPLE 5-1: Pressure Gradients in Fully Developed Flow

In this example, both mathematical and physical proofs are offered to show that the pressure gradient in fully developed flow must be linear.

Mathematical

This can be proven from a mathematical point of view by differentiating Equation (5.6) with respect to x.

$$-\frac{\partial}{\partial x}\left[\frac{d\,\tau_{yx}}{dy}\right] - \frac{\partial}{\partial x}\left[\frac{\Delta P}{L_x}\right] = 0$$

    For fully developed flow, the velocity gradients in the y-direction cannot change with x and so the differentiation of the first term must be identically zero. Since the flow is laminar and one-dimensional, this means that the second term in Equation (5.6) must be a constant, or that $\frac{\Delta P}{L_x}$ is a constant and thus, the pressure varies linearly with x.

Physical

The physical proof can be found by viewing the shear stresses acting at both parallel planes of Figure 5-4 as forces. The net difference between accelerating forces and retarding forces is then

$$\Sigma F = P_{x|_x}(B)(L_z) + \tau_{xy|_0}(L_x L_z) - P_{x|_{x+L_x}}(B)(L_z) - \tau_{xy|_B}(L_x L_z)$$

where the subscripts 0 and B, denote the bottom and top parallel planes in Figure 5-4.

    From Newton's second law we know that the sum of the forces must equal the acceleration. However, in fully developed flow, there is no acceleration or else the flow would certainly change in the flow direction. In other words, the sum of the forces must be zero and, since the shear stresses at the wall will not vary with x in fully developed flow (because the velocity gradients do not vary with x), then

$$\Delta P = P|_x - P|_{x+L_x} = constant$$

    Figure 5-6 shows a sketch of the velocity profiles as a function of V. Note that the shape of the profile can vary anywhere from a near-straight line to a near-parabolic shape depending on the relative magnitudes of ΔP and V.

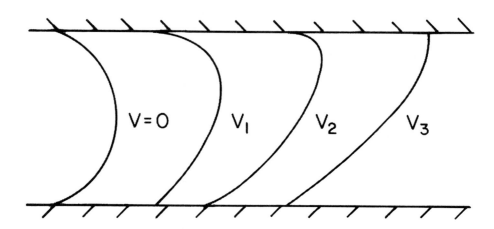

Figure 5-6 Velocity Profiles in Couette Flow

5-3 FILM FLOW OVER A SOLID SURFACE

Now let's turn to another type of flow which is often encountered; that of a liquid film falling freely under the pull of gravity down a cylindrical tube. A practical example of such a situation would be the flow of condensate in a shell-and-tube condenser. Figure 5-7 shows a sketch of the system including details of the differential volume element necessary to describe the momentum transport which is occurring. In this case, the gravitational force acts as a SOURCE of momentum since it acts uniformly on the entire fluid volume and in the positive z-direction. Forces which act on fluids in this manner are sometimes referred to as <u>body forces</u>. The corresponding differential momentum balance is then

$$\tau_{rz|_r} 2\pi r|_r L_z \delta t - \tau_{rz|_{r+\delta r}} r|_{r+\delta r} L_z \delta t + \rho g_z 2\pi r \delta r L_z \delta t = 0 \tag{5.10}$$

which results in the differential equation given in Equation (5.11)

$$-\frac{1}{r}\frac{d}{dr}(r\tau_{rz}) + \rho g_z = 0 \tag{5.11}$$

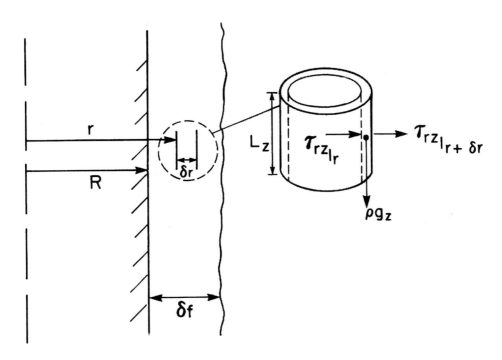

Figure 5-7 Film Flow over a Cylindrical Tube

The no slip condition at the liquid-film interface gives one boundary condition in terms of velocity, $v_z = 0$ at $r = R$; whereas the second boundary condition is in terms of the momentum flux at the liquid-gas interface. Specifically, if the gas is stagnant, then it offers little resistance to the flow of the liquid and the applicable boundary condition is (see Table 3-1)

$$at\ r = R + \delta_f \ ; \ \tau_{rz} = 0 \qquad (5.12)$$

Integrating (5.11) once and applying Equation (5.12)

$$\tau_{rz} = \frac{\rho\, g_z\, r}{2} + \frac{C_1}{r} \qquad (5.13)$$

$$0 = \frac{\rho\, g_z}{2}(R + \delta_f) + \frac{C_1}{(R + \delta_f)}$$

so that

$$C_1 = -\frac{\rho g_z}{2}(R+\delta_f)^2 \qquad (5.14)$$

If Newton's Law of Viscosity is now substituted into Equation (5.13), one more integration yields

$$v_z = -\frac{\rho g_z r^2}{4\mu} - \frac{C_1}{\mu}\ln(r) + C_2 \qquad (5.15)$$

The second constant of integration, C_2, is found by applying the no-slip boundary condition at R and the resulting expression for the velocity profile is

$$v_z = \frac{\rho g_z R^2}{4\mu}\left(1 - (\frac{r}{R})^2\right) + \frac{\rho g_z}{2\mu}(R + \delta_f)^2 \ln\left(\frac{r}{R}\right) \qquad (5.16)$$

The volumetric flow rate of the liquid can be calculated from

$$\dot{V} = \int_A v_z\, dA = 2\pi \int_R^{\beta R} v_z\, r dr$$

where $\beta = \dfrac{(R + \delta_f)}{R}$

When Equation (5.16) is substituted into the integral, the integrated expression for the volumetric flow rate becomes

$$\dot{V} = \frac{\rho g_z R^4}{16\mu} F(\beta)$$

where

$$F(\beta) = \beta^4(4\ln\beta - 3) + 4\beta^2 - 1$$

5-4 LAMINAR PIPE FLOW

The final example of molecular momentum transport to be considered in this chapter is steady-state laminar pipe flow. Figure 5-8 is a sketch of a pipe inclined at an angle θ, with respect to the horizontal with a flow under the influence of both an applied pressure gradient (from a pump, for example) and gravity. The differential volume element shown in Figure 5-8 is appropriate for this type of flow as long as we are sufficiently removed from the entrance and exit of the flow field. Under these conditions ("fully developed flow") it can be shown that the pressure gradient is a constant (pressure varies linearly with z), and the application of a differential momentum balance results in

$$\left[(r\,\tau_{rz})\,|_r - (r\,\tau_{rz})|_{r+\delta r}\right] 2\pi\,L_z\,\delta t + [P_{|z} - P_{|z+L_z}]\,2\pi r\,\delta r\,\delta t + \rho\,g\sin\theta\;2\pi r\,\delta r\,L_z\,\delta t = 0$$

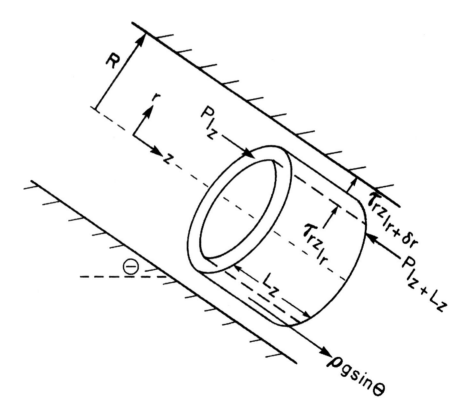

Figure 5-8 Laminar Flow in an Inclined Pipe

Note that, whereas the pressure forces act on the cross-sectional area of the volume element, the gravitational force acts on the entire volume of the element. Note also that the pressure terms can be written as a negative pressure gradient

$$-\frac{\Delta P}{L_z} = \left[\frac{P\mid_z - P\mid_{z+L_z}}{L_z}\right]$$

Dividing through by the volume of the differential element and δt,

$$-\frac{1}{r}\frac{d}{dr}(r\,\tau_{rz}) - \frac{\Delta P}{L_z} + \rho\,g\sin\theta = 0 \tag{5.17}$$

It should be pointed out that Equation (5.17) could also have been derived by considering a static force balance on the differential element (see <u>Alternative Derivation</u>, in the box below).

In order to obtain an expression for the velocity profile, we will need to substitute Newton's Law of Viscosity for τ_{rz} in Equation (5.17) resulting in a second-order differential equation and requiring two boundary conditions for a solution. Before deciding on the appropriate boundary conditions, let us first integrate Equation (5.17) once

$$\tau_{rz} = \frac{\wp r}{2} + \frac{C_1}{r} \tag{5.18}$$

where

$$\wp = -\frac{\Delta P}{L_z} + \rho\,g\sin\theta \tag{5.19}$$

From a strictly mathematical point of view we can see that there is a singularity at r = 0 unless $C_1 = 0$. Since the point at r = 0 is certainly within the region of the pipe being analyzed, and we expect such a mundane physical problem to yield a continuous solution, we must conclude that $C_1 = 0$. Note that we would reach the same conclusion if we used the boundary condition: at $r = 0$, $\tau_{rz} = 0$. The latter gives some insight into a physical rationale for such a boundary condition. Since the pipe is symmetrical with respect to the longitudinal axis, the radial velocity profile should also be symmetrical. In other words, $\frac{d v_z}{dr}$ should equal zero at $r = 0$; which, by virtue of Newton's Law of Viscosity, leads to the conclusion that τ_{rz} should be zero at that point. Viewed in another way, if τ_{rz} were not zero at the pipe center, we would have to imagine some point source or sink at that location, which is clearly inconsistent with the physical problem.

Alternative Derivation of Equation (5.17)

    As shown in the sketch, below, the τ_{rz} terms can be viewed as shear forces acting on the surfaces of the differential volume element. When the problem is viewed in this way, the differential momentum balance is actually a static force balance. Using as a convention, that the shear force acting on the input surface of the differential volume element acts in the positive direction of the coordinate system, then the accelerating forces (SOURCES) are:

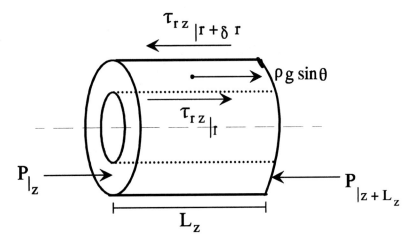

$$\Sigma F_{acc} = \tau_{rz}\,|_r\, 2\pi\,r\,|_r\, L_z + P\,|_z\, 2\pi r \delta r + \rho g \sin\theta\; 2\pi r \delta r\; L_z$$

and the retarding forces (SINKS) are:

$$\Sigma F_{ret} = \tau_{rz}\,|_{r\,+\,\delta r}\, 2\pi\,r\,|_r\, L_z + P\,|_{z\,+\,L_z}\, 2\pi r \delta r$$

From Newton's second law we can now state

$$\Sigma(\vec{F}_{acc} - \vec{F}_{ret}) = \frac{d}{dt}(m\vec{v}) = 0$$

$$\left[(r\,\tau_{rz})\,|_r - (r\,\tau_{rz})\,|_{r\,+\,\delta r}\right]2\pi\,L_z + [P\,|_z - P\,|_{z\,+\,L_z}]2\pi\,r\,\delta r + \rho g \sin\theta\; 2\pi\,r\,\delta r\,L_z = 0$$

which leads to the same differential equation as Equation (5.17).

With $C_1 = 0$ and $\tau_{rz} = -\mu \dfrac{dv_z}{dr}$, we can integrate Equation (5.18) once more to yield

$$v_z = -\frac{\wp r^2}{4\mu} + C_2 \tag{5.20}$$

C_2 can be evaluated by using the no-slip condition at the wall; i,e., *at* $r = R$, $v_z = 0$ which results in the well known parabolic velocity profile

$$v_z = \frac{\wp R^2}{4\mu}\left[1 - \left(\frac{r}{R}\right)^2\right] \tag{5.21}$$

A knowledge of the velocity profile also allows for the computation of the volumetric flow rate in the pipe

$$\dot{V} = \int_0^R v_z \, 2\pi \, rdr$$

which, upon substitution for v_z from Equation (5.21) and subsequent integration, yields

$$\dot{V} = \frac{\pi \wp R^4}{8\mu} \tag{5.22}$$

For a horizontal pipe or a pipe flow problem where the gravitational force is small relative to the applied pressure gradient, $\wp = -\dfrac{\Delta P}{L_z}$ and Equation (5.22) can be written as

$$\dot{V} = \frac{\pi R^4 (-\Delta P)}{8\mu L_z} \tag{5.23}$$

which is known as the <u>Hagen-Poiseuille Law</u> and which states that, in laminar pipe flow, the pressure drop in the pipe varies linearly with the volumetric flow rate.

The velocity profile predicted by Equation (5.21) can also be used to determine the average velocity in the pipe, either by noting that

$$\bar{v} = \frac{\dot{V}}{\pi R^2}$$

or calculating it directly via

$$\overline{v} = \frac{\int_0^R v_z \, 2\pi \, rdr}{\int_0^R 2\pi \, rdr}$$

which, after the substitution for v_z from (5.21) and integration, gives

$$\overline{v} = \frac{\wp R^2}{8\mu} \tag{5.24}$$

Since the maximum velocity in the pipe, v_{max} occurs at $r=0$, we can see from Equation (5.21) that $v_{max} = \dfrac{\wp R^2}{4\mu}$ and thus, comparison with (5.24) produces the interesting result that

$$v_{max} = 2\,\overline{v} \tag{5.25}$$

so that the velocity distribution (Equation (5.21)) can also be expressed as

$$v_z = 2\,\overline{v}\left[1 - \left(\frac{r}{R}\right)^2\right] \tag{5.26}$$

EXAMPLE 5-2: Velocity Distribution in an Extruder

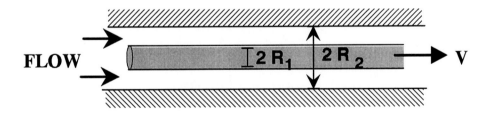

The barrel of an extruder can be modeled as if a solid rod is moving with a velocity, V, through a fluid inside a horizontal cylindrical tube as shown in the sketch. There is also a pressure gradient imposed on the Newtonian fluid in the annulus. We wish to derive an expression for the velocity distribution of the fluid in the annulus under steady-state, fully developed laminar-flow conditions.

<u>**Solution**</u>

The DVE corresponding to this problem is shown in the sketch below. The sketch is not drawn to scale in order to accent the characteristics of the DVE and to illustrate how it fits with the physical system (the annular gap in an extruder is typically small). Notice that the DVE is identical to that shown in Fig 5-8 (with $\theta = 0$) for flow in an pipe.

Applying the general conservation equation for momentum to the DVE, we get

$$\tau_{rz_r} 2\pi\, r_{|rL}\, \delta t - \tau_{rz_{r+\delta r}} 2\pi\, r_{|r+\delta r}\, L\delta t + (P_{|z} - P_{|L})2\pi r\ \delta r\ \delta t = 0$$

After dividing through by the product of the volume of the DVE ($2\pi r\delta r L$) and δt , the appropriate differential equation is

$$-\frac{1}{r}\frac{d}{dr}(r\,\tau_{rz}) - \frac{\Delta P}{L} = 0$$

where $\Delta P = P_{|z+L} - P_{|z}$. As pointed out earlier, ΔP is a constant in fully developed flow and so the differential equation can be separated and integrated once to get

$$r\,\tau_{rz} = -\frac{\Delta P}{L}\frac{r^2}{2} + C_1$$

where C_1 is a constant of integration. Unlike the case of fully developed flow in a tube, we do not have a boundary condition for τ_{rz} since the point, r = 0, is NOT part of the physical system. The next step is to substitute Newton's Law of Viscosity and to integrate once more to obtain v_z

$$-\mu\frac{dv_z}{dr} = \frac{\Delta P}{L}\frac{r}{2} + \frac{C_1}{r}$$

$$v_z = -\frac{\Delta P}{\mu L}\frac{r^2}{4} - \frac{C_1}{\mu}\ln r + C_2$$

The two constants of integration can be evaluated by using the two boundary conditions on velocity:

$$\text{at } r = R_1,\ \ v_z = V$$
$$\text{at } r = R_2,\ \ v_z = 0$$

Upon substitution of the two boundary conditions into the expression for v_z, the two resulting algebraic equations can be solved simultaneously for C_1 and C_2

$$C_1 = -\frac{\mu}{\ln \dfrac{R_1}{R_2}} \left[V - \frac{\Delta P}{4\mu L} (R_1^2 - R_2^2) \right]$$

$$C_2 = -\frac{\Delta P}{4\mu L} R_2^2 + \frac{C_1}{\mu} \ln R_2$$

so that the final expression for the velocity distribution becomes

$$v_z = -\frac{\Delta P R_2^2}{4\mu L} \left[1 - \left(\frac{r}{R_2}\right)^2 \right] + \frac{\left[V + \dfrac{\Delta P R_2^2}{4\mu L} \left(1 - \left(\dfrac{r}{R_2}\right)^2\right) \right]}{\ln \dfrac{R_1}{R_2}} \ln \frac{r}{R_2}$$

Even though this solution looks complex, it is apparent that the first term on the right hand side is the same solution as would be obtained for fully developed flow in a tube of radius, R_2. The second term is then a consequence of the extruder barrel. In fact, if $\dfrac{R_1}{R_2} \to 0$, the second term drops out of the solution. Note also that if $V = 0$, then the solution is identical to fully developed flow in the annulus of two concentric tubes (see Problem 5-14).

RERFERENCES

[1] Schlicting, H., *Boundary Layer Theory*, 4th ed., McGraw-Hill, N.Y., 1960.

PROBLEMS

5-1

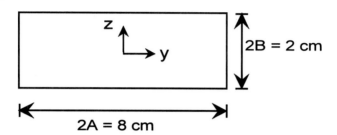

2B = 2 cm

2A = 8 cm

Given the steady-state, laminar fully developed flow of a Newtonian fluid in the rectangular duct shown in the sketch, an approximation to the velocity distribution in the duct is

$$v_x = \frac{9}{5}\bar{v}\left[1-\left(\frac{y}{A}\right)^6\right]\left[1-\left(\frac{z}{B}\right)^2\right]$$

where $\bar{v}$ is the average velocity in the duct.

(a) Derive expressions for the shear stress at the top and side walls of the duct; i.e., $\tau_{zx}|_{z=B}$ and $\tau_{yx}|_{y=A}$.

(b) Obtain expressions for the <u>average</u> values of the shear stresses at the top and side walls.

(c) If water is flowing in the duct at an average velocity of 4.5 cm/s, calculate the <u>total</u> viscous force/length of duct; that is, the integral of the shear stress over the surface area of the duct. Remember that force = (shear stress) x (surface area)

(d) Use an overall force balance to relate the total viscous force calculated in (c) to the pressure gradient in the duct, $\dfrac{\Delta P}{L}$, and then calculate $\dfrac{\Delta P}{L}$ in units of cm -H_2O/m.

5-2 A river is flowing downhill on a calm, windless day (at an angle of θ with respect to the horizontal) and is in fully developed, laminar flow. The depth of the river is H, and it may be assumed that the river bed is rectangular of width, W.

(a) Define a coordinate system and then use a differential momentum balance to derive the differential equation which describes the velocity distribution in the river.

(b) State the boundary conditions needed to solve the differential equation in (a).

(c) What, if anything, would change in parts (a) and (b) if the wind were blowing against the river flow direction at a velocity, v_w ?

5-3 Consider fully developed, steady-state, laminar flow between two stationary, horizontal parallel plates separated by a distance, B.

(a) Provide a sketch of this system, label all parameters pertaining to the physical flow system, and show your coordinate system.

(b) Draw the appropriate DVE showing all dimensions and illustrate the molecular momentum transport as both fluxes and shear stresses.

5-4

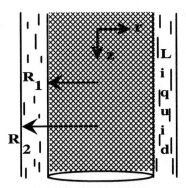

The sketch above shows a liquid film flowing down the outside of a cylindrical tube in fully developed flow. Derive the differential equation (in terms of fluid velocity) which describes this flow using a <u>static force balance</u>.

5-5 Derive an equation for the velocity profile corresponding to the Couette flow problem shown in Figure 5-4 but with the bottom plane also moving with a velocity, V_1 in the <u>negative</u> x-direction.

5-6   Using the solution given by Equation 5.9 for Couette flow with a pressure gradient, derive an expression for the locus of points where the momentum flux is zero.

5-7   A Newtonian fluid is flowing downward under the influence of gravity between two infinite, stationary parallel, plates separated by a distance, δ. If the flow is steady-state and fully developed, solve for both the velocity distribution and the average velocity between the plates.

5-8

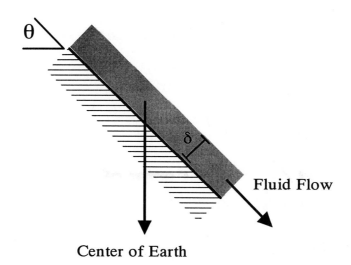

Center of Earth

A Newtonian fluid of viscosity, μ, and density, ρ, is flowing as a uniform film (thickness, δ_f) down a vary wide inclined plane of width, W, under the pull of gravity. Derive an expression for the steady-state velocity profile and for the average velocity within the film.

5-9   Derive expressions for the velocity profile and the volumetric flow rate for the falling liquid film shown in the sketch associated with Problem 5-4. The fluid is Newtonian and the flow is steady-state, laminar, and is fully developed.

5-10   (a) Reanalyze the film-flow problem in Section 5-3 by assuming that the thickness of the film is much less than the tube radius. That is, assume that the film has no curvature.

(b) Compare the solution obtained in (a) with that given in the text [Equation (5.19)] and, realizing that $\alpha \equiv \dfrac{\delta_f}{R}$ is small, determine the criterion for $\dfrac{\delta_f}{R}$ so that the velocity prediction at the edge of the film is within 5% of the value calculated by Equation (5.19).

5-11

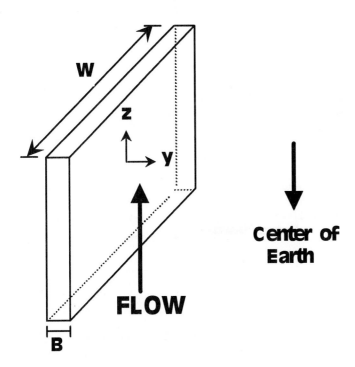

A fluid is being pumped up a vertical duct ($\dfrac{W}{B} >> 1$) as shown above. If the fluid is Newtonian and the flow is laminar and steady-state:

(a) Solve for the fully developed velocity distribution in the fluid.

(b) Solve for both the average velocity and the volumetric flow rate of the fluid.

5-12 Derive the differential equation and state the boundary conditions which describe the steady-state, laminar, fully developed flow of a Newtonian fluid in the duct shown in Problem 5-1.

(a) Does the approximate velocity distribution given in 5-1 satisfy the boundary conditions?

(b) Does the approximate velocity distribution satisfy the differential equation at all points in the duct? **HINT:** Use the results from Problem 5-1(b and d) to relate the pressure gradient in the differential equation to the shear stresses at the wall and, hence to the velocity gradients.

5-13

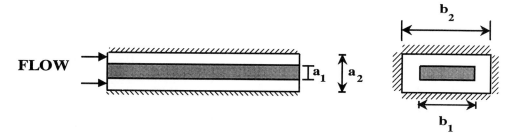

Water is flowing in the rectangular annulus of the horizontal duct shown in the above sketch. If the flow is laminar, fully developed and steady-state, derive the differential equation and state the boundary conditions necessary to solve for the velocity distribution in the annulus.

5-14

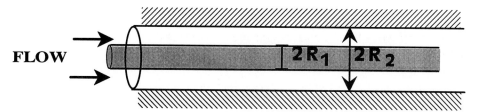

The sketch shows flow of a Newtonian fluid in the annulus of a horizontal cylindrical tube. If the flow is laminar, steady and fully developed with a pressure gradient, $\dfrac{\Delta P}{L}$ derive:

(a) An expression for the velocity distribution of the fluid in the annulus.

(b) An expression for the volumetric flow rate of the fluid in the annulus.

5-15

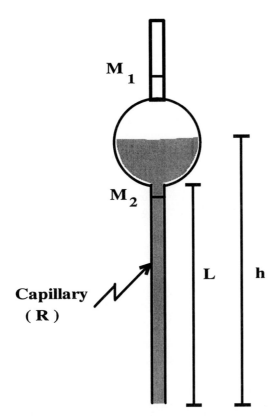

The sketch shows a portion of a <u>capillary viscometer</u>. The fluid, initially at position M_1, slowly falls under gravitational forces through a capillary of radius, R. The flow in the capillary is essentially laminar and the viscosity is measured by knowing the density of the fluid, ρ, and measuring the time it takes for a known volume of the fluid to pass between M_1 and M_2.

(a) Assuming that the flow is fully developed and at "quasi" steady-state, show that the viscosity is related to the measured time via $\dfrac{\mu}{\rho} = A t$ and then show how A is related to the parameters of the physical system.

(b) For a particular viscometer, the capillary radius is 0.23 mm, the fluid volume is 13.5 cc and the manufacturer claims that $A = .0031 \dfrac{\text{mm}^2}{\text{s}^2}$. Compare your calculation of A with the manufacturer's value and discuss why they might be different.

5-16 A particular non-Newtonian fluid is a <u>Bingham Plastic</u> where, in one-dimensional transport, the momentum flux is related to the velocity gradient by

$$\tau_{ij} = \tau_0 - \mu \frac{dv_j}{dx_i}$$

where τ_0 is the *yield stress*. Obtain an expression for the velocity distribution for a Bingham Plastic in Couette Flow between horizontal parallel plates (width, H) where the top plate has a velocity V and the bottom plate is stationary.

5-17 The "no-slip" boundary condition is typically assumed for most problems in momentum transport. However, if slip should occur, then Jackson (J. Rheol., <u>30</u>, p. 907, 1986) suggests that the following boundary condition be used

$$v\,|_{wall} = V + \beta \left(\frac{dv}{dz} \right)_{|wall}$$

where V is the velocity of the wall and v is the velocity of the fluid.

(a) Apply this boundary condition at the upper wall of the Couette flow problem discussed in Section 5-2 in the text in the <u>presence</u> of a pressure gradient and obtain an expression for the velocity profile between the walls.

(b) Show that this solution is equivalent to Equation (5.8) in the text when there is no slip at the wall.

5-18 In <u>Magneto-Hydrodynamic (MHD) Flows</u> an electrically charged fluid flows between the poles of a magnet. If an electric field is externally applied at right angles to both the flow and the magnetic field, then the system behaves as an electromagnetic pump. If there is no external electrical field, then the induced potential can be measured and the system becomes an electromagnetic flow meter.

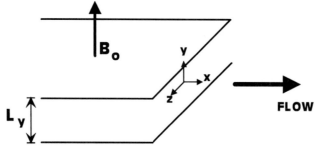

In either case, a <u>body force</u> (i.e., a force/vol) is generated according to the vector-cross product of the current density, $\vec{J}$, and the magnetic field, $\vec{B}$; i.e., $\vec{J}$ x $\vec{B}$. For an electrically conducting Newtonian fluid flowing under the influence of a pressure gradient between two stationary parallel plates as depicted in the sketch, it can be shown from Ohm's Law that the induced current, J_z , is related to the fluid velocity by, $J_z = \sigma \, v_x \, B_o$, where σ is the electrical conductivity of the fluid. Derive expressions for both the velocity profile and the average velocity in this situation if the flow is fully developed and at steady-state.

5-19 Another example of a non-Newtonian fluid is a <u>power law</u> fluid; i.e., a fluid where the relation between the momentum flux and the velocity gradient is of the form $N_{By} = J_{By}$. Derive an expression for the velocity profile of a power law fluid flowing in fully developed, steady-state, laminar flow in a horizontal tube.

5-20 Two immiscible Newtonian fluids (`A and `B) are flowing in stratified flow in a horizontal, rectangular duct. The flow is fully developed, steady-state and laminar. The duct is H units high and W units wide ($\dfrac{W}{H} >> 1$). The interface between the two fluids is smooth (no ripples) and is located at the center of the duct. Derive expressions for the velocity profiles for each fluid.

5-21

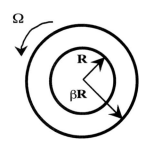

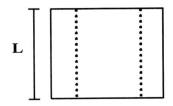

The accompanying sketch shows a concentric-cylinder viscometer, commonly used to measure the viscosities of liquids. The outer cylinder is rotated at a constant angular velocity, Ω, and the torque, T, required to turn the outer cylinder is measured. For this physical system, and for a Newtonian fluid, it can be shown (see Problem 8-11) that the differential equation describing v_θ is $\dfrac{d}{dr}\left(\dfrac{1}{r}\dfrac{d}{dr}(r\, v_\theta) \right) = 0$

For this geometry, $\tau_{r\theta}$ can be related to the velocity gradient of a Newtonian fluid by

$$\tau_{r\theta} = -\mu\left[r\frac{d}{dr}(\frac{v_\theta}{r}) \right]$$

Using the above equations together with the appropriate boundary conditions, solve for the torque at the outer cylinder and state how you would use this relationship to measure the viscosity of a Newtonian fluid.

CHAPTER 6

The Transport Coefficients

Up to this point, we have been using the transport coefficients in our analyses without really saying much more than they are properties which can be found in a handbook. However it has already been pointed out that this is not always the case since diffusivity values depend on the properties of other species in the media and, for fluids which do not follow Newton's Law of Viscosity (non-Newtonian fluids), viscosity values will depend on the existing velocity gradients. In this chapter we address the prediction and behavior of the transport properties in more detail. The first three sections deal with the prediction of the transport equations in gases, liquids, and solids, while the last section of the chapter specifically discusses the complications introduced by non-Newtonian fluids. It should be pointed out that these discussions are only meant to be introductory; entire text books have been written in each of these subject areas. Nevertheless, exposure to these subjects at this level should be sufficient to allow for more comprehensive studies by consulting any of the appropriate references.

6-1 TRANSPORT PROPERTY PREDICTIONS IN DILUTE, PURE GASES

The prediction of physical properties, whether thermodynamic or transport, is an extremely important field. The subject is so vast and there are so many new results being published that to attempt to do justice to it would be impossible. However, an excellent starting point is the reference text by Reid, Prausnitz and Sherwood. [1] While the prediction of the transport properties of liquids is complicated by our fundamental lack of a consistent theoretical model for the liquid state, the same is not true of gases, particularly at low densities (dilute). The manner in which this is approached is to use an appropriate model for the gas behavior and then to analyze the mechanisms by which momentum, energy, and mass are transported. Since we are dealing with the actual gas molecules, the net transport fluxes are due to molecular transport and thus the phenomenological laws will apply. By comparing the net transport fluxes to the phenomenological laws, we can then arrive at relationships to predict the viscosity, thermal conductivity, and diffusivity of a gas.

6-1.a Predictions Based on the Elementary Kinetic Theory

One model we can use for gas behavior is the so-called "billiard ball" model or the elementary kinetic theory (EKT) of gases. Here all molecules are considered to be rigid spheres and are assumed to be unaffected by the presence of other molecules until they actually collide. As a result, this theory is restricted to dilute gases where the influence of neighboring molecules is reduced since the average distance between molecules is many times their molecular diameter. One way to determine whether a gas is sufficiently dilute or not is to see if the ideal gas law is valid. If it is, then the requisite conditions are met. However, it is also possible for the gas density to be so low (high vacuum, for example) that we lose the ability to use statistical averaging. In such cases, the probability that a gas molecule collides with another gas molecule is less than the probability that the molecule will collide with the surrounding walls. This state of affairs is called <u>free molecule flow</u> and is characterized by the Knudsen number, Kn being greater than 1.0, or

$$Kn \equiv \frac{\lambda}{L} >> 1$$

where λ is the "mean free path" of a molecule (the average distance traveled between collisions) and 'L' is a characteristic dimension of the physical system. Note that Kn will be greater than one when λ is large (high vacuum) or when the dimensions of the physical system are quite small (in micropores, for example). Assuming then, that $Kn << 1$ but that the gas density is sufficiently low that molecule - molecule interactions are not significant,[*] we can use the results of the kinetic theory of gases [2] to decide the random motion of the gas molecules.

In such a gas, the EKT predicts that the average velocity (relative to any bulk fluid velocity) due to molecular motion is given by

$$\bar{u} = \sqrt{\frac{8KT}{\pi m}} \tag{6.1}$$

---

[*] Generally speaking, this occurs when the reduced pressures ($P_r = P/P_c$) are less than about 0.6 and when the reduced temperatures are greater than about 1.3.

where K is the Boltzmann constant, T is the absolute temperature of the gas, and m is the mass of one molecule. There are also some other useful mathematical relations which derive from the EKT; specifically:

$$Z = number\ of\ molecules\ per\ unit\ area\ per\ unit$$
$$time\ crossing\ a\ plane\ from\ one\ side \qquad\qquad (6.2)$$
$$= \frac{1}{4} n\bar{u}$$

where n is the *number density* of the molecules; i.e., the number of molecules per unit volume of gas. The mean free path in the gas can be calculated from

$$\lambda = avg.\ distance(in\ all\ directions)\ travelled$$
$$by\ a\ molecule\ between\ collisions \qquad\qquad (6.3)$$
$$= \frac{1}{\sqrt{2\pi}\,d^2\,n}$$

where d is the diameter of the molecule, and, more specifically

$$a = average\ distance(in\ one\ direction)$$
$$travelled\ between\ collisions \qquad\qquad (6.4)$$
$$= \frac{2}{3}\lambda$$

At this point, it is well to digress and to consider that in any fluid there are basically two sources of motion for the molecules. In simple molecular motion, the molecules move because they contain internal energy, part of which is manifested as microscopic kinetic energy. In a tank of compressed gas,, all the molecules are moving in the molecular mode but the total mass is not moving at all. If the valve on the tank is opened, the molecules leaving are subject to two types of motion. One is the molecular motion they had in the tank and the other is motion due to the fact that the exiting gas is moving as a unit (or in bulk). This latter movement is referred to as bulk flow. The distinction between these two is extremely important.

To examine the phenomena which occur during gas motion, consider a plane at y as shown in Figure 6-1. Molecules at both planes $y+a$ and $y-a$ experience random collisions and we can track those molecules which happen to cross plane y after the collisions at $y \pm a$. Furthermore, we can also tabulate their velocities, energy content, and species identity as they cross. By so doing, we can then calculate the momentum and kinetic energy associated with the

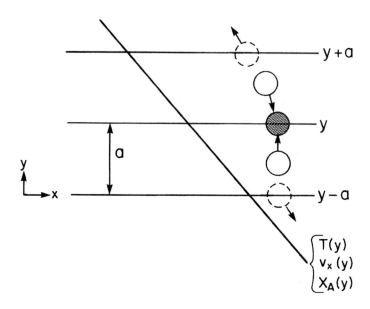

Figure 6-1 Molecular Collisions in a Dilute Gas

molecules crossing plane y as well as the number of any particular species. Assume that the gas is also being subjected to bulk motion such that there is a bulk velocity profile in the gas; that is, the bulk velocity at plane $y-a$ is greater than that at y. Similarly, we can superimpose both a bulk temperature and a bulk concentration gradient on the gas. In other words, not only do the gas molecules at $y-a$ have a higher velocity, but they also have a higher temperature and a higher species concentration of A at $y-a$ than they do at $y+a$. We will deal with the net quantities of energy and species A crossing plane y shortly; but first let's tabulate the net momentum crossing y.[#]

    Because of the bulk velocity gradient, molecules that collide at $y-a$ and do not suffer another collision until they arrive at y will have a greater momentum than those molecules that are still at y. If now a molecule arriving at plane y (from plane $y-a$) undergoes a collision with one of the molecules at plane y, there will be an exchange of momentum; i.e., momentum will be <u>added</u> to molecules at plane y. The net effect is that, due to random molecular collisions, there is a transport of momentum from regions of higher bulk velocity to regions of lower bulk velocity. Note also, that even though there are always molecular collisions, there

---

[#]   This is one of the few occurrences in transport phenomena where momentum transport is easier to deal with than energy or mass transport.

would be no momentum transport if adjacent planes were all at the same bulk velocity. Of course, it is important to keep in mind that to talk about particular molecules is very simplistic, in reality we are discussing the behavior of particular molecules, <u>on the average.</u>

Since Newton's law of viscosity is really the defining equation for viscosity, a logical approach would be to formulate the equivalent of Equation (2.13) in terms of the behavior of the gas molecules. The first requirement is to be able to express the <u>net</u> momentum transport at y in terms of a momentum flux (to be equivalent to τ_{yx}). Thus if the momentum carried (on the average) by individual molecules is known in addition to the number of molecules per unit time, per unit area which cross plane y, then the multiplication of these two quantities yields the net momentum flux at y. The momentum carried per molecule is simply mv_x and the average number of molecules per unit time per unit area which cross a plane from one side, is given by Z Equation (6.2). Thus the net momentum flux at y is given by

$$\tau_{yx} = Zm\,v_{x\,|\,y\text{-}a} - Zm\,v_{x\,|\,y\text{+}a} \tag{6.5}$$

Note that we have written (6.5) so that τ_{yx} is positive since $v_{x\,|\,y\text{-}a} > v_{x\,|\,y\text{+}a}$. Over the short distance of $2a$ * it is a good assumption that the velocity profile is linear, or

$$v_{x\,|\,y\text{-}a} = v_{x\,|\,y} - a\frac{d\,v_x}{dy}$$

$$\tag{6.6}$$

$$v_{x\,|\,y\text{+}a} = v_{x\,|\,y} + a\frac{d\,v_x}{dy}$$

since $\dfrac{d\,v_x}{dy}$ is inherently negative. Substituting Equations (6.6) and (6.2) into Equation (6.5) we obtain

$$\tau_{yx} = -\frac{1}{2}n\bar{u}m\,a\frac{d\,v_x}{dy} \tag{6.7}$$

If (6.7) is compared to Newton's Law of Viscosity, (Equation (2.13)

$$\tau_{yx} = -\mu\frac{d\,v_x}{dy} \tag{2.13}$$

---

* For example, in air at atmospheric pressure and room temperature, a ~ 4x10⁻⁶ cm.

then it is apparent that

$$\mu = \frac{1}{2} \, numa$$

or

$$\mu = \frac{1}{3} \, \rho \bar{u} \lambda \qquad (6.8)$$

where we have substituted for a in terms of λ (Equation 6.4) and recognized that $\rho \equiv nm$. In order to get μ in terms of measurable variables, Equations (6.1) and (6.3) can be substituted into (6.8) for $\bar{u}$ and λ, so that

$$\mu = \frac{2}{3\pi^{3/2}} \frac{\sqrt{mKT}}{d^2} \qquad (6.9)$$

To be able to use (6.9), a value of the collision diameter, d, is required which must be obtained from an experimental value of μ. Even so this equation does not, in general, give quantitatively accurate values for $\mu,$. However this simplified treatment does give a qualitatively correct picture of the variation of the viscosity of gases with temperature and pressure. The independence of viscosity with pressure is verified by experimental data up to about 10 atm (where the "dilute" assumption ceases to be valid), and its dependence on $T^{.5}$ is close to actual observations of $T^{.7}$. Note also that the viscosity of a gas <u>increases</u> with <u>increasing temperature</u>, a consequence of the mechanism by which momentum is transported in a gas. That is, higher temperatures result in a higher value of $\bar{u}$ and thus there are proportionately more collisions (higher values of Z) capable of transporting momentum.

We can use the same type of molecular model to predict the <u>thermal conductivity of a dilute gas</u> and, while there is a clear analogy with the molecular transport of momentum in such a gas, there are also some differences. Since we are dealing with energy transport here, we require one additional piece of information from the kinetic theory of gases - the mean translational energy associated with a molecule. The Elementary Kinetic Theory predicts that this is given by

$$\frac{1}{2} m u^2 = \frac{3}{2} KT \qquad (6.10)$$

Note that this is strictly for translational energy; i.e., the kinetic energy associated with rectilinear motion. Note also that this is an average of u^2, not the square of $\bar{u}$. Of course when dealing with the Elementary Kinetic Theory, we are assuming that all molecules are monatomic and thus they can exhibit energy only in terms of translational energy. As we shall see, polyatomic molecules can also absorb energy by changing vibrational and rotational modes; this is a complication in molecular energy transport that was not encountered in the model to predict viscosities of dilute gases.

Referring back to Figure 6-1, consider the net energy flux crossing the plane at y. Molecules which have undergone their last collision at $y - a$ will carry kinetic energy $\frac{1}{2} m u^2 \big|_{y-a}$ in the positive y-direction and molecules from the plane at $y + a$ will carry kinetic energy $\frac{1}{2} m u^2 \big|_{y+a}$ in the negative y-direction.

Consequently the net energy flux crossing the y-plane in the positive y-direction is

$$q_y = net\ energy\ flux = \left[\frac{collisions}{area\text{-}time}\right] \circ \left[\frac{net\ energy}{collision}\right]$$

(6.11)

$$q_y = Z \circ m \left[\frac{1}{2} u^2 \big|_{y-a} - \frac{1}{2} u^2 \big|_{y+a}\right]$$

Equation (6.11) is not in a useful form to extract the thermal conductivity since it is not expressed in terms of the temperature gradient as is the case with Fourier's Law. Therefore to convert this equation into the desired form, we substitute for the kinetic energy in terms of temperature from Equation (6.10) to obtain

$$q_y = Z \left[\frac{3}{2} KT \big|_{y-a} - \frac{3}{2} KT \big|_{y+a}\right]$$

(6.12)

Once again, we can assume that over the small distance, 2a, the temperature gradient is linear, and since $a = 2/3\lambda$

$$T\big|_{y-a} = T\big|_y - a \frac{dT}{dy} = T\big|_y - \frac{2}{3} \lambda \frac{dT}{dy}$$

(6.13)

$$T_1 \big|_{y+a} = T_{1_y} + a \frac{dT}{dy} = T_{1_y} + \frac{2}{3} \lambda \frac{dT}{dy}$$

Substituting (6.13) into (6.12) and simplifying, we obtain

$$q_y = -2KZ\lambda \frac{dT}{dy} \tag{6.14}$$

Comparing Equation (6.14) term-by-term with FoFthermodynamic urier's Law, Equation (2.10),

$$q_y = -k \frac{dT}{dy} \tag{2.10}$$

we see that

$$k = 2KZ\lambda \tag{6.15}$$

Substituting for Z and λ from the results of the EKT, Equations (6.2) and (6.3), k is given by

$$k = \frac{1}{d^2} \sqrt{\frac{K^3 T}{\pi^3 m}} \tag{6.16}$$

Note that the results predict a square root dependency on temperature and no effect of pressure which is the same dependency predicted for viscosity as a function of temperature and pressure.

Due to the similarity of the results for viscosity and thermal conductivity, it is interesting to examine the relationship between the two as given by the EKT. In the process of making this comparison it will be useful to write the results in terms of the heat capacity of the gas at constant volume, $\tilde{c}_V$. By definition

$$\tilde{c}_V \equiv \left(\frac{\partial \tilde{Q}}{\partial T} \right)_V$$

But, from the first law of thermodynamics at constant volume and no "shaft" work, $d\tilde{E} = d\tilde{Q}$ so that

$$\tilde{c}_V = \left(\frac{\partial \tilde{E}}{\partial T} \right)_V \tag{6.17}$$

$\tilde{E}$ can be expressed in terms of temperature from the EKT

$$\tilde{E} = \frac{molecules}{mole} \frac{energy}{molecule} = \tilde{N} \left(\frac{1}{2} mu^2 \right) = \frac{3}{2} \tilde{N}KT = \frac{3}{2} R_g T \tag{6.18}$$

Since the gas constant, R_g, is equal to the product of Avogadro's number and the Boltzmann constant, $\tilde{N}K$, Equation (6.18) can be differentiated according to Equation (6.17) to yield

$$\tilde{c}_V = \frac{3}{2} R_g \tag{6.19}$$

or

$$\hat{c}_V = \frac{3}{2} \frac{R_g}{M}$$

With this in mind, the EKT prediction for thermal conductivity, Equation (6.16), can be compared with the prediction for viscosity, Equation (6.9), by forming the ratio, k/μ, which gives

$$\frac{k}{\mu} = \frac{3}{2} \frac{K}{m} = \frac{3}{2} \frac{\tilde{N}K}{\tilde{N}m} = \frac{3}{2} \frac{R_g}{M} = \hat{c}_V \tag{6.20}$$

So, the EKT predicts that the thermal conductivity and viscosity are simply related by a thermodynamic property, $\hat{c}_V$, and estimates for one can be directly calculated from measurements of the other.

Now let's turn our attention to the prediction of <u>diffusion coefficients of dilute gases</u> by means of the EKT. As pointed out in Chapter 4, one of the major difficulties in mass transport is that we are always dealing with mixtures. This poses major difficulties when trying to predict diffusion coefficients in dilute gases since the kinetic theory of gases can barely give an adequate representation of single component gases. Consequently, the best we can do is to consider the "self" diffusion of a particular species, (two isotopes, for example). Recalling Equation (2.11), the y-component of the diffusion flux of A with respect to fixed coordinates in a mixture of A and its isotope, A^*, is given by

$$J_{A_y} = -C D_{AA^*} \frac{d X_A}{dy} \tag{6.21}$$

In order to predict D_{AA^*} we need to take a molar balance of A crossing the y-plane in Figure 6-1 and then express the balance in terms of a molar flux so that we can compare the results with (6.21). Given that the flux of species A in one direction is simply the volumetric concentration of A multiplied by its velocity of transport (e.g., $(n\,X_A)\,\bar{u}$), then the net molecular flux of A in one direction and at y is

$$Net\;molecular\;flux\;of\;A\;at\;y = Z\left(X_{A_{|\,y-a}} - X_{A_{|\,y+a}}\right) \tag{6.22}$$

Note that the units of Equation (6.22) are molecules A/area-time. Thus, in order to compare the molecular model with Equation (6.21) we need to express the molecular flux in terms of moles and this can be done by dividing Equation (6.22) by Avogadro's number. Then, with Z given by Equation (6.2), we arrive at

$$J_{A_y} = \frac{n\bar{u}}{4\,\tilde{N}}\left(X_{A_{|\,y-a}} - X_{A_{|\,y+a}}\right) \tag{6.23}$$

Once again, if we use a linear approximation for the mole fraction gradient

$$X_{A_{|y-a}} = X_{A_{|y}} - a\frac{dX_A}{dy}$$

$$X_{A_{|y+a}} = X_{A_{|y}} + a\frac{dX_A}{dy}$$

Equation (6.23) becomes, after employing Equation (6.4) and noting that $\dfrac{n}{\tilde{N}} = C$,

$$J_{A_y} = -\frac{\overline{Cu}\lambda}{3}\frac{dX_A}{dy}$$

which, when compared with Equation (6.21), predicts that

$$D_{AA^*} = \frac{\bar{u}\lambda}{3} \tag{6.24}$$

Before proceeding with the substitution for $\bar{u}$ and λ from the kinetic theory we must first confront the problem of molecular properties. Since the simplest system we will encounter is a binary system composed of A & B, D_{AA^*} is not of much use. In other words, since the quantities we need from the kinetic theory are all functions of the molecular diameter and we have two molecular species present, we must decide on how to calculate these properties in order to predict D_{AB}. The approach which has been surprisingly successful is to assume uniform molecular diameters, d_{AB}, and molecular masses, m_{AB}, which are calculated according to

$$d_{AB} \equiv \frac{1}{2}(d_A + d_B)$$

(6.25)

$$\frac{1}{m_{AB}} \equiv \frac{1}{2}\left(\frac{1}{m_A} + \frac{1}{m_B}\right)$$

(6.26)

When this kind of a molecular model is combined with the results for the EKT,[#] we arrive at

$$D_{AB} = \frac{2}{3 d_{AB}^2} \sqrt{\frac{K^3}{\pi^3 m_{AB}}} \frac{T^{\frac{3}{2}}}{P}$$

(6.27)

where the pressure, P, is given by the ideal gas law,[##] $P = C R_g T = nKT$.

Notice that the binary diffusivity depends on pressure whereas both viscosity and thermal conductivity were found to be independent of pressure. The reason for the appearance of pressure is that here we are dealing with the net <u>movement</u> of molecules. Increased pressure means more molecules/volume and hence increased difficulty for one species to move through another. With the molecular transport of momentum and energy, we were not concerned with the movement itself, but rather with the <u>results</u> of molecular collisions (momentum and energy exchanges). Since in all cases we assumed a dilute gas, the collision frequency was independent of pressure and consequently so were μ and k. However, in diffusional mass transport we are interested in the <u>penetration</u> of one species through the other and increased concentrations make it all the more difficult for penetration to occur.

6-1.b Predictions Based on the Chapman-Enskog Theory

Whereas the results obtained with the EKT give reasonable predictions for simple monatomic gases and give an approximately correct temperature dependence for most gases, there is a refinement which improves the predictive capability in dilute gases - the <u>Chapman-Enskog</u>

[#]   See Problem 6-1 at the end of the chapter.

[##]   Note that the ideal gas law is valid under the same assumptions required by the EKT.

Theory. This theory improves on one of the most restrictive assumptions connected with the elementary kinetic theory - the assumption that there are no intermolecular forces. To improve on this, the theory assumes a long-range interaction potential between two colliding molecules so that the molecules do not actually collide before influencing one another. The key to this interaction is the potential function and the most commonly used function is the Lennard-Jones Potential

$$\varphi \ (r) = 4\varepsilon \left[\left(\frac{\sigma}{r} \right)^{12} - \left(\frac{\sigma}{r} \right)^{6} \right]$$

where r is the radial distance between the two particles. Note that the magnitude of the potential, $\varphi(r)$, is dependent on two parameters: ε and σ (the collision diameter). As might be expected, these two parameters influence the prediction of the transport coefficients. The detailed derivations of the dependency of the transport coefficients on ε and σ are given by Hirschfelder, Curtis and Bird [3], but it will serve our purposes to merely state the results.

$$\mu = (2.669) \ 10^{-8} \ \frac{\sqrt{MT}}{\sigma^2 \Omega_\mu} \tag{6.28}$$

$$k = (8.322) \ 10^4 \ \frac{\sqrt{T/M}}{\sigma^2 \Omega_k} \tag{6.29}$$

$$D_{AB} = \frac{(1.8583)10^{-7} \sqrt{T^3 \left(\frac{1}{M_A} + \frac{1}{M_B} \right)}}{P \ \sigma_{AB}^2 \Omega_D} \tag{6.30}$$

In these equations, T is in degrees K, P in kPa and σ in nm; which produces values of μ in Pa-s, k in W/m-K and D_{AB} in m²/s.

If Equations (6.28 - 6.30) are compared with the results obtained with the Elementary Kinetic Theory, Equations (6.9), (6.16) and (6.27), the only differences lie in the presence of the "collision integrals" ($\Omega_{\mu,k,D}$) " and the fact that the molecular diameter, d, is replaced by the collision diameter, σ. Since the collision integrals are functions of temperature (and ε), the dependency of the transport coefficients on temperature is different than that obtained from the EKT and, in fact, is closer to that which is observed experimentally. The collision integrals for

μ and k are identical and are given along with Ω_D, as a function of $\dfrac{KT}{\varepsilon}$ in Appendix C. A word of caution is in order here. Although the Chapman-Enskog theory is technically sound, it is by no means a thorough description of intermolecular forces. So, in many respects, it is an improvement on the results of elementary kinetic theory simply because it has an additional parameter, ε, with which to fit experimental data.[*] Thus it is important to use <u>pairs</u> of ε and σ values from the same source since there are a large combinations of ε and σ which fit the experimental data to the same degree of confidence. Appendix C also lists sets of ε and σ for some of the more common gases. For other gases, ε and σ can be estimated from empirical correlations of the critical properties [1]. Finally, as was necessary in the EKT, we need to modify the ε and σ parameters for binary diffusivity calculations; specifically

$$\varepsilon_{AB} = \sqrt{\varepsilon_A \varepsilon_B} \qquad\qquad (6.31)$$

and

$$\sigma_{AB} = \frac{\sigma_A + \sigma_B}{2} \qquad\qquad (6.32)$$

EXAMPLE 6-1: Estimating the Diffusion Coefficients for HCN-In-Air

Use the Chapman-Enskog Theory to estimate the diffusivity of HCN in air at $0\,^{\circ}\mathrm{C}$.

In order to use Equation (6.30), it is necessary to compute values of the collision diameter, σ_{AB} and the collision integral, Ω_D. The collision integral, in turn, depends on the parameter $\dfrac{KT}{\varepsilon}$. The appropriate parameters for a binary system are given by Equations (6.31) and (6.32), and the individual values for HCN and air are given in Table C-1 in Appendix C.

$$\sigma_{AB} = \frac{1}{2}(36.3 + 36.17) = 36.235 \text{ nm}$$

$$\frac{\varepsilon_{AB}}{K} = \sqrt{(301)(97)} = 170.87$$

Interpolating in Table C-2, gives a value for the collision integral, $\Omega_D = 1.167$ and then the diffusivity can be estimated from Equation (6.30)

---

[*] In fact, the ε and σ parameters are normally obtained by comparing the results of the Chapman-Enskog Theory to experimental values (usually for μ).

$$D_{AB} = 1.8583(10)^{-7} \frac{\sqrt{(273)^3 (\frac{1}{27} + \frac{1}{28.97})}}{101.3 \ (36.235)^2 \ 1.167} = 2.43(10)^{-5} \frac{m^2}{s}$$

This value is about 40% higher than the experimental value of 0.173 cm² /s (Perry's Handbook of Chemical Engineering, 5ᵗʰ ed. p. 3-222). This large a discrepancy would normally cause one to investigate the source of the experimental value, since experimental values are usually quite old and subject to criticism (see Problem 6.12)

6-1.c Prandtl Number Predictions.

A useful dimensionless parameter which relates molecular momentum transport to molecular energy transport is the Prandtl number. It is defined by[#]

$$Pr \equiv \frac{\hat{c}_p \mu}{k} \tag{6.33}$$

Substituting for $\frac{\mu}{k}$ from Equation (6.20), we see that the EKT predicts that

$$Pr = \frac{\hat{c}_p}{k/\mu} = \frac{\hat{c}_p}{\hat{c}_V} \tag{6.34}$$

At this point we can use still another result from the kinetic theory of gases which states that, for a monatomic gas,

$$\hat{c}_p = \frac{5}{2} \frac{R_g}{M} \tag{6.35}$$

Substituting Equations (6.19) and (6.35) into (6.34), we find that the EKT predicts that the Prandtl Number is a constant; specifically

$$Pr = \frac{5/2 \ R/M}{3/2 \ R/M} = 1.67$$

which predicts that all monatomic, ideal gases have the same Prandtl number which is independent of temperature.

[#] Note that it can also be written in terms of "diffusivities"; i.e., $Pr = \nu/\alpha$.

On the other hand, if we use the Chapman Enskog theory to predict the relationship between conductivity and viscosity, it can be shown that (see Problem 6-3)

$$\frac{k}{\mu} = \frac{5}{2}\hat{c}_v \tag{6.36}$$

and the predicted Prandtl number is

$$Pr = \hat{c}_p \frac{\mu}{k} = \frac{2}{5}\frac{\hat{c}_p}{\hat{c}_v}$$

Substituting for $\hat{c}_v$ and $\hat{c}_p$ for a monatomic gas (Equations 6.19 and 6.35),

$$Pr = \frac{2}{5}\frac{5/2\ R_g\ /M}{3/2\ R_g\ /M} = \frac{2}{3}$$

a value which is still independent of temperature but which is 40% of the value predicted by the Elementary Kinetic Theory.

Thus, even though the individual results predicted by the EKT for viscosity and thermal conductivity are reasonable, the results for the ratio, k/μ, differ from the more accurate Chapman-Enskog Theory by a factor of 2.5 [compare Equation (6.36) with Equation (6.20)]. Stated another way, serious error could arise if we attempted to predict k from μ (or vice versa) on the basis of the Elementary Kinetic Theory.

Given the large discrepancy in Prandtl Number predictions with the two theories, one additional point should be made concerning the manner in which molecules absorb energy. Although it is true that the viscosity predictions based on the Chapman-Enskog Theory hold reasonably well for polyatomic molecules (remember this theory also assumes only monatomic molecules), the same is not true for thermal conductivity predictions. This is because it has been assumed that the energy exchanged during a collision was manifested only by a change in translational energy. Since polyatomic molecules can also vibrate and rotate, it is possible for a portion of the energy exchange to appear as changes in the vibrational and rotational characteristic of the molecules. To handle this problem, Euken [4] developed a semi-empirical correction to Equation (6.36) (the *Euken Correction*) in the form of

$$\frac{k}{\mu} = \left(\hat{c}_p + \frac{5}{4}\frac{R_g}{M}\right) \tag{6.37}$$

Note that if a molecule is monatomic, $\hat{c}_p = \dfrac{5}{2}\dfrac{R_g}{M}$ and Equation (6.36) results.

To summarize, the three methods give the following Prandtl Number predictions

Elementary Kinetic Theory: $\quad\quad\quad Pr = 1.667$

Chapman-Enskog (Monatomic): $\quad\quad Pr = .667$

Euken Correction (Polyatomic): $\quad\quad Pr = \dfrac{\tilde{c}_p}{\tilde{c}_p + 1.25\,R_g}$

Note that only the Euken Correction predicts a Prandtl Number which is a function of temperature (due to the dependency of $\tilde{c}_p$, on temperature). Table 6-1 shows the Prandtl Number variation with temperature for a number of gases. As can be seen, the temperature variation is slight and the numerical values are close to those predicted by the Chapman-Enskog Theory even though all the molecules are polyatomic and some are polar.

6-2 TRANSPORT PROPERTIES IN LIQUIDS AND SOLIDS

6-2.a Liquids

Unfortunately, despite the overwhelming presence of liquid water on the planet, our ability to model the liquid state lags far behind our ability to do so in the gaseous and solid states. As a result, the prediction of liquid transport properties is largely empirical and texts such as Reid, Prausnitz and Sherwood [1] should be consulted in order to determine the most appropriate method for a given situation.

With respect to the *viscosity of liquids* we refer here only to Newtonian fluids where the viscosity is an inherent property of the species and independent of the shear rate (velocity gradient) to which the fluid is exposed. The subject of non-Newtonian fluids is discussed in Section 6-4. Even in Newtonian fluids, there are only empirical relations with which to predict viscosity values and these are generally valid over specific temperature ranges. The most common means of correlating viscosity data of liquids is given in Equation 6.38,

$$\mu = A\exp\frac{B}{T} \tag{6.38}$$

where T is the absolute temperature and A and B are parameters chosen to fit the experimental data. This equation is valid above a <u>reduced temperature</u> ($T_r = \dfrac{T}{T_c}$) of about .75 and up to the normal boiling point. At lower reduced temperatures, the same type of logarithmic relationship holds but the parameters, A and B, will differ. In general, the parameter, B, is a positive number and the viscosity of Newtonian liquids <u>decrease with increasing temperature</u>. Equation (6.38) is best used with two experimental values of the viscosity but, failing that, Reid, Prausnitz and Sherwood list values of A and B for a number of liquids [5].

In discussing predictive methods for the *thermal conductivities of liquids*, it should be borne in mind that experimental values are not easy to obtain. Because liquids are almost always susceptible to natural convection currents whenever there is a temperature differential, errors in experimental measurements are often on the same order as differences in the various predictive methods. For <u>organic</u> liquids, one predictive relationship which has stood the test of time is

$$k = (4.184)10^{-5}\left[\frac{(88.0-4.94H)}{\Delta \tilde{S}^*}\right]\left(\frac{.55}{T_r}\right)^N \tilde{c}_p C^{4/3} \tag{6.39}$$

where k is in W/m-K, C is the molar density in kg moles/m^3, T is the reduced temperature and

$$\Delta S^* = \frac{\Delta \tilde{H}_{vb}}{T_b} + R_g \ln\frac{273}{T_b}$$

with: T_b the normal boiling point ($^\circ$K), $\Delta \tilde{H}_{vb}$ the molal heat of vaporization at T_b

Reid, Prausnitz and Sherwood [6] list values of the parameters, H and N. Equation (6.39) is applicable for all liquids except highly polar, multihydroxyl and multi-amino compounds. The thermal conductivities of most liquids <u>increase with increasing temperature</u> whereas they <u>decrease</u> with increasing temperature in the latter compounds. In either case, it is reasonable to assume a linear relationship between the thermal conductivities of liquids with temperature; i.e.,

$$k = k_o[1+\alpha(T-T_o)] \tag{6.40}$$

where $\alpha = \dfrac{dk}{dT}$ and is a constant for a given compound, typically varying between $-5 \times (10)^{-4}$ and $-2 \times (10)^{-3}$ K^{-1}.

Table 6-1 Variation of Prandtl Number with Temperature for Various Gases

GAS	TEMPERATURE (°F)	PRANDTL NO.
Air	0	.721
	100	.703
	200	.654
	300	.686
	600	.680
H_2O	212	1.06
	250	.994
	300	.963
	500	.922
	800	.912
N_2	0	.719
	200	.690
	600	.686
O_2	0	.718
	200	.703
	600	.688
CO_2	0	.792
	200	.730
	600	.700

As far as the *diffusivity of liquids* is concerned, a simple hydrodynamic model has been found to be useful. This model visualizes diffusion as spherical molecules of A moving through a liquid B, where the diffusion coefficient is given by

$$D_{AB} = KT \frac{u_A}{F_A} \tag{6.41}$$

Here u_A is the steady state velocity of A and F_A is the force acting on A. Since molecules move very slowly in liquids, their motion can be described in terms of *creeping flow* (i.e., very low Reynold's number flows) for which there are accurate mathematical solutions to describe the momentum transport [7]. If F_A is calculated from the creeping flow solution over a sphere, then Equation (6.41) becomes

$$D_{AB} = \frac{KT}{6\pi \, \mu_B \, R_A} \tag{6.42}$$

where μ_B is the viscosity of the surrounding liquid (solvent) and R_A is the radius of molecule A. It is interesting that the viscosity of the solvent is a factor in this prediction scheme. One word of caution is in order, Equation (6.42) was derived by assuming that F_A was a symmetrical hydrodynamic force. If species A is a polar molecule this will not be the case and corrections to (6.42) are necessary. A semi-empirical approach which has been found useful for those situations is the Wilke-Chang Equation [8]

$$D_{AB} = (1.17) \, 10^{-16} \left[\frac{T \sqrt{\psi \, M_B}}{\mu_B \, \tilde{V}_A^{.6}} \right] \tag{6.43}$$

where D_{AB} is given in m²/s, μ_B is the viscosity of the solvent in Pa-s, $\tilde{V}_A$ is the specific molar volume of the diffusing species in m³/kg mol and ψ is an association parameter for the solvent (B) which takes on values of: 1 for non-associated solvents, 1.5 for ethanol, and 2.6 for water. For comparison purposes, Table 6-2 lists some diffusion coefficients for common gases (in air) and liquids (in water).

Table 6-2 Diffusivities for Common Gases and Liquids

GASES (IN AIR)		LIQUIDS (IN H₂0)	
273 K, 1 atm	m²/s x 10⁴	293 K	m²/s x 10⁴
Acetone	.0816	O_2	1.8×10^{-5}
Ammonia	.213	CO_2	1.77×10^{-5}
Benzene	.0765	H_2	5.13×10^{-5}
CO_2	.136	CH_3OH	1.28×10^{-5}
CCl_4	.0614	NaCl	1.35×10^{-5}
H_2	.605	NaOH	1.51×10^{-5}
CH_4	.155	NH_3	1.76×10^{-5}
CH_3OH	.131	Cl_2	1.44×10^{-5}
n-octane	.05	HNO_3	2.98×10^{-5}
SO_2	.102	SO_2	1.75×10^{-5}
H_2O	.217		

6-2.b Transport Properties of Solids

It is obviously difficult to utilize the concept of viscosity in a solid. Most solids transport momentum very efficiently- a force applied to one end of a steel rod is immediately transmitted to the other end. In this sense, the viscosity can be thought of as an infinite value. While there are some solids that are somewhat deformable, these systems are usually described in terms of structural mechanics rather than in terms of viscosity. Fluids which have high viscosities are usually non-Newtonian; this is covered in Section 6-4.

The *thermal conductivity of solids* is relatively easy to measure and thus predictive methods are not as critical as they are with liquids. Nevertheless, it is worthwhile to discuss the topic in order to appreciate the wide range of values which may be encountered. For example, crystalline copper at very low temperatures has a thermal conductivity on the order of 1.2×10^4 W/m-K while cryogenic insulation can have values as low as 3×10^{-5} W/m-K. Table 6-3 lists the range of thermal conductivities for various solid materials.

The fact that metals have the highest values of thermal conductivity and are also good electrical conductors, leads to the conclusion that electron transfer is involved in the molecular transport of energy through solids. In fact, the <u>Weideman-Franz-Lorenz Correlation</u> states that $\dfrac{k}{\sigma_e\,T} = Constant$, where σ_e is the electrical conductivity. While this correlation only holds for pure metals, Equation (6.44) is a method of predicting the thermal conductivity of metals and their alloys

$$k = (2.61)10^{-4}\frac{T}{\rho_e} - \frac{(8.37)10^{-8}}{\hat{c}_p\rho}\left(\frac{T}{\rho_e}\right)^2 + (2.32)10^{-3}\frac{\hat{c}_p\rho^2}{MT} \tag{6.44}$$

where k is expressed in W/m-K, $\hat{c}_p$ in KJ/kg-K, ρ in kg/m^3, T in K, and ρ_e is the electrical resistivity in ohm-m.

Table 6-3 Thermal Conductivities of Solids

Substance	k (W/m-K)
Metals	3 - 450
Ceramics	.5 - 5
Insulators	.02 - .2

Whereas diffusion in gases occurs by means of molecular collisions and is visualized in liquids as occurring via hydrodynamic motion, *diffusion in solids* is vastly different. In general, solid-state diffusion is due to intra-crystalline movement of ionic species, usually along defects or via vacancies in the crystalline structure. In polycrystalline solids, diffusion can be much faster if it occurs along grain boundaries. As a result the entire subject is exceedingly complex and consequently we will not spend much time discussing it here. Because of the type of mechanisms involved, solid state diffusion is "activated." That is, the diffusion coefficients vary exponentially with temperature in an <u>Arrhenius</u> manner; namely

$$D_S = D_o \exp\left[\frac{-E_D}{RT}\right] \qquad (6.45)$$

Not only are the diffusion coefficients in solids much lower than in gases and liquids (values typically range from 10^{-10} to 10^{-18} m^2/s), but they are extremely sensitive to temperature with E_D values on the order of 250 KJ/mol. However, if grain boundary diffusion is prevalent, the apparent diffusion coefficients can be as high as 10^{-6} m^2/s.

6-3 TRANSPORT COEFFICIENTS IN MULTICOMPONENT MIXTURES

6-3.a Viscosity and Thermal Conductivity

Generally speaking, the composition of a liquid mixture does not have a strong effect on the transport properties of the mixture, since it is most common that one particular component in the species (the "solvent") will dominate and the specific coefficient will be largely determined by that particular component. The same is not true for gaseous mixtures however, and so we will examine methods useful for predicting the transport coefficients in those mixtures.

For dilute gases (low pressures), the Chapman-Enskog theory can be approximated and the viscosity of mixtures can be predicted by

$$\mu_{mix} = \sum_{i=1}^{n}\left[\frac{X_i\mu_i}{\sum_{j=1}^{n}X_j\phi_{ij}}\right] \qquad (6.46)$$

Wilke's method [9] can be used to estimate the parameters, ϕ_{ij}, (although it is not too accurate when H$_2$ is present):

$$\phi_{ij} = \frac{\left[1 + \left(\dfrac{\mu_i}{\mu_j}\right)^{1/2} \left(\dfrac{M_j}{M_i}\right)^{1/4}\right]^2}{\left[8 \left(1 + \left(\dfrac{M_i}{M_j}\right)\right)\right]^{1/2}} \tag{6.47}$$

where it should be noted that $\phi_{ii} = 1$, and ϕ_{ji} can be calculated from ϕ_{ij} by

$$\phi_{ji} = \frac{\mu_j\, M_i}{\mu_i\, M_j}\, \phi_{ij} \tag{6.48}$$

Identical relationships can also be used to predict the thermal conductivities of gas mixtures at low pressures. That is,

$$k_{mix} = \sum_{i=1}^{n} \left[\frac{X_i k_i}{\displaystyle\sum_{j=1}^{n} X_j \phi_{ij}} \right] \tag{6.49}$$

with the ϕ_{ij} parameters given by Equations (6.47) and (6.48).

6-3.b Diffusivity

If mixtures of two components cause problems with predicting diffusion coefficients, the difficulty with multicomponent mixtures can easily be appreciated. Add this to the complications presented by liquid systems and the situation is rapidly out of hand. Thus we will discuss, only briefly, multicomponent gaseous mixtures.

First of all, in the general gaseous mixture, multicomponent diffusion is a complicated function of the concentrations of all the species present. Rigorous application of the molecular theory of gases will lead to a set of equations relating the concentration gradients of the various species to the corresponding concentrations and molar fluxes. These equations are in terms of the individual binary diffusivities of all the various species and must be solved simultaneously.

While this method is feasible, it is not as simple as dealing with single transport coefficients. Thus the concept of the effective binary diffusivity, D_{im}, which is based upon analogy with Equation (4.11), is often used.

$$\vec{N}_i = -C D_{im} \vec{\nabla} X_i + X_i \sum_{j=1}^{n} \vec{N}_j \qquad (6.50)$$

where D_{im} is the diffusivity of component i in the mixture.

Actually, for an ideal gas mixture, a rigorous expression for the effective binary diffusivity can be obtained [10]

$$\frac{1}{C D_{im}} = \frac{\sum_{j=1}^{n} \left(\frac{1}{C D_{ij}}\right)(X_j N_i - X_i N_j)}{N_i - X_i \sum_{j=1}^{n} N_j} \qquad (6.51)$$

For the particular case where species i is diffusing into a stagnant gas mixture (i.e., $N_j = 0 \ all \ j \ne i$), then the rigorous Equation (6.51), simplifies to

$$D_{im} = \frac{(1 - X_i)}{\left[\sum_{\substack{j=1 \\ j \ne i}}^{n} \frac{X_j}{D_{ij}}\right]} \qquad (6.52)$$

There are some special cases that produce a further simplification of Equation (6.52); namely,

(a) For <u>small concentrations</u> of trace components mixed with a large concentration of species B

$$D_{im} \sim D_{iB}$$

and we can use the binary diffusivities of species i with B.

(b) For a mixture where all the binary diffusivities are equal[*]

$$D_{im} = D_{ij}$$

and any of the binary diffusivities are directly applicable.

---

[*]   Depending on the species present in the mixture, this is often a reasonable assumption.

6-4 Non-Newtonian Fluids

While the most commonly encountered fluids, water and air, are Newtonian, there are also a great number of industrial fluids which are not. Polymer solutions, slurries, pastes, and suspensions are all examples of fluids which exhibit non-Newtonian behavior. Basically, non-Newtonian fluids are those that do not follow Equation (2.13) and are often characterized by an "apparent" viscosity which varies with either the velocity gradient or the momentum flux. Since most of the work in this field has been in the hands of rheologists a good deal of the nomenclature and jargon is somewhat different than is common in engineering usage. Thus τ_{yx} is often referred to as a <u>shear stress</u> while the velocity gradient is termed the <u>shear rate</u>. Consequently, if the shear stress is plotted versus shear rate as shown in Figure 6-2, Newtonian fluids will plot in a straight line while non-Newtonian fluids will be nonlinear.

Shear-thinning fluids are characterized by an apparent viscosity (the slope of the curve in Figure 6-2) which decreases with increasing shear rate while *shear-thickening* fluids have an apparent viscosity which increases with increasing shear rate. The former fluids are sometimes referred to as <u>pseudoplastics</u> and the latter as <u>dilatants</u>. While these categories describe molecular momentum transport in most fluids, there is still one other category; <u>Bingham plastics</u>. As shown in Figure 6-2, Bingham plastics are characterized by a shear-shear rate behavior which is Newtonian once a <u>yield stress</u>, τ_0 is reached. The rheology of "slurries" (solid-liquid suspensions) is often described by a Bingham plastic model.

The curves sketched in Figure 6-2 can be misleading unless it is appreciated that the characteristics of these fluids can change substantially over large ranges of shear rate. Thus attempts to present a general equation [comparable to Equation (2.13)] for non-Newtonian fluids have met with only limited success. There have been many non-Newtonian models developed but in view of their limitations, probably the most useful is the <u>Ostwald-de Waele</u> model which is most commonly referred to as the <u>power-law</u> model. This model, given in Equation (6.53), is a two-parameter model

$$\tau_{yx} = -m \left| \frac{d\,v_x}{dy} \right|^{n-1} \frac{d\,v_x}{dy} \tag{6.53}$$

where m and n are parameters which result from the fitting of the shear - shear rate data. Note that the power law is written in the form given so that the molecular momentum flux will have the proper direction, dependent on the sign of the velocity gradient. The power law model is also convenient in the sense that it is compatible with Newtonian fluids ($n = 1$), as well as with dilatants ($n > 1$) and psuedoplastics ($n < 1$).

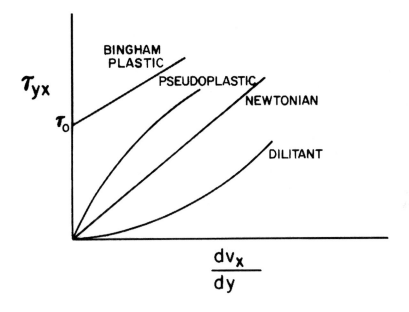

Figure 6-2 Shear Stress Vs. Shear Rate For Various Fluids

REFERENCES

[1] Reid, R.C., Prausnitz, J.M and T.K. Sherwood, *The Properties of Gases and Liquids*, McGraw-Hill, N.Y., 1977.

[2] Moore, W.J., *Physical Chemistry*, 4th ed., Prentice-Hall, Englewood Cliffs, N.J., 1972, p.116

[3] Hirschfelder, J.O., Curtis, C.F. and R.B. Bird, *Molecular Theory of Gases and Liquids*, John Wiley & Sons, N.Y., 1954

[4] Euken, A., PHYSIK. Z., 14, p.324, 1913

[5] Reid, R.C., Prausnitz, J.M and T.K. Sherwood, *The Properties of Gases and Liquids*, McGraw-Hill, N.Y., 1977 , p.629

[6] Reid, R.C., Prausnitz, J.M and T.K. Sherwood, *The Properties of Gases and Liquids*, McGraw-Hill, N.Y., 1977, p.520

[7] Bird, R.B., Stewart, W.E. and E.N. Lightfoot, *Transport Phenomena*, John Wiley & Sons, N.Y., 1960

[8] Wilke, C.R. and P. Chang, AIChE J., 1, p.264, 1955

[9] Wilke, C.R., J. CHEM. PHYS., <u>18</u>, p.517, 1950

[10] Perry, R.H. and C.H. Chilton, *Chemical Engineer's Handbook*, 5th ed., McGraw-Hill, N.Y., 1973

PROBLEMS

6-1   Derive, in detail, the expression for D_{AB} (Equation 6.30), starting with Equation (6.27).

6-2   Use Equation (6.15) and the results from the elementary kinetic theory of gases to show that the thermal diffusivity ($\alpha = \dfrac{k}{\rho \hat{C}_p}$) of a monatomic ideal gas is equal to $\dfrac{1}{3} \bar{u} \lambda \dfrac{\hat{C}_v}{\hat{C}_p}$.

6-3   Using Equations (6.28) and (6.29), show that Equation (6.36) is valid.

6-4   Estimate the viscosity of nitric oxide at 473 K by <u>both</u> the elementary kinetic theory and the Chapman-Enskog theory. Compare your answers to the measured value.

6-5   Use the elementary kinetic theory, the Chapman-Enskog theory and the Euken correction to estimate the thermal conductivity of ethane at:

   (a) 300 K from a measured value of its viscosity at that temperature.

   (b) 900 K from a measured value of its viscosity at that temperature.

6-6   Estimate the viscosity of a gaseous mixture containing 45% N_2 , 35% CO_2 , and 20% H_2 at 473 K.

6-7   Estimate:

   (a) The diffusivity of benzene-in-air at 273 K and atmospheric pressure. Compare with experimental data.

(b) The diffusivity of benzene in liquid heptane at 298 K and 10 atmospheres pressure. Compare with experimental data.

(c) The diffusivity of the system in (b) but at 400 K.

6-8   Use the Chapman-Enskog theory to estimate the diffusivities of the following binary systems:

(a) Carbonyl sulfide-in-nitrogen at 125 C and 2 atmospheres pressure.

(b) CO-in-H_2 at 400 °C and 700 kPa pressure.

(c) Carbon tetrachloride-in-air at 70 °C and 20 psia.

6-9   Using the results of the Chapman-Enskog theory and assuming that the properties of the gas are essentially due to species B, show that the *Schmidt Number* ($\dfrac{v}{D_{AB}}$) can be estimated from

$$Sc = 0.1436\, R_g \left(\frac{\sigma_{AB}}{\sigma_B} \right)^2 \left(\frac{M_A}{M_A + M_B} \right)^{\frac{1}{2}} \frac{\Omega_D}{\Omega_\mu}$$

that is, independent of pressure and only weakly dependent on temperature.

6-10   Starting with Equation (6.51), derive Equation (6.52) for the diffusion of species i into a stagnant gas mixture.

6-11   The viscosity of ethylene at 50 °C and 200 °C is 110.3 and 154.1 micropoise, respectively [*Handbook of Chemistry and Physics*, Chemical Rubber Co., Cleveland, OH].

(a) Extrapolate from each of these data points, using the Chapman-Enskog theory to predict the viscosity of ethylene at 400 °C.

(b) Using the experimental value of μ, predict a value of the thermal conductivity of ethylene at 100 °C, using both the Chapman-Enskog theory and the Euken correction.

(c) Compare your answers with measured data (e.g., CRC *Handbook of Chemistry and Physics*)

6-12 Perry's <u>Handbook of Chemical Engineering</u> (5th ed., McGraw-Hill Companies, N.Y., 1973, p. 222) lists an experimental value for the diffusivity of methane in oxygen at 500 $^{\circ}$C at 1.1 cm^2/s. Use the Chapman-Enskog Theory to estimate this diffusivity and compare it to the experimental value.

6-13 Use Equation (6.42) and the data given in Table 6-2 to predict the diffusivity of NH$_3$-in-water at 340 K.

6-14 Dudley and Tyrrell [J. Chem. Soc., Fara. Trans. I, <u>69</u>, 2200 (1973)] present data on the effect of concentration on the binary diffusivities of triethylamine-in-water at 15 $^{\circ}$C and give the following values as the mole fraction of triethylamine is varied from 0.0005 to 0.9630:

0.589 x 10^9 m^2/s	at X = 0.0005
0.201 x "	" = 0.4935
1.893 x "	" = 0.9630

Wilke has proposed [Chem. Engr. Progr., <u>45</u>, 218, 1949] that the effect of concentration on the binary diffusivities of ideal solutions can be estimated by linearly combining the diffusivities predicted by Equation (6.42) at high dilutions; i.e.,

$$\frac{D_{12}\mu_{mix}}{T} = X_1 \left(\frac{D^0_{21}\,\mu_1}{T} \right) + (1-X_1) \left(\frac{D^0_{12}\,\mu_2}{T} \right)$$

Use the data given above in conjunction with this equation to predict the diffusivity of triethylamine-in-water at 15 $^{\circ}$C at a triethylamine mole fraction of 0.4935 if the viscosities of pure triethylamine and the 49.35% mixture are 0.42 x 10^{-3} and 1.348 x 10^{-3} Pa-s, respectively. What is the percent error and the likely reason for the error?

6-15

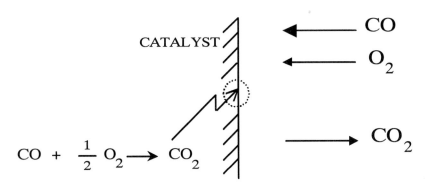

As shown in the sketch, CO is being burned at the surface of a catalyst at 600 C and 1 atm pressure under steady-state conditions. The gas composition is CO = 1.0%, CO_2 = 5.0%, N_2 = 78%, O_2 = 16%. Compare the predictions of Equations (6.51) and (6.52) for the diffusivity of CO in the mixture with the value for the binary diffusivity of CO-in-N_2. Use the Chapman-Enskog theory to calculate the various binary diffusivities.

6-16 Boersma, Laven and Stein [AIChE J., 36, 321 (1990)] have measured the viscosities of polystyrene in a 86.1% glycerol/water mixture as a function of temperature and shear rate. The data below give the *relative viscosities* (viscosity of solution relative to viscosity of glycerol/water) as a function of temperature for a shear rate of 10 s^{-1} and as a function of shear rate at a temperature of 20 C.

(a) Does the relative viscosity at $\dot{\gamma}$ = 10 s^{-1} vary with temperature as predicted by Equation (6.38)? Comment!

(b) Do the relative viscosity data at 20 $^\circ$C correspond to a power law model over the entire range of $\dot{\gamma}$? Is there a region of $\dot{\gamma}$ where the fluid behaves as a Newtonian fluid? Note that the glycerol/water mixture is Newtonian.

VISCOSITY DATA - PS/86.1% Glycerol Solution

$\dot{\gamma}$ (s^{-1})	T (C)	$\dfrac{\mu}{\mu_{86\%Gly}}$
10	20	1050.
"	25	530.
"	30.	280.

"	35.	105.
"	40.	92.
"	50.	62.
1.0	20.	82.
2.0	"	100
5.0	"	300.
7.0	"	680.
20.	"	1300.

6-17   Shah and Lord [AIChE J., 37, 863 (1991)] have collected rheological data on the fully developed, laminar flow of a 4.8 kg/m³ solution of cross linked hydroxypropyl guar (HPG) in tubes. They related the pressure drop, ΔP, and the volumetric flow rate, $\dot{V}$, to the wall shear stress, τ_w, and the "nominal" shear rate, $\dfrac{8\bar{v}}{D}$, respectively. Use their data from the table below to determine the two parameters in the power law model for non-Newtonian fluids.

τ_w x 10³ (kPa)	$\dfrac{8\bar{v}}{D}$ (s⁻¹)
2.3	15
4.8	50
5.84	80
6.6	100
13.9	500
17.0	700

CHAPTER 7

Similarity Analyses

The similarities and analogies between the three transport processes have been emphasized throughout the first six chapters. However, similarity is more than an interesting observation and *similarity analysis* can often be used to obtain a great deal of information about a system without actually solving the mathematics. In other situations, this approach can sometimes result in a transformation of complex mathematics to simpler mathematics; to the point where closed-form mathematical solutions will result. The key to this approach is to formulate the mathematical description of physical problems in such a way that similar mathematical equations will result from the analyses of very different physical phenomena. The way in which this is achieved is through the use of <u>dimensionless analysis</u>. Here in Chapter 7, dimensionless analysis is first introduced via a reexamination of the phenomenological laws, followed by placing differential equations into dimensionless form and then showing how this approach can produce variable transformations to change partial differential equations into ordinary differential equations. Finally, the utility of similarity analysis is illustrated by its application to one-dimensional, unsteady-state transport processes.

7-1 DIMENSIONLESS GROUPS IN MOLECULAR TRANSPORT

Recall that in Chapter 2, it was shown that the phenomenological laws could all be expressed in terms of "diffusion" processes. The one-dimensional forms of these relationships are

$$J_{Ax} = -D_{Am}\frac{dC_A}{dx} \tag{2.12}$$

$$q_x = -\alpha\frac{d}{dx}(\rho\hat{C}_p T) \tag{2.15}$$

$$\tau_{xy} = -v\frac{d}{dx}(\rho v_y) \tag{2.16}$$

In each of these equations, the parameter in front of the differential has identical units, l^2/t, and is referred to as the molar, thermal or momentum diffusivity, respectively. If Equation (2.16) is divided by Equation (2.15), then

$$\frac{\tau_{xy}}{q_x} = \frac{v}{\alpha}\frac{d(\rho\,v_y)}{d(\rho\,\hat{H})} \tag{7.1}$$

Equation (7.1) relates the relative "ease" of molecular momentum transport to molecular energy transport. Stated in another way, for equivalent changes in the concentrations of momentum and enthalpy, the magnitude of the ratio of the molecular momentum and energy fluxes will depend only on the parameter, $\dfrac{v}{\alpha}$. Of course this parameter is nothing but the *Prandtl Number*,

$$Pr = \frac{v}{\alpha} = \frac{\hat{C}_p\mu}{k} \tag{7.2}$$

Similarly, Equation (2.16) can be divided by Equation (2.12) so that

$$\frac{\tau_{xy}}{J_{Ax}} = \frac{v}{D_{Am}}\frac{d(\rho\,v_y)}{d(C_A)}$$

In this case, the relative ease of molecular momentum versus molecular mass transport is governed by the *Schmidt Number, Sc*

$$Sc = \frac{v}{D_{Am}} = \frac{\mu}{\rho\,D_{Am}} \tag{7.3}$$

As might be expected, a third ratio can also be formed, comparing molecular energy and mass transport. The applicable dimensionless group here is the *Lewis Number*, Le

$$\text{Le} = \frac{\alpha}{D_{Am}} = \frac{k}{\rho \hat{C}_p D_{Am}} \tag{7.4}$$

Because convection phenomena necessarily involves momentum transport, the rates of convective energy and mass transport will depend on the Prandtl and Schmidt numbers, respectively, and this will be discussed in greater detail in Chapter 8. Simultaneous energy and mass transport problems (e.g., "drying" processes), on the other hand, will depend on the Lewis number.

7-2 DIMENSIONLESS DIFFERENTIAL BALANCES

In order to illustrate the utility of placing the equations which result from differential balances into dimensionless form, let us reconsider the unsteady heating of a sphere which was discussed in 3-4a (Figure 7-1). In this case, the partial differential equation which describes the unsteady-state transport of energy is Equation (3.30)

$$\frac{\partial T}{\partial t} = \frac{\alpha}{r^2} \frac{\partial}{\partial r} (r^2 \frac{\partial T}{\partial r}) \tag{3.30}$$

where α is the thermal diffusivity, $\dfrac{k}{\rho \hat{c}_p}$

An initial condition and two boundary conditions are required in order to solve Equation (3.30) and these are:

$$at\ t = 0,\ T = T_0,\ \text{all}\ r$$

$$at\ r = 0,\ \frac{\partial T}{\partial r} = 0,\ \text{all}\ t \tag{7.5}$$

$$at\ r = R,\ -k\frac{\partial T}{\partial r} = h(T - T_a),\ \text{all}\ t$$

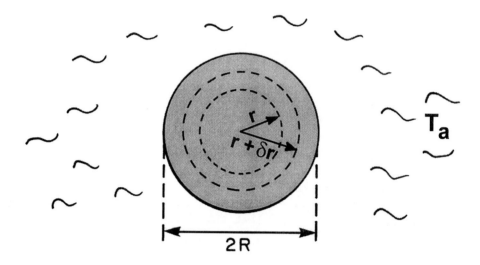

Figure 7-1 Unsteady Heating of a Sphere

Note that, since internal temperature gradients are being accounted for, the uniform SOURCE term which appeared in the more approximate analysis of Example 3-2 [see Equation (3.31)] appears here as a flux boundary condition.

Since this is the first time such a boundary condition has been encountered, it is worthy of additional discussion. The observation that the energy flux delivered to or from a fluid to a solid is observed to be proportional to the temperature difference between the two is sometimes referred to as <u>Newton's Law of Cooling</u>. Strictly speaking, it is a phenomena which occurs at a surface and, in that sense, it is more properly a boundary condition.[#] The last condition in Equation (7.5) is actually a statement of continuity of flux at an interface, as was discussed in connection with energy transport through a furnace wall (see Section 3-2b). Thus Fourier's Law applies at the solid side of the fluid-solid interface and Newton's Law of Cooling applies at the fluid side of the interface.[##] This type of a boundary condition can actually cover two other conditions; viz., either a constant temperature or a zero flux condition. To show this, we can rewrite the boundary condition as

---

[#]   Mathematicians refer to this type of a boundary condition as a <u>Neumann boundary condition</u>.

[##] Actually, Fourier's Law also applies on the fluid side of the interface. Its replacement by Newton's Law of Cooling eliminates the need for solving the mathematics in that phase.

$$T_{|R} - T_a = -\frac{k}{h}\left[\frac{\partial T}{\partial r}\right]_{|R}$$

Thus, when $h \to \infty$ (vigorous mixing in the fluid), $T_{|R} \to T_a$ and the constant surface temperature boundary condition emerges. Similarly, when $h \to 0$, there is no energy transported between the fluid and the solid and we have a perfectly insulated boundary condition, or $\left[\dfrac{\partial T}{\partial r}\right]_{|R} = 0$.

In order to place Equation (3.30) into dimensionless form, it is first necessary to define appropriate dimensionless variables for both the dependent as well as the independent variables. In general, this is done by dividing each of the variables by a characteristic value which, of course, must have the same units as the variable being rendered dimensionless. However, it is equally important that the characteristic value be appropriately related to the system being analyzed. For example, we can define a dimensionless radius, η, corresponding to Equation (3.30) by

$$\eta = \frac{r}{R}$$

Here, R, the radius of the sphere being analyzed, serves as the characteristic value for r. Not only does it have the proper units and an appropriateness for the physical system, but it conveniently varies between 0 and 1. In fact, the characteristic values are often chosen so that the numerical values of the dimensionless variables take on such well-defined ranges. For example, we could choose either T_a or T_0 as the characteristic value for a dimensionless temperature. Both have the proper units, and both are appropriately characteristic of the physical system. However we can also define a dimensionless temperature by

$$\theta = \frac{T - T_a}{T_0 - T_a}$$

With this definition, θ takes on a value of 1 at t = 0, and approaches 0 as the surface temperature approaches the bulk fluid temperature, T_a.

While the choice of convenient characteristic values often requires some trial-and-error, it is sometimes possible to leave the choice until the dependent and independent variables have been placed into dimensionless form. At that point, the choice of a convenient characteristic value may be obvious. To illustrate both the placing of a differential equation into dimensionless form as well as this latter point, let's define the following dimensionless variables relative to Equation (3.30)

$$\theta = \frac{T - T_a}{T_0 - T_a}, \quad \eta = \frac{r}{R}, \quad t^* = \frac{t}{t_c} \tag{7.6}$$

In order to place the differentials in Equation (3.30) into dimensionless form, we employ the chain rule. Thus the second-order differential becomes

$$\frac{\partial^2 T}{\partial r^2} = \frac{\partial}{\partial \eta} \frac{\partial \eta}{\partial r} \left[\frac{\partial \theta}{\partial \eta} \frac{\partial \eta}{\partial r} \frac{\partial T}{\partial \theta} \right] \tag{7.7}$$

and the first-order differential becomes

$$\frac{\partial T}{\partial t} = \frac{\partial \theta}{\partial t^*} \frac{\partial T}{\partial \theta} \frac{\partial t^*}{\partial t} \tag{7.8}$$

Since

$$\frac{\partial \eta}{\partial r} = \frac{1}{R}, \quad \frac{\partial t^*}{\partial t} = \frac{1}{t_c}, \quad \frac{\partial T}{\partial \theta} = T_0 - T_a$$

the dimensionless form of Equation (3.30) becomes

$$\frac{\partial \theta}{\partial t^*} = \frac{\alpha t_c}{R^2} \frac{1}{\eta^2} \frac{\partial}{\partial \eta} (\eta^2 \frac{\partial \theta}{\partial \eta}) \tag{7.9}$$

Notice that Equation (7.9) was purposely written so that there were no dimensions in front of one of the differential terms. This insures that <u>all</u> of the terms in the equation are now dimensionless and the individual dimensional properties and parameters will subsequently collect as one or more dimensionless parameters. For example, here we can make the choice of t_c so that all the dimensional properties and parameters are removed from the differential equation. That is, if $t_c = \frac{R^2}{\alpha}$, then the dimensionless equation becomes

$$\frac{\partial \theta}{\partial t^*} = \frac{1}{\eta^2} \frac{\partial}{\partial \eta} (\eta^2 \frac{\partial \theta}{\partial \eta}) \tag{7.10}$$

Of course we must also place the initial and boundary conditions in dimensionless form as well. With the definition of the dimensionless variables in Equations (7.6) and (7.7) and the chosen value of t_c, the dimensionless initial and boundary conditions become

$$\text{at } t^* = 0; \ \theta = 1, \text{ all } \eta$$

$$\text{at } \eta = 0; \ \frac{\partial \theta}{\partial \eta} = 0, \text{ all } t^* > 0 \qquad (7.11)$$

$$\text{at } \eta = 1; \ \frac{\partial \theta}{\partial \eta} = -\frac{1}{m}\theta, \text{ all } t^* > 0$$

where $m = \dfrac{k}{hR}$ [#]

Because this problem has been placed in dimensionless form, it can be solved for θ as a function of η, t^* and m and independent of the magnitude ("scale") of the particular problem being analyzed. The solution to this problem as well as similar problems in rectangular and cylindrical geometries were derived a number of years ago [1] and have been presented in many different graphical forms. Figures 7-2 - 7-4 give these solutions in the form of the "Gurney-Lurie" charts [2] for each of the three regular geometries. A summary of the differential equations and the accompanying boundary conditions are given in Table 7-1 for each geometry.

Even though most practical problems are not likely to fit these assumptions very exactly, reasonably close estimates can be obtained for many engineering problems if the Gurney-Lurie charts are applied judiciously. However, it is very important to keep in mind that both the differential equation and the boundary conditions must match those described by the Gurney-Lurie charts in Figures 7.2 - 7.4. Quite often the dimensionless variables themselves must be carefully defined in order to obtain the applicable boundary conditions (see Example 7.1).

7-3 SIMILARITY TRANSFORMS

Similarity analysis can also be a proven asset in certain problems where molecular transport takes place for only a very short duration, since it is sometimes possible to transform partial differential equations into ordinary ones. These so called short contact problems can often be representative models for complex physical problems and, as will be seen in Chapter 9, can also be used to model turbulent transport at an interface. In unsteady-state problems, short contact

---

[#] The reciprocal of m, hR/k, is known as the Biot Modulus and can be interpreted as a measure of the relative rates of convective-to-molecular transport.

corresponds to short periods of time whereas, in a steady-state problem, it would represent a situation where the flux of energy, mass, or momentum can only penetrate a very short distance (relative to the dimensional scale of the physical system).

Table 7-1 Equations for use with Figures 7-2 through 7-4

Geometry	Equations	Boundary Conditions
Spherical	$$\frac{\partial \theta}{\partial t^*} = \frac{1}{\eta^2}\frac{\partial}{\partial \eta}\left(\eta^2 \frac{\partial \theta}{\partial \eta}\right)$$	at $t^* = 0$; $\theta = 1$, all η at $\eta = 0$; $\dfrac{\partial \theta}{\partial \eta} = 0$, all $t^* > 0$ at $\eta = 1$; $\dfrac{\partial \theta}{\partial \eta} = -\dfrac{1}{m}\theta$, all $t^* > 0$
Cylindrical	$$\frac{\partial \theta}{\partial t^*} = \frac{1}{\eta}\frac{\partial}{\partial \eta}\left(\eta \frac{\partial \theta}{\partial \eta}\right)$$	at $t^* = 0$; $\theta = 1$, all η at $\eta = 0$; $\dfrac{\partial \theta}{\partial \eta} = 0$, all $t^* > 0$ at $\eta = 1$; $\dfrac{\partial \theta}{\partial \eta} = -\dfrac{1}{m}\theta$, all $t^* > 0$
Rectangular	$$\frac{\partial \theta}{\partial t^*} = \frac{\partial^2 \theta}{\partial \eta^2}$$	at $t^* = 0$; $\theta = 1$, all η at $\eta = 0$; $\dfrac{\partial \theta}{\partial \eta} = 0$, all $t^* > 0$ at $\eta = 1$; $\dfrac{\partial \theta}{\partial \eta} = -\dfrac{1}{m}\theta$, all $t^* > 0$

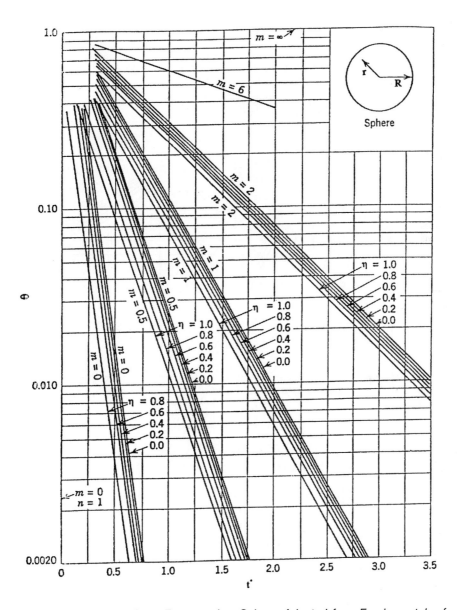

Figure 7-2 Unsteady-State Transport in a Sphere, Adapted *from Fundamentals of Momentum, Heat and Mass Transfer*, Welty, Wicks & Wilson, (1969), by permission from John Wiley & Sons,

$$(\eta = \frac{r}{R}, \quad t^* = \frac{\alpha t}{R^2}, \quad m = \frac{k}{hR})$$

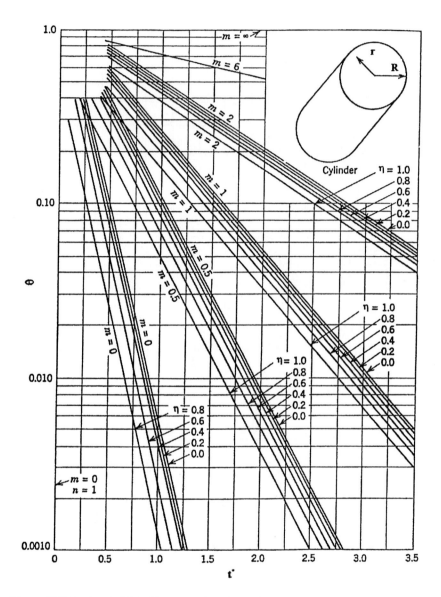

Figure 7-3 Unsteady State Transport in a Long Cylinder, Unsteady-State Transport in a Sphere, Adapted *from Fundamentals of Momentum, Heat and Mass Transfer*, Welty, Wicks & Wilson, (1969), by permission from John Wiley & Sons, $(\eta = \dfrac{r}{R},\ t^{*} = \dfrac{\alpha t}{R^{2}},\ m = \dfrac{k}{hR})$

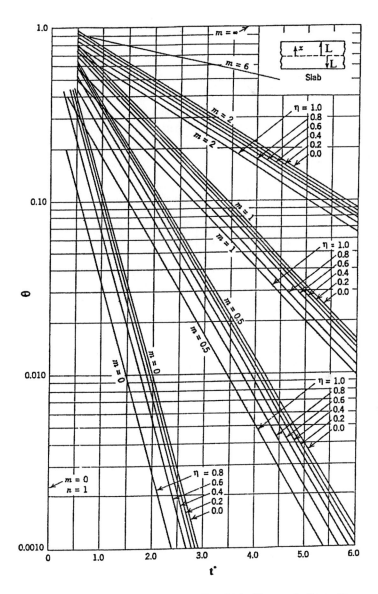

Figure 7-4 Unsteady State Transport in a Flat Slab, Unsteady-State Transport in a
Sphere, Adapted *from Fundamentals of Momentum, Heat and Mass
Transfer*, Welty, Wicks & Wilson, (1969), by permission from John Wiley &
Sons, $\qquad (\eta = \dfrac{x}{L}, \; t^* = \dfrac{\alpha t}{L^2}, \; m = \dfrac{k}{hL})$

EXAMPLE 7-1: Unsteady Heating of a Sphere

A 5-cm diameter stainless steel sphere, initially at $0\,^{\circ}C$, is immersed in a hot ($57\,^{\circ}C$) bath (similar to the problem described in Chapter 3.4b). Use the Gurney-Lurie charts to determine the temperature <u>difference</u> between the center and the surface of the sphere at 60, 90, and 120 seconds after it is immersed in the bath. Assume that the heat transfer coefficient is equal to the value determined by neglecting internal temperature gradients in the aluminum sphere described in Example 3.2. The thermal conductivity and thermal diffusivity of the stainless steel are: 19 J/m-s-K and 5.6×10^{-6} m^2/s, respectively.

Solution

The first decision to be made is how to define the dimensionless temperature, θ. Since the Gurney-Lurie chart (Figure 7-2) is based on $\theta = 1$ at $t = 0$, a compatible definition for θ would be

$$\theta = \frac{T_H - T}{T_H - T_o}$$

which has a value of 1.0 at time zero since, initially, $T = T_o$. With the conventional definition of η, a check of the other boundary conditions shows that the symmetry boundary condition at the center of the sphere is satisfied and, placing the surface boundary condition

$$-k \left(\frac{\partial T}{\partial r}\right)_{|R} = h\left(T_H - T_{|R}\right)$$

into dimensionless form gives

$$-k \frac{(T_H - T_o)}{R} \frac{\partial \theta}{\partial \eta} = h\left[\theta\left(T_H - T_o\right)\right]$$

$$\frac{\partial \theta}{\partial \eta} = \frac{hR}{k}\theta = \frac{1}{m}\theta$$

which is identical to the other dimensionless boundary condition in Figure 7-2.

Using the dimension and the thermal properties of the stainless sphere and the heat transfer coefficient from Example 3.2 which was 0.079 J/cm^2-s-K,

$$m = \frac{19 \text{ J/m-s-K}}{(.079) \text{ J/cm}^2\text{-s-K} \ (10^4) \text{cm}^2/\text{m}^2 \ (.025) \text{m}} = .96$$

$$t^* = \frac{5.6(10)^{-5} \ \text{m}^2/\text{s}}{(.025)^2 \ \text{m}^2} t = .009 \, t$$

The surface and center of the sphere correspond to $\eta = 1$ and $\eta = 0$. From Figure 7-2, the values of θ at $t = 60$ s ($t^* = .54$) are approximately 0.38 and 0.25 at the center and surface of the sphere, which correspond to temperatures of 35 °C and 43 °C, respectively. At 120 s, the temperatures are 51 °C and 53 °C. As would be expected, the temperature difference between the surface and center of the sphere starts off high and becomes negligible as the sphere approaches the temperature of the bath.

Now let us consider a one-dimensional, unsteady-state problem, represented by the sketch shown Figure 7-5. A medium which is initially at temperature T_0, is brought into contact with a surface which is at a higher temperature, T_s. During this contact, molecular energy transport takes place. Furthermore, if the duration of contact is short enough then the region far from the surface will remain at T_a during the entire contact time. Thus for practical purposes, the distance between the surface and this position may as well be infinite. Mathematically, this problem can be expressed as

$$\frac{\partial T}{\partial t} = \alpha \frac{\partial^2 T}{\partial z^2} \tag{7.12}$$

with initial and boundary conditions

$$\begin{aligned}
&\text{at } t = 0; \ T = T_0 \text{, all } z \\
&\text{at } z = 0; \ T = T_s \text{, all } t > 0 \\
&\text{at } z = \infty; \ T = T_0 \text{, all } t > 0
\end{aligned} \tag{7.13}$$

In order to transform this partial differential equation into an ordinary one, it is necessary to reduce the number of independent variables by one. However we still have to account for three initial and boundary conditions, which means that two of the original boundary conditions must also be able to be combined into one. Note that if we form a ratio of the two independent variables, z and t; specifically, $\zeta = \dfrac{z}{t}$, then we can see that the first and third boundary conditions

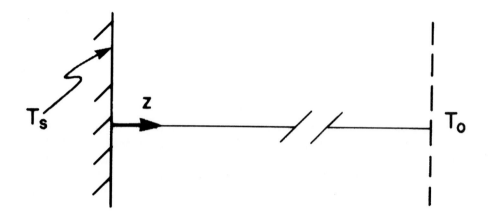

Figure 7-5 Unsteady State Transport in an Infinite Media

in Equation (7.13) can be combined into one. Namely; at $\zeta = \infty$, $T = T_0$. With this requirement met, we next define a dimensionless combined variable[*]

$$\eta = \left(\frac{z^2}{4\alpha t}\right)^{\frac{1}{2}} \tag{7.14}$$

along with the dimensionless temperature

$$\theta = \frac{T - T_0}{T_s - T_0}$$

Using the chain rule, the individual terms in Equation (7.12) can now be placed in dimensionless form; specifically

$$\frac{\partial T}{\partial t} = \left[\frac{\partial T}{\partial \theta}\right]\left[\frac{\partial \eta}{\partial t}\right]\frac{\partial \theta}{\partial \eta} = \left[(T_s - T_0) \right]\left[-\frac{1}{2}t^{-\frac{3}{2}}\left(\frac{z^2}{4\alpha}\right)^{\frac{1}{2}}\right]\frac{\partial \theta}{\partial \eta} = -[T_s - T_0]\frac{1}{2}\frac{\eta}{t}\frac{\partial \theta}{\partial \eta} \tag{7.15}$$

---

[*]   For this reason, the similarity transform technique is sometimes referred to as the <u>combination of variables</u> method

$$\frac{\partial^2 T}{\partial z^2} = \frac{\partial}{\partial z}\frac{\partial T}{\partial z} = \left[\frac{\partial \eta}{\partial z}\right]\frac{\partial}{\partial \eta}\left(\left[\frac{\partial T}{\partial \theta}\right]\left[\frac{\partial \eta}{\partial z}\right]\frac{\partial \theta}{\partial \eta}\right) = \left[\left(\frac{\partial \eta}{\partial z}\right)^2\right]\left[\frac{\partial T}{\partial \theta}\right]\frac{\partial}{\partial \eta}\frac{\partial \theta}{\partial \eta}$$

and since

$$\frac{\partial \eta}{\partial z} = \left(\frac{1}{4\alpha t}\right)^{\frac{1}{2}}, \quad \frac{\partial T}{\partial \theta} = T_s - T_o$$

then

$$\frac{\partial^2 T}{\partial z^2} = \left(\frac{1}{4\alpha t}\right)(T_s - T_0)\frac{\partial^2 \theta}{\partial \eta^2} \tag{7.16}$$

Substituting Equations (7.15) and (7.16) into (7.12), we obtain a second-order, ordinary differential equation

$$\frac{d^2 \theta}{d\eta^2} + 2\eta\frac{d\theta}{d\eta} = 0 \tag{7.17}$$

with boundary conditions

$$\text{at } \eta = 0;\, \theta = 1$$

$$\tag{7.18}$$

$$\text{at } \eta = \infty;\, \theta = 0$$

If we temporarily let $p = \dfrac{d\theta}{d\eta}$, then Equation (7.18) can be written

$$\frac{dp}{d\eta} + 2\eta p = 0$$

which can be integrated once, to give

$$p = \frac{d\theta}{d\eta} = C_1 \exp(-\eta^2)$$

and then once more, so that the solution for theta is

$$\theta = C \int_0^\eta \exp(-\eta^2) \, dn + C_2 \tag{7.19}$$

Applying the first boundary condition in Equation (7.18), we see that $C_2 = 1$, and applying the second boundary condition, we obtain

$$C_1 = -\frac{1}{\int_0^\infty \exp(-\eta^2) \, d\eta} \tag{7.20}$$

The integral in the denominator is $\dfrac{\sqrt{\pi}}{2}$ and the solution becomes

$$\theta = 1 - \frac{2}{\sqrt{\pi}} \int_0^\eta \exp(-\eta^2) \, d\eta \tag{7.21}$$

The second term on the right-hand side of Equation (7.21) is known as the <u>Error Function</u> (Erf(η)) and is tabulated in Appendix D as a function of η. Thus the solution can be written as[#]

$$\theta = 1 - Erf(\eta) = Erfc(\eta)$$

Figure 7-6 is a sketch showing how θ varies with η.

[#] Erfc(η) is called the <u>Complementary Error Function</u>.

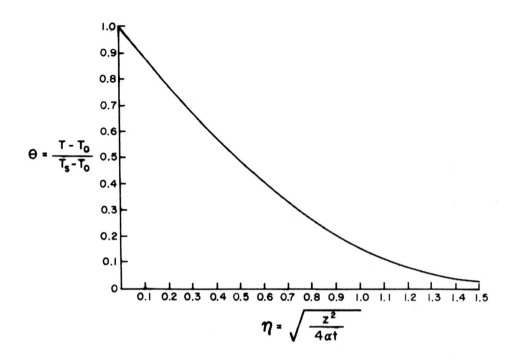

$$\Theta = \frac{T - T_0}{T_s - T_0}$$

$$\eta = \sqrt{\frac{z^2}{4\alpha t}}$$

Figure 7-6 Solution from Equation 7.21

REFERENCES

[1] Carslaw, H.S. and J.C. Jaeger, *Conduction of Heat in Solids*, 2nd ed., Oxford Univ. Press, Oxford, UK (1959).

[2] Gurney, H.P. and J. Lurie, Ind. Eng. Chem., 15, p. 1170 (1923).

PROBLEMS

7-1   (a) Describe, for *mass transport*, the equivalent <u>physical</u> problem to the unsteady-state conduction problem discussed in Section 7-2 but *in cylindrical coordinates*.

       (b) Place the appropriate differential equations corresponding to the situation in (a) and state <u>how</u> you could use the Gurney-Lurie charts to obtain concentration as a function of time and distance.

7-2   If a generalized momentum balance is taken (accounting for unsteady-state and both molecular and convective transport in all directions), then, for a newtonian fluid with constant density, the result in vector form is[*]

$$\rho \frac{\partial}{\partial t}(\vec{v}) = -\rho \nabla \bullet (\vec{v}\,\vec{v}) + \mu \nabla^2 \vec{v} - \nabla \vec{P} + \rho \vec{g}$$

Place this equation in dimensionless form. Use D, $\bar{v}$ and $\dfrac{D}{\bar{v}}$ as characteristic values for position, velocity and time, respectively and use $\rho\,\bar{v}^2$ as the characteristic pressure. To be perfectly general, also place the gravitational force in dimensionless form by using the gravitational acceleration at sea level, g_0, as its characteristic value. Write the resulting equation so that only the derivative of dimensionless velocity with respect to dimensionless time appears on the left-hand side and examine the dimensionless parameters in front of each term in the resulting equation.

7-3   If molecular transport in the flow direction is neglected, the partial differential equation describing energy transport in steady-state, laminar flow in a tube is

$$\rho \hat{c}_P v_z \frac{\partial T}{\partial z} - \frac{k}{r}\frac{\partial}{\partial r}\left(r\frac{\partial T}{\partial r}\right) = 0$$

(a) Place this equation in dimensionless form by choosing the tube diameter as the characteristic dimension for both r and z and the average temperature, $\bar{T}$, and average velocity, $\bar{v}$, for the characteristic temperature and velocity.

(b) Group all of the constant terms in front of the <u>molecular transport</u> term and rearrange in terms of the Reynolds and Prandtl number. Note that the product of the Reynolds and Prandtl numbers is sometimes referred to as the <u>Peclet Number</u> for heat transfer.

7-4   The differential equation describing the transport of A in a <u>plug flow</u> reactor (i.e., a reactor with uniform concentration in the radial direction) with a first order reaction and with <u>dispersion</u> in the flow direction is

$$D_t \frac{d^2 C_A}{dz^2} - \bar{v}\frac{dC_A}{dz} - kC_A = 0$$

---

[*]   This is another form of Equation (A.4) which is derived in Appendix A.

where the dispersion phenomena is modeled as though it were molecular transport and D_t is the Dispersion Coefficient.

Using the inlet concentration of A (C_{A_0}) and the reactor length (L) as the characteristic dimensions for the dependent and independent variables, place this equation in dimensionless form and show that the solution will be dependent on the Peclet Number ($\frac{\bar{v}L}{D_t}$) and the ratio of the reactor residence time ($\theta_{Reac} = \frac{L}{\bar{v}}$) to the "reaction time" ($\theta_{rxn} = \frac{1}{k}$)

7-5

Hot (T_1)

(a) Derive a second-order, nonhomogeneous ordinary differential equation which describes one-dimensional, steady-state energy transport in the *thin* rectangular slab shown in the sketch if temperature gradients in the z-direction can be neglected and convective cooling at the upper and bottom surfaces of the slab is accounted for by means of Newton's Law of Cooling (heat transfer coefficient, h) with the temperature driving force being equal to the difference between the temperature of the slab and, T_a, the ambient temperature.

(b) Place this equation in dimensionless form by choosing characteristic dimensions which will result in a homogeneous equation and dependent and independent variables which will vary between 1 and 0 and 0 and 1, respectively, as the slab cools.

7-6   If molecular transport in the flow-direction is ignored, the partial differential equation describing momentum transport of a Newtonian fluid in steady-state, developing laminar flow between two parallel plates is

$$\rho \frac{\partial}{\partial z}(v_z{}^2) = \mu \frac{\partial^2 v_z}{\partial y^2} - \frac{\partial P}{\partial z}$$

Choosing the length and distance between the plates for the characteristic dimensions of the independent variables and the average velocity, $\bar{v}$, and $\rho \bar{v}^{-2}$ as the characteristic dimensions for velocity and pressure, place this equation in dimensionless form and show that the solution will depend on only <u>one</u> dimensionless parameter; viz., the product of the Reynolds Number and the ratio of the characteristic lengths.

7-7

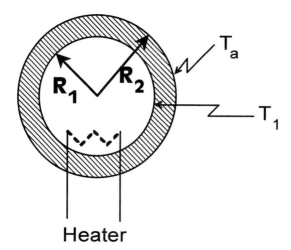

The above sketch depicts steady-state molecular energy transport through the walls of a spherical shell (constant thermal conductivity, k,) where the inner shell wall is maintained at a constant temperature, T_1, by virtue of a self-contained electrical heater, The outer shell wall is effectively at the ambient temperature, T_a.

(a) Place the applicable differential equation <u>and</u> boundary conditions into dimensionlesss form and solve the equation to obtain expressions for both the dimensionless temperature as a function of dimensionless radial distance and the energy rate at the outer wall ($\dot{Q}_{|R_2}$).

(b) Describe, in words, an analogous problem in molecular mass transport, defining the appropriate dimensionless variables which will result in the <u>identical</u> dimensionless solution obtained in (a).

(c) Use the solutions in (a) and your analogy in (b) to directly write the corresponding expressions for the mass transport problem.

7-8

$$\frac{D_e}{r^2}\frac{d}{dr}\left(r^2\frac{dC_A}{dr}\right) - kC_A = 0$$

at $r = R_s$, $-D_e\frac{dC_A}{dr} = k_{mc}(C_A - C_{A_a})$

at $r = 0$, $\frac{dC_A}{dr} = 0$

(a) Give a <u>physical</u> description of the mass transport problem described by the above equation and boundary conditions.

(b) Place this equation and its boundary conditions into dimensionless form and then sketch the dimensionless solution(s).

(c) Provide both a physical and mathematical description of an energy transport problem which is analogous to the above.

7-9   In Chapter 5, the solution for the velocity distribution in fully developed, steady-state laminar flow was derived to give

$$v_z = \frac{\wp R^2}{4\mu}\left[1 - \left(\frac{r}{R}\right)^2\right]$$

with an average velocity, $\bar{v} = \frac{1}{2}\frac{\wp R^2}{4\mu}$. Describe an analogous problem in energy transport and then derive expressions for both the temperature distribution and the average temperature.

7-10 Referring to Problem 5-18, place the differential equation into dimensionless form (use $\bar{v}$ and L_y as the characteristic dimensions) and then obtain a dimensionless solution that depends on the

Hartmann Number, $Ha \equiv \left(\dfrac{B_o^2 L_y^2 \sigma}{\mu} \right)^{\frac{1}{2}}$, which represents the ratio of magnetic to viscous forces.

7-11 The two sets of equations given below describe various physical problems in transport processes. For each set, describe a physical situation (include a sketch) in <u>both</u> energy and mass transport, which could be described by the equations.

(a)

$$a_1 \frac{1}{x} \frac{\partial}{\partial x} \left(x \frac{\partial \phi}{\partial x} \right) - a_2 \frac{\partial \phi}{\partial y} - a_3 \phi = 0$$

$\phi = 1$ at $y = 0$, $0 \leq x \leq 1$

$\dfrac{\partial \phi}{\partial x} = 0$ at $x = 0$, all y

$\phi = 0$ at $x = 1$, all $y \geq 0$

(b)

$$\frac{\partial \phi}{\partial t} = b_1 \frac{\partial^2 \phi}{\partial y^2} - b_2 \phi$$

$\phi = 0$ at $t = 0$, all y

$\phi = 1$ at $y = 0$, all $t > 0$

$\phi = 0$ at $y = \infty$, all $t > 0$

7-12 (a) Derive the steady-state differential energy balance equation for a fluid in a plug-flow reactor being heated <u>uniformly</u> by a first-order, exothermic reaction (which takes place at the surface of a catalyst; i.e., at the catalyst surface temperature, T_s) and which is being cooled at the reactor walls. The reactant concentration is taken to be constant at the entering concentration, C_{A0} , and the reaction rate constant has an Arrhenius dependency ($k = A \exp\left[\dfrac{-E}{R_g T_s}\right]$).

(b) Balakotaiah and Luss [AIChE J., 37, 1780 (1991)], have analyzed the same situation in order to determine "runaway" criteria in packed bed catalytic reactors. They present the following dimensionless energy balance:

$$\frac{d\theta}{d\xi} = \Delta \exp(\theta_c) - St\,(\theta - \theta_w)$$

where the <u>Stanton Number</u>, is defined by $St = \dfrac{4hL}{Dv\rho\hat{c}_P}$ and

$$\Delta \equiv \frac{k_{|_{T_0}}}{v}\frac{L}{\rho\hat{c}_p T_0}\frac{|\Delta\tilde{H}_r|C_{A0}}{R_g T_0}\frac{E}{R_g T_0}. \quad \text{In these equations,}$$

$k, E, \Delta\tilde{H}_r$ and C_A correspond to the rate constant, activation energy, heat of reaction and reactant concentration corresponding to the reaction and $h, L,$ and D correspond to the fluid-wall heat transfer coefficient, the reactor length and diameter respectively. The subscripts o, s, and w refer to the reactor inlet conditions, the catalyst surface and the reactor wall, respectively. They also assume that the Arrhenius dependence of k can be approximated by

$$k_{|_{T_s}} \sim k_{|_{T_0}}\; Exp\left[\frac{E}{R_g T_0^2}(T_s - T_0)\right]$$

From these equations, determine T_c if the dimensionless dependent variable is defined in the form $\theta = \dfrac{T - T_0}{T_c}$.

7-13 Referring to the discussion related to the unsteady heating of an aluminum sphere (Section 3-4a), decide if the neglect of internal temperature gradients justified. [NOTE: $k_{Al} = 0.49$ cal / cm-s-K].

7-14 Referring to Problem 3-26, use the heat transfer coefficient determined from the experiment (h = 546 J/m²-s-K) together with the Gurney-Lurie charts to determine:

(a) the difference between the surface and centerline temperatures at $t = 25\ s$.

(b) If the stirring speed in the bath is increased to the point where the heat transfer coefficient increases to 2048 J/m²-s-K, determine the time below which the surface and center line temperatures will differ by more than 2 °C.

7-15 Compare the solutions for unsteady-state conduction as given by Equation (7.21) and the Gurney-Lurie charts (for rectangular geometry, assuming an infinitely large heat transfer coefficient) by doing the following:

(a) First show that the definitions of dimensionless temperature in the two solutions are related by $\theta_{GL} = 1 - \theta_{SA}$ (where "GL" is the Gurney-Lurie solution and "SA" is the similarity analysis solution).

(b) Then plot the values of distance (from the <u>surface</u>) vs. time at $\theta_{GL} = .1$ and $.4$ for the two solutions for a 30-cm long, 310 stainless steel ($\alpha = .13\quad ft^2 /hr$) slab.

7-16

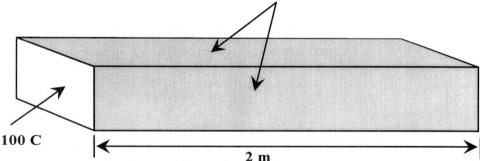

Insulated (4 surfaces)

100 C

2 m

If a <u>very</u> long (2 m) aluminum slab is originally at 25 C and has one surface suddenly raised to 100 C, how long will it take for the point 2 cm from the heated surface to reach 85 °C if energy transport is one-dimensional? <u>Al Properties:</u> $k = 118$ BTU/hr-ft-°F, $\alpha = 2$ ft²/hr, $\hat{c}_P = .36$ BTU/lb$_m$-°F

7-17

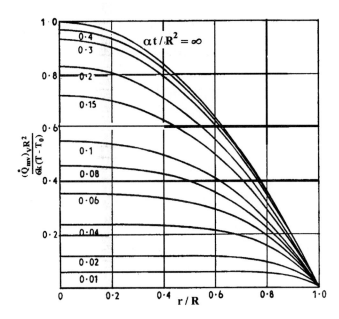

The figure is a graphical representation of the solution (H.S. Carslaw & J.C. Jaeger, "*Conduction of Heat in Solids*", Oxford University Press, 2nd ed., p. 244, 1959, reproduced, courtesy of Oxford University Press) for unsteady-state energy transport in a sphere with zero initial and surface temperatures and internal generation, $\dot{Q}_{mwv}$. Reanalyze the microwave heating of a squash (Problem 3-12) and decide whether the neglect of internal temperature gradients was justified for a 10-cm diameter squash.

[Squash Properties: $\rho = 980 \, \text{kg}/\text{m}^3, \hat{c}_P = 3.52 \, \text{KJ/kg-K}, k = .55 \, \text{w/m-s-K}$].

HINT: Calculate the values of t^*

7-18 Referring to the steady-state problem described in Problem 3-27,

(a) Derive the differential equation and state the boundary conditions applicable to the unsteady-state situation; i.e., when the sphere at temperature, T_s, is initially immersed in the stagnant fluid. Assume that the surface temperature of the sphere remains at T_s.

(b) By means of two successive transformations; viz.,

(i) $\phi = r\theta$

(ii) $s = r - R_I$

show that the differential equation and its boundary conditions are identical to the situation described in Section 7-3 and then give the solution for the unsteady-state temperature distribution in the fluid.

PART II

Convective Transport

Being able to quantitatively analyze the complexities introduced by convective flow fields is one of the real challenges in the transport phenomena approach to engineering problems. Chapters 8-11 discuss this aspect of transport phenomena by introducing the various methods which have been successfully applied over the years. However it should be kept in mind that this is an active area of research and new approaches and techniques are being continuously developed. Thus Chapter 8 illustrates how convective transport can be mathematically handled in some simple laminar flow systems, Chapter 9 introduces some of the concepts which have been used to predict velocity, temperature and concentration profiles in turbulent flow and Chapter 10 describes the use of *transfer coefficients* and "film theory" to handle complex systems which are not yet tractable from a mathematical point of view. Finally, Chapters 11- 14 illustrate how these concepts can be used in macroscopic calculations to analyze and size the equipment required to accomplish the movement, energy exchange and component separation of fluids.

CHAPTER 8

Convective Transport in Laminar Flow

Although we have yet to specifically analyze convective transport problems, convection has been previously encountered in discussions of both bulk diffusion (Section 4-1) and laminar pipe flow (Section 5-4). However, bulk diffusion was considered as a component of molecular mass transport, and there was no need to deal with convection in laminar pipe flow since the direction of the transport was <u>not</u> in the direction of the convection. As we shall see, it is easy to separate molecular from convective transport in laminar flow, but it is not so easy in turbulent flow where there is convective transport in <u>all</u> directions. Thus, convective transport in laminar flow is a matter of dealing with more complex mathematics (usually partial differential equations), but turbulent problems leave us with the difficult problem of being able to mathematically describe the problem at all[*]. Consequently, we are usually forced to rely on semi-empirical methods in an attempt to be able to predict turbulent transport rates. In this chapter, we only address some simple laminar flow convective transport problems which <u>can</u> be solved mathematically and which illustrate how to set up such problems in momentum, energy, and mass transport situations. In addition, a more generalized approach to analyzing laminar transport problems is also introduced. This approach starts with equations that allow for gradients in all directions (the "Equations of Change") and which must then be simplified for the specific problem at hand. While, in principle, these equations can be solved, they are so complex that semi-empiricism is usually employed here as well. This will be demonstrated in Chapter 13 in the context of laminar-flow heat exchangers.

---

[*] Actually, the advent of supercomputers has allowed for some progress in this regard and has sparked an entire subdiscipline — <u>computational fluid dynamics</u>. It is just that the ratio of computing time to real time is still a number much greater than one!

In this chapter, one specific topic in each of the transport areas is chosen to illustrate solutions to laminar-flow problems: developing flow in a pipe, for momentum transport, energy transport in a laminar-flow heat exchanger and a laminar flow chemical reactor, for mass transport.

8-1 DEVELOPING FLOW IN A PIPE

In Chapter 5, we analyzed steady-state, fully developed, laminar pipe flow and showed how the velocity profile has a parabolic shape which does not change as we proceed down the pipe. We now turn our attention to the entrance region of the pipe where the velocity profile is in the process of "developing" into its final, parabolic shape. Figure 8-1 shows an idealized flow at the entrance to a smooth pipe where the velocity profile at $z = 0$ is perfectly "flat." As the fluid immediately adjacent to the wall encounters friction, it experiences a sudden deceleration so that its velocity at the wall is zero (no-slip condition). Since total mass is conserved and under steady-state conditions there can be no accumulation of mass, fluid further away from the wall must accelerate in order to compensate for the deceleration at the wall. As we move down the pipe, momentum is continually transported (by molecular mechanisms) to the wall so that more of the fluid close to the wall also decelerates. Again, total mass conservation dictates that the fluid near the center of the pipe accelerate even more. Eventually, we reach fully developed conditions so that there is no longer any deceleration or acceleration with distance downstream; i.e., the "fully developed" situation which was analyzed in Chapter 5.

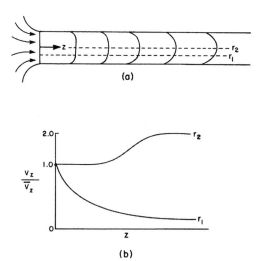

Figure 8-1 Velocity Profiles in Developing Laminar Flow

8-1.a The Continuity Equation

While this description has been purely qualitative, we can use a differential mass balance to illustrate these same principles from a mathematical point of view. For the moment, consider a rectangular geometry and a general flow system where there are velocity components in all directions. Recall that convective fluxes can always be expressed as the product of the volumetric concentration of the species being transported, multiplied by the velocity of transport in a particular direction (see Section 1-2). Since, in this case, we are dealing with total mass, the convective flux is simply

$$convective\ flux\ of\ mass\ =\ \rho\ \vec{v}$$

Applying the general conservation balance for total mass to the three-dimensional, rectangular geometry shown in Figure 8-2, we obtain, for the general case of unsteady-state conditions:

$$[\rho\,v_{z\,|\,z}\ -\ \rho\,v_{z\,|\,z+\delta z}]\delta x\,\delta y\,\delta t+[\rho\,v_{x\,|\,x}\ -\ \rho\,v_{x\,|\,x+\delta x}]\delta z\,\delta y\,\delta t$$

$$+[\rho\,v_{y\,|\,y}\ -\ \rho\,v_{y\,|\,y+\delta y}]\delta x\,\delta z\,\delta t=[\rho_{\,|\,t+\delta t}-\rho_{\,|\,t}]\delta x\,\delta y\,\delta z$$

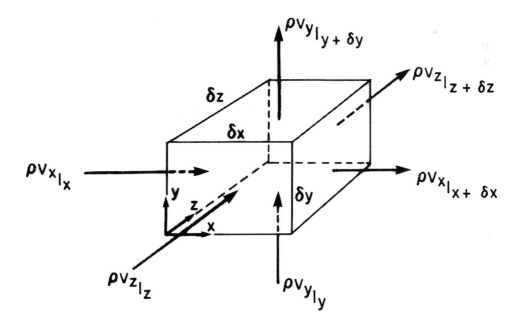

Figure 8-2 Mass Transport across a 3-Dimensional DVE

Taking the limit as $\delta x\ \delta y\ \delta z\ \delta t \rightarrow 0$ we obtain the general conservation of mass equation, or, what is known as the continuity equation

$$\frac{\partial \rho}{\partial t} = -[\frac{\partial(\rho v_x)}{\partial x} + \frac{\partial(\rho v_y)}{\partial y} + \frac{\partial(\rho v_z)}{\partial z}] \qquad (8.1)$$

In vector notation, this can be written as

$$\frac{\partial \rho}{\partial t} = -\vec{\nabla} \bullet (\rho \vec{v}) \qquad (8.2)$$

Of course our pipe problem is not a rectangular geometry and thus, a coordinate transformation to cylindrical coordinates is necessary [1], with the general result[#]

$$\frac{\partial \rho}{\partial t} = -[\frac{1}{r}\frac{\partial}{\partial r}(r\rho v_r) + \frac{1}{r}\frac{\partial}{\partial \theta}(\rho v_\theta) + \frac{\partial(\rho v_z)}{\partial z}] \qquad (8.3)$$

Since we have axial symmetry in developing pipe flow, there are no changes with respect to θ, and for steady-state conditions and constant density, Equation (8.3) becomes

$$\frac{1}{r}\frac{\partial}{\partial r}(r v_r) + \frac{\partial v_z}{\partial z} = 0 \qquad (8.4)$$

Since the flow is in the process of developing its fully developed profile, we know that $\dfrac{\partial v_z}{\partial z} \neq 0$ and thus we can see from Equation (8.4) that, in developing flow, it is necessary to have a radial component of velocity. In other words, as the fluid near the wall decelerates, additional fluid must be transported towards the center of the pipe in order to conserve total mass.

With this knowledge we can now take a differential momentum balance (z-component) around the DVE shown in Figure 8-3. Because there is a velocity gradient in the direction of flow, there is a convective transport as well as a molecular transport of z-momentum in the z-direction. Although Figure 8-3 only shows a molecular transport flux of z-momentum in the r-direction, strictly speaking there is also a convective transport in that direction as well (since $v_r \neq 0$). However, except for regions very close to the wall, this velocity component is small and can usually be neglected. Using the methods of Chapters 2 and 5, the differential equation which describes the steady-state momentum transport in developing laminar flow is

---

[#]   The general continuity equation is given in Appendix A in all three standard coordinate systems.

$$-\frac{\partial}{\partial z}(\rho v_z v_z) - \frac{\partial \tau_{zz}}{\partial z} - \frac{1}{r}\frac{\partial}{\partial r}(r\,\tau_{rz}) - \frac{\partial P}{\partial z} = 0 \tag{8.5}$$

Since the point velocities are changing in the flow direction, there is both acceleration (near the center of the pipe) and deceleration (regions near the wall). Consequently it is not surprising that Equation (8.5) is related to Newton's Second Law (see EXAMPLE 8.1). In this context, the first term on the left hand side of Equation (8.5) is actually an *inertial* term and, in its absence (fully developed flow), the momentum balance becomes a static force balance; or, $\Sigma(F_V)_z = 0$.

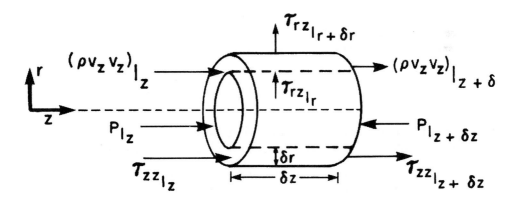

Figure 8-3 DVE Momentum Balance in Developing Flow

EXAMPLE 8-1: Comparison of Equation (8.5) with Newton's Second Law

Show that Equation (8.5) can be written as a form of Newton's Second Law.

Solution:

Newton's Second Law can be generally written as

$$\Sigma \vec{F} = \frac{d}{dt}(m\vec{v}) \tag{8.6}$$

If we consider only the z-component of Equation (8.6) and express it on a per unit volume basis then, for constant density

$$\Sigma(F_V)_z = \rho \frac{dv_z}{dt} \tag{8.7}$$

Furthermore, the chain rule can be used so that the time derivative is converted to a spacial derivative; viz.,

$$\frac{d}{dt}(v_z) = \frac{dz}{dt}\frac{d}{dz}(v_z)$$

$$= v_z\frac{d}{dz}(v_z)$$

so that Equation (8.7) becomes

$$\Sigma(F_V)_z = \rho v_z\frac{d}{dz}(v_z) \tag{8.8}$$

Noting that the last three terms in Equation (8.5) are force/volume terms (remember that τ can be considered either as a momentum flux or a shear stress) and that the first term can be rewritten as $2\rho v_z\frac{d}{dz}(v_z)$, the comparison of Equation (8.5) with Newton's Second Law, Equation (8.8) is readily evident.

In general, τ_{zz} can usually be neglected when compared in magnitude to the convective transport in the z-direction [the first term in Equation (8.5)]. While this is a simplification, the problem is still difficult due to the nonlinearity of the convective transport term. It should also be noted that, unlike fully developed flow, the pressure gradient term in Equation (8.5) is not a constant. Physically, this is due to the additional energy losses associated with the acceleration/deceleration of the fluid. Mathematically, it means that we require another equation (the simplified continuity equation, Equation (8.4)) in order to be able to solve for the additional dependent variable. Actually, we need two more equations, since employment of the continuity equation introduces still another dependent variable (v_r), thus we must also utilize a momentum balance for r-momentum in order to be able to solve for the three dependent variables, v_z, v_r, and P.

Once a problem becomes this complex, it is easier to make use of a more generalized approach rather than attempt to take a momentum balance around a specific DVE. Up to this point, all of the problems have been analyzed by first assessing the physical situation and then developing the mathematics so that they apply specifically to the given problem. An alternate method of solution is to formulate the general <u>Equations of Change</u> and then simplify them so that they apply to the given situation. The derivation of the Continuity Equation is an example of this approach and the same can be done to derive general momentum, energy, and species conservation equations. The derivations of these equations in rectangular coordinates and their presentation in all three standard coordinate systems are given in Appendix A. Example 8.2 illustrates the application of these equations to the developing pipe-flow problem.

EXAMPLE 8-2: Use of the Equations of Change – Developing Pipe Flow

Use the Equations of Change in Appendix A to determine the differential equations which apply to the developing pipe flow problem discussed above.

Solution:

From Table A-1, the form of the continuity equation needed here is Equation [A-1].1, which is also given as Equation (8.3). For this specific case, the continuity equation simplifies to Equation (8.4) and since this equation contains both axial and radial velocity components, we must look to both the axial and radial components of the Momentum Equation[#] in Table A-2b. The physical system allows for the following simplifications:

Steady-state: $\dfrac{\partial}{\partial t} = 0$

No circumferential flow: $v_\theta = 0$,

Flow is symmetric about the z axis: $\dfrac{\partial}{\partial \theta} = 0$

On the basis of these factors, Equations [A.2b]. and [A.2b].2 in Appendix A simplify to:

$$\rho\left[v_r\right] = -\frac{\partial P}{\partial z} + \mu\left[\frac{1}{r}\frac{\partial}{\partial r}\left(r\frac{\partial v_z}{\partial r}\right) + \frac{\partial^2 v_z}{\partial z^2}\right]$$

$$\rho\left[v_r\frac{\partial v_r}{\partial r}\right] = -\frac{\partial P}{\partial r} + \mu\left[\frac{\partial}{\partial r}\left(\frac{1}{r}\frac{\partial}{\partial r}(rv_r)\right) + \frac{\partial^2 v_z}{\partial z^2}\right]$$

Comparing these equations to Equation (8.5) we can see that Equation (8.5) neglects v_r as well as the entire r-component of momentum. The following discussion of boundary layer theory (below), will show that there are good reasons for neglecting these terms.

[#] For a Newtonian fluid with constant density and viscosity, this equation is commonly known as the Navier-Stokes equations.

Thus the set of differential equations describing a flow as simple as laminar convective transport in two dimensions, are exceedingly complex. There are three dependent variables (v_z, v_r, and P) and three differential equations [the two above plus the simplified continuity equation, (8.5)] Of course turbulent convective problems are even more intractable since the nature of turbulence introduces time variations into the Equations of Change.

8-1.b The Boundary Layer

Although solutions to this complex set of equations describing developing flow are really beyond the scope of this introductory text, it is important to be knowledgeable of a particular strategy which was designed to specifically analyze two-dimensional laminar flows — boundary layer theory[#]. The strategy involves the separation of the flow into two distinct regions: flow near the wall in the absence of a pressure gradient (i.e., within the boundary layer), and *inviscid flow* (i.e., "frictionless" flow) away from the wall which is driven by pressure gradients. The two regions are then linked via boundary conditions at the edge of the boundary layer. In this section we will deal with the boundary layer near the wall and the next section, will discuss flow near the center of the pipe.

To illustrate the approach to laminar boundary layer analysis, let us consider the flow <u>very</u> close to the pipe wall; i.e., where pipe curvature can be neglected. In this case the flow becomes similar to flow over a flat plate and the developing velocity profiles as we proceed from the pipe entrance, are shown in Figure 8-4. As can be seen, we can define a boundary layer thickness, δ, as that distance from the plate where the velocity equals the unchanging, free-stream velocity, v_∞. For pipe flow, the boundary layer thickness will grow until it is equal to the pipe radius and the flow is fully developed.

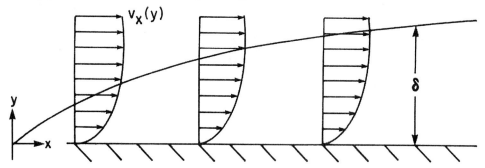

Figure 8-4 The Developing Boundary Layer

[#]   The classical development of this theory is generally attributed to Blasius [2], but a thorough analysis of all aspects of boundary layer theory can be found in the text by Schlichting [3].

In complex problems such as this it is best to utilize the generalized equations of change (given in Appendix A) and to first apply the continuity equation in order to see what information it might give which could simplify the analysis. For an incompressible fluid, Equation (A-1).1 from Table A-1 simplifies to

$$\frac{\partial v_x}{\partial x} + \frac{\partial v_y}{\partial y} = 0 \tag{8.9}$$

Note that since $\frac{\partial v_x}{\partial x} \neq 0$, there <u>must</u> be a vertical component of velocity in order to satisfy mass conservation. This of course is a direct consequence of the vertical growth of the boundary layer thickness as we move downstream. The next step in the analysis is to simplify the generalized momentum equation for this situation. From Table A-2, the x- and y-components of momentum, i.e., Equations (A-2a).1 and (A-2a).2 become

X-COMPONENT

$$v_x \frac{\partial v_x}{\partial x} + v_y \frac{\partial v_x}{\partial y} = \nu \left[\frac{\partial^2 v_x}{\partial x^2} + \frac{\partial^2 v_x}{\partial y^2} \right] \tag{8.10}$$

Y-COMPONENT

$$v_x \frac{\partial v_y}{\partial x} + v_y \frac{\partial v_y}{\partial y} = \nu \left[\frac{\partial^2 v_y}{\partial x^2} + \frac{\partial^2 v_y}{\partial y^2} \right] \tag{8.11}$$

The next step in the analysis is to attempt further simplification on the basis of the relative magnitudes of each of the terms in Equations (8.9) through (8.11)[*]. To do this we first estimate the magnitudes of v_x, δx, δy as

$$v_x \approx v_\infty$$

$$\delta x \approx L$$

$$\delta y \approx \delta$$

---

[*] This is sometimes referred to as <u>order-of-magnitude analysis</u>.

With these estimates, we can now obtain an estimate of the magnitude of v_y from the simplified continuity equation, Equation (8.9); viz.,

$$v_y \approx \left[v_\infty \frac{\delta}{L} \right]$$

Using these estimated magnitudes, the relative magnitudes of each of the terms in Equations (8.10) and (8.11) can also be estimated, so that

$$v_\infty \left[\frac{v_\infty}{L} \right] + \left[v_\infty \frac{\delta}{L} \right] \left[\frac{v_\infty}{\delta} \right] = \nu \left(\left[\frac{v_\infty}{L^2} \right] + \left[\frac{v_\infty}{\delta^2} \right] \right) \tag{8.12}$$

and

$$v_\infty \left[\frac{v_\infty \frac{\delta}{L}}{L} \right] + \left[v_\infty \frac{\delta}{L} \right] \left[\frac{v_\infty \frac{\delta}{L}}{\delta} \right] = \nu \left(\left[\frac{v_\infty \frac{\delta}{L}}{L^2} \right] + \left[\frac{v_\infty \frac{\delta}{L}}{\delta^2} \right] \right) \tag{8.13}$$

Since $\frac{\delta}{L} << 1$, the first term on the right-hand side of both Equations (8.12) and (8.13) can be neglected relative to the last terms on the right-hand side of these equations. Furthermore, the magnitude of Equation (8.13) is $\frac{\delta}{L} \frac{v_\infty^2}{L}$, which is much less than $\frac{v_\infty^2}{L}$, the magnitude of Equation (8.12). In other words, the entire y-component of momentum can be neglected and the x-component simplifies to

$$v_x \frac{\partial v_x}{\partial x} + v_y \frac{\partial v_x}{\partial y} = \nu \frac{\partial^2 v_x}{\partial y^2} \tag{8.14}$$

which must be solved simultaneously with Equation (8.10). Blasius [2] obtained a solution to these equations using a similarity transform, η, (see Section 7-3), so that

$$\frac{v_x}{v_\infty} = f(\eta)$$

where

$$\eta = y\sqrt{\frac{V_\infty}{\nu x}}$$

and a plot of his solution, is given in Figure 8-5. Note from Figure 8-5 that, if the boundary layer thickness, δ, is defined as the value of y when $\frac{V_x}{V_\infty} \sim 0.99$, the Blasius solution predicts that this occurs at $\eta = 5$, or

$$y = \delta = \frac{5}{\sqrt{\frac{V_\infty}{\nu x}}} = 5\frac{x}{\sqrt{N_{Re_x}}} \qquad (8.15)$$

where the Reynolds number is defined in terms of the length along the plate; i.e.,

$$Re_x = \frac{x V_\infty}{\nu}.$$

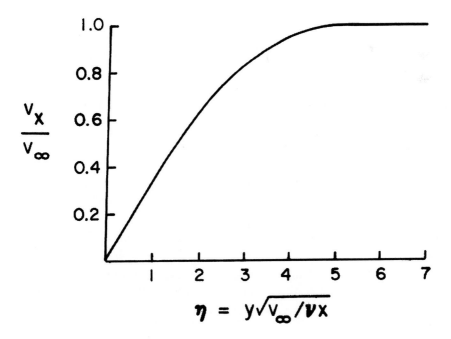

Figure 8-5 Velocity Distribution: Flat Plate Boundary Layer

To further illustrate the power of order-of-magnitude analyses, refer back to Equation (8.12) where it was shown that the order of magnitude of the simplified x-component of momentum was

$$\left[\frac{v_\infty^2 \delta}{L^2}\right] = v\left[\frac{v_\infty}{L\delta}\right] \tag{8.16}$$

Now, in order for the equality in Equation (8.16) to hold, it is necessary that

$$\delta^2 = \frac{v\,L}{v_\infty} = \frac{L^2}{\left[\dfrac{L v_\infty}{v}\right]}$$

or, for any x,

$$\delta = \frac{x}{\sqrt{N_{Re_x}}}$$

which is certainly within an order of magnitude of the Blasius prediction.

EXAMPLE 8.3: Calculation of the Entrance Length in Laminar Flow

Use the Blasius solution for boundary layer thickness as a function of distance to obtain an expression for L/D as a function of Reynolds number; where L is the length required to reach fully developed flow and D is the diameter of the tube.

Solution:

Starting with the Blasius solution for boundary layer growth along a flat plate, Equation (8.15),

$$\delta = 5\,\frac{x}{\sqrt{Re_x}}$$

the flow in a tube will become fully developed when the boundary layer equals ½ the tube diameter. Calling L, the length where this occurs and recognizing that, for fully developed laminar flow in a tube, $v_\infty = 2\bar{v}$, Equation (8.15) becomes

$$\frac{D}{2} = \frac{5L}{\left[\dfrac{2L\bar{v}}{\nu}\right]^{1/2}}$$

Rearranging

$$\frac{L}{D} = \frac{1}{10}\left[\frac{2L\bar{v}}{\nu}\frac{D}{D}\right]^{1/2}$$

or

$$\left(\frac{L}{D}\right)^{1/2} = \frac{\sqrt{2}}{10}Re^{1/2}$$

$$\frac{L}{D} = 0.02\ Re$$

The accepted correlation for the entrance length in laminar pipe flow is $\dfrac{L}{D} = 0.035\,Re$

8-1.c Inviscid Flow

Having concentrated on the flow very near the wall, let's now examine the second region of flow, the flow at the center of the pipe. In this region of the pipe, we make the rather idealized assumption that molecular momentum transport in the radial direction is still very small and τ_{zz} can be neglected ($\tau_{zz} < \tau_{rz} \sim 0$). This is called <u>inviscid flow</u> since the fluid will have no viscosity under these assumptions. For the more general case of a vertical pipe, Equation (8.5) simplifies to

$$-\frac{d}{dz}(\rho v_z v_z) - \frac{dP}{dz} - \rho g_z = 0 \tag{8.17}$$

In this region of flow, the continuity equation states that

$$\frac{d}{dz}(\rho\,v_z) = 0$$

and therefore, the first term in Equation (8.17) can also be written as

$$\frac{d}{dz}(\rho v_z v_z) = \rho v_z \frac{dv_z}{dz} + v_z \frac{d}{dz}(\rho v_z) = \rho v_z \frac{dv_z}{dz} = \frac{\rho}{2}\frac{d}{dz}(v_z^2)$$

and so Equation (8.17) becomes

$$-\frac{\rho}{2}\frac{d}{dz}(v_z^2) - \frac{dP}{dz} - \rho g_z = 0 \qquad (8.18)$$

Equation (8.18) is sometimes called the point form of the Bernoulli Equation but it is usually employed in its integrated form and applied in a macroscopic manner to practical piping calculations. To do this, v_z is taken to be the average velocity in the pipe and then Equation (8.18) is integrated between points 1 and 2 in a piping system to give

$$\frac{\rho}{2}(\bar{v}_2^2 - \bar{v}_1^2) + (P_2 - P_1) + \rho g_z (h_2 - h_1) = 0 \qquad (8.19)$$

where h_2 and h_1 correspond to the <u>vertical</u> height of points 2 and 1 in the pipe. If the pipe were vertical, then h_2 and h_1 are also equal to the distance along the pipe between the two points, z_2 and z_1. Note that Equation (8.19) has units of energy/volume and thus it can also be derived from the First Law of Thermodynamics for an open system. The first term is a kinetic energy term, the second is "work" (force-distance) term and the last is potential energy.

This equation can be solved in conjunction with the boundary layer equations to give a complete picture of momentum transport in the pipe. However, it has its most useful application in problems of flow over solid objects, flow over an airplane wing, for example. Because Equation (8.19) is a result of integration, it is a macroscopic equation, applying to the system as a whole. In Chapter 11, we will deal with macroscopic momentum calculations and will return to this equation and modify it to account for the friction losses in the system.

For now, lets see how Equation (8.19) can be used in conjunction with the boundary layer equations. Figure 8-6 shows the fluid impinging on the leading edge of the flat plate (point 1) where it has zero velocity and a pressure, P_{sat}, which is called the <u>stagnation pressure</u>. Beyond the boundary layer, the fluid can be taken to be inviscid and thus Equation (8.18) applies. At some distance downstream, point 2, the pressure will be less than P_{sat} and the free-stream velocity, $v_\infty(z)$, will have increased. Solving Equation (8.18) for $v_\infty(z)$ along the horizontal plate, we have

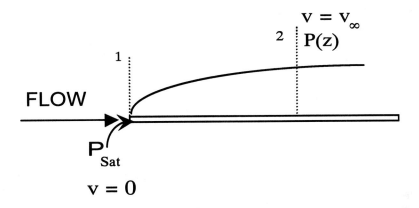

Figure 8-6 Inviscid Flow along a Flat Plate

$$\frac{v_\infty^2(z)}{2} - 0 = -\frac{1}{\rho}[P_{Sat} - P(z)]$$

(8.20)

$$v_\infty(z) = \sqrt{\frac{2[P_{Sat} - P(z)]}{\rho}}$$

This calculated value of $v_\infty(z)$ can then be used in the Blasius solution (Figure 8-5) to obtain a complete momentum analysis of developing flow along a flat plate, or of the entrance sections of pipe flow.

EXAMPLE 8-4: Pitot Tube Analysis

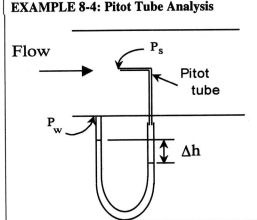

The velocity of a fluid at a point in a duct can be measured by using a *pitot tube*, as shown in the accompanying sketch. The pressure at the head of the pitot tube is P_s and it is measured relative to the pressure at the wall, P_w. The pressure difference between the two readings is measured with a manometer containing a fluid of density, ρ_M and it registers a height difference, Δh. Derive an equation that expresses the velocity of the fluid in terms of ρ and Δh.

Solution

If we apply Equation (8.19) between the free-flowing fluid, where the velocity is v and the tip of the probe, where the velocity is zero, then

$$\frac{\rho_f}{2}\,(v^2 - 0) + [P_w - P_s] = 0$$

where ρ_f is the density of the fluid flowing in the duct, P_w is the pressure of the fluid (measured at the wall) and P_s is the pressure at the tip of the pitot probe (the stagnation pressure). The pressure difference is related to the manometer reading by

$$P_S - P_w = \rho_M\,g\,\Delta h$$

and so,

$$v = \sqrt{\frac{2\,\rho_M\,g\,\Delta h}{\rho_f}}$$

The above equation assumes inviscid flow; thus a correction factor is usually applied and is related to the design ("shape") of the probe itself. In this case it is more generally written as

$$v_S = C_P\sqrt{\frac{2\,\rho_M\,g\,\Delta h}{\rho_f}}$$

where C_p is the correction factor.

8-2 ENERGY TRANSPORT IN A SHELL AND TUBE CONDENSER

Figure 8-7 shows a shell and tube condenser where a vapor is condensing on the shell side due to heat removal by a coolant fluid which flows through the tubes. To analyze the steady-state energy transport in this situation, we can focus on a single tube and apply the differential balance techniques of Chapter 2. Since the coolant is heated as it flows through the tube, we have both molecular and convective energy transport in the flow (z) direction. Furthermore, if we have laminar flow (Reynolds number less than 2,100), then the only mode of energy transport in the r-direction is molecular. A differential energy balance over the DVE shown in Figure 8-7, yields

$$(q_{z|_z} - q_{z|_{z+\delta z}}) \, 2\pi r\delta r + \left[(\rho\hat{H}v_z)_{|_z} - (\rho\hat{H}v_z)_{|_{z+\delta z}} \right] 2\pi r \delta r$$

$$(8.21)$$

$$+ \left[(rq_r)_{|_r} - (rq_r)_{|_{r+\delta r}} \right] 2\pi\delta z = 0$$

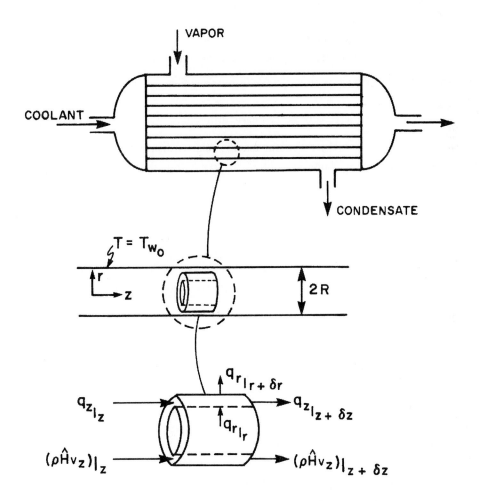

Figure 8-7 Shell and Tube Condenser

If we once again assume that the magnitude of the molecular transport in the flow direction is significantly less than the convective transport, then the first grouping of terms in Equation (8.21) can be dropped. While this simplifies the differential energy balance, it still leaves us with a partial differential equation to be solved. Specifically, substituting for $\hat{H}$ in terms of the heat capacity at constant pressure and using Fourier's Law, the applicable differential equation becomes, for constant properties

$$\rho \hat{c}_p v_z \frac{\partial T}{\partial z} - \frac{k}{r} \frac{\partial}{\partial r} \left(r \frac{\partial T}{\partial r} \right) = 0 \tag{8.22}$$

However, this equation can be reduced to an ordinary equation if it is assumed that the temperature profiles are fully developed; that is, that their "shape" does not change in the flow direction. If the shape of the temperature profile is invariant, then $\frac{\partial T}{\partial z}$ is constant everywhere[#],

($\frac{\partial T}{\partial z} = \frac{\Delta T}{\Delta z}$) and we are left with an ordinary differential equation with non-constant coefficients (because $v_z = f(r)$). Assuming that the velocity distribution is fully developed, we can use the solution for v_z from Equation (5.26) so that Equation (8.22) becomes

$$\frac{2\bar{v}}{\alpha} \frac{\Delta T}{\Delta z} \left[1 - \left(\frac{r}{R} \right)^2 \right] - \frac{1}{r} \frac{d}{dr} \left(r \frac{dT}{dr} \right) = 0 \tag{8.23}$$

Of course the boundary conditions must be stipulated before we can solve Equation (8.23). While the usual symmetry condition prevails at $r = 0$ ($\frac{dT}{dr} = 0$), the appropriate boundary condition at the wall is not so obvious. Because of the physical problem and the stated assumptions, the wall is not at constant temperature and a constant wall flux condition will lead to some inconsistencies [4]. Instead we use

$$\text{at } r = R, \ T = T_{w0} + \left(\frac{\Delta T}{\Delta z} \right) z \tag{8.24}$$

The temperature distribution can now be obtained by integrating Equation (8.22) twice[##], using the stated boundary conditions, to give

$$T - T_{w0} = \frac{\Delta T}{\Delta z} \left[z + \frac{\bar{v} R^2}{8\alpha} \left(4\eta^2 - \eta^4 - 3 \right) \right] \tag{8.25}$$

[#]   Note that this assumption is tantamount to assuming that the heat input to the tube is a constant; thus, the magnitude of the temperatures do change.

[##]   See problem 8-2.

where $\eta = \dfrac{r}{R}$

8-3 PLUG-FLOW CHEMICAL REACTOR

An example of a simple convective mass transport problem is depicted by the sketch in Figure 8-8 which shows a tubular chemical reactor in which species 'A' decomposes homogeneously and irreversibly to species B. The fluid entering the reactor is in laminar flow and thus the axial velocity will vary with r according to the parabolic equation, Equation (5.26). This means that fluid near the center of the tube will spend less time in the reactor and will have a conversion less than fluid near the wall. To eliminate this problem, a series of "re-distributors" are placed periodically along the length of the reactor, thus promoting radial mixing and a "flat" velocity profile. In such a situation, the fluid can be thought of as moving through the tube as a "plug"; that is, with perfect radial mixing but no axial mixing. This somewhat idealistic situation is commonly referred to as <u>plug flow</u>.

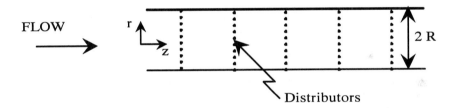

Figure 8-8 Plug-Flow Reactor

A differential mass balance for species A must account for the axial transport of A by both molecular and convective mechanisms since there will be a concentration gradient of 'A' in the axial direction. If we distinguish between molecular and convective transport[#], then the molecular flux is J_{A_z} and the z-component of the convective transport term is $C_A v_z$ (see Chapter 2-1), so that the differential mass balance describing the transport of species A in the reactor results in

$$-\frac{dJ_{AZ}}{dz} - \frac{d(C_A v_z)}{dz} - R_{Av} = 0 \tag{8.26}$$

[#] This is not necessary if we work with the molecular flux of A with respect to fixed coordinates; see Problem 8-3.

where R_{Av} represents the sink term due to the chemical reaction rate. In most situations the molecular flux is much smaller than the convective flux and so, $J_{Az} \ll C_A v_z$. With this assumption, the fact that the velocity is constant (equal to $\bar{v}$), and assuming, as we did in Chapter 4-2, that the reaction rate is first order with respect to A, the differential equation and accompanying boundary condition is simply

$$-\bar{v}\,\frac{dC_A}{dz} - k_r\,C_A = 0$$

$$\text{at } z=0; \; C_A = C_{A0}$$

where k_r is the rate constant. The solution is given by

$$C_A = C_{A_o}\,\text{Exp}\left[-\frac{k_r\,z}{\bar{v}}\right] \tag{8.27}$$

Note that $\dfrac{z}{\bar{v}}$ can be thought of as a *residence time*, τ, and thus this solution is identical to that obtained in an unsteady-state <u>batch</u> reactor[5].

It should be pointed out that while the neglect of molecular transport relative to convective transport is almost always reasonable, the presence of the packing (or catalyst) in the reactor can produce deviations from plug flow behavior. That is, irregularities in the packing can result in additional axial mixing; due to <u>channeling</u>, for example. Since the net effect appears very similar to a flux superimposed on the bulk motion, mathematics analogous to molecular mass transport[#] have been used to describe the phenomena. In this case, Equation (8.26) applies except that J_{A_z} is a dispersion flux and is analogously related to the concentration gradient by means of a <u>dispersion coefficient</u>; i.e.,

$$J_{A_z} = -D_L\,\frac{dC_A}{dz}$$

The dispersion coefficient will of course be related to the flow situation and thus empirical correlations are used to obtain its value. The differential equation describing this situation (the so called <u>dispersion model</u>) is

---

[#] Since molecular mass transport, i.e., diffusion, is also superimposed on the bulk transport.

$$D_L \frac{d^2 C_A}{dz^2} - \bar{v} \frac{dC_A}{dz} - k_r C_A = 0 \tag{8.28}$$

The boundary conditions corresponding to this model can be quite involved and they are discussed, along with the solution to Equation (8.28) by Fogler [6].

REFERENCES

[1] Bird, R.B., Stewart, W.E., and Lightfoot, E.N., *Transport Phenomena*, Appendix A, John Wiley & Sons, N.Y., 1960.

[2] Blasius, H., NACA Tech. Memo No. 1256 (1956)

[3] Schlichting, H., *Boundary Layer Theory*, McGraw-Hill Co., N.Y., 1960

[4] Bird, R.B., Stewart, W.E., and E.N.Lightfoot, , *Transport Phenomena*, John Wiley & Sons, N.Y., 1960, p. 296

[5] Moore, W.J., *Physical Chemistry*, 4th ed., Prentice Hall, Englewood Cliffs, NJ, 1972, p. 333

[6] Fogler, H.S., *Elements of Chemical Reaction Engineering*, Prentice Hall, Englewood Cliffs, NJ, 1986, p. 701

PROBLEMS

8-1 Calculate the reading of a manometer (in inches of water) of a pitot tube placed in the center of a 1" ID tube with air flowing at 25 °C and 1 atm pressure, at a Reynolds number of 1,000. Assume that the pitot tube reading is referenced to the static pressure at that point.

8-2 Starting with Equation (8.23), derive the solution given by Equation (8.25). Then take an overall energy balance and relate $\dfrac{\Delta T}{\Delta z}$ to the flux at the wall, q_w.

8-3 Re-derive Equation (8.26) by considering only the flux of A relative to fixed coordinates (N_{A_z}) and then using Equations (4.10), (4.8), and (4.5).

8-4   Consider fully developed laminar flow between parallel plates where a newtonian fluid is being heated by the walls which are maintained at a constant wall flux, If the distance between the walls is 2H, derive and solve the differential energy balance corresponding to this flow situation to obtain solutions for both the temperature distribution between the walls and the average temperature in the duct, $\overline{T}$. It may be assumed that molecular energy transport in the z-direction can be neglected and that the axial temperature gradient, $\dfrac{dT}{dz}$, is constant at $\dfrac{\Delta T}{L}$. Use the boundary condition given by Equation (8.24) and note that it is necessary to first solve for the velocity profile between the walls.

8-5

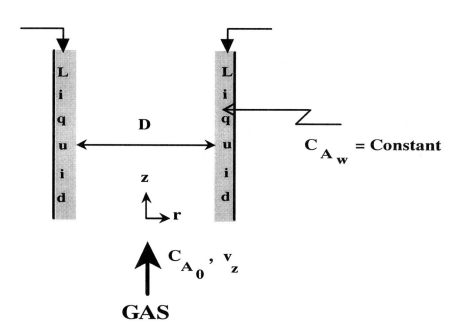

A gas containing species `A` enters the bottom of a <u>Wetted Wall Column</u> at a concentration C_{A0} and a velocity, v_z. As the gas flows up the column, A is absorbed by the film of liquid running down the inside wall of the column. Solve for the concentration of A in the gas, as a function of z if the following assumptions can be made: (1) the gas concentration profile and the gas velocity across the column are "flat;" (2) the radial transport of A can be accounted for by a uniform SINK with a rate proportional to the interfacial surface area and the difference between its concentration in the gas and its concentration in the liquid, C_{A_w} (proportionality coefficient is k_{mc}); (3) the only mode of mass transport in the z-direction is convective.

8-6   Reconsider Problem 3-21, this time with a reaction rate which is first-order reaction with respect to species A. Assuming that the reaction rate constant is only a weak function of temperature, solve for the temperature distribution in the reactor if the reactor behaves as a plug-flow reactor.

8-7

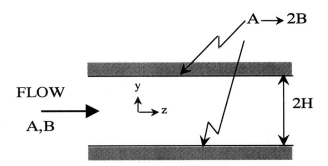

A gas containing species A and B is flowing between the parallel walls of a wide ($W/2H \gg 1$) duct in isothermal, steady-state, laminar flow, as shown in the above sketch. A catalytic reaction takes place at the walls with a zero-order <u>surface</u> reaction rate, (r_{A_s} , moles/area-time). If the entering flow contains only A at a concentration, C_{A0} , and the molecular transport in the z-direction can be neglected

(a) Derive the differential equation (in terms of X_A) and state the boundary conditions necessary to solve for the concentration of `A` in the duct. [NOTE: the total concentration, C , is <u>not</u> constant].

(b) Assuming that both the velocity and concentration profiles are flat with a velocity, $\bar{v}$, and a concentration, $C_A(z)$, solve for the concentration of `A` as a function of z .

8-8

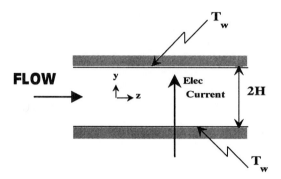

A Newtonian, electrolytic solution is flowing in the positive z-direction between two parallel electrodes. An electrical current flows uniformly through the fluid and the two electrodes are cooled to maintain the wall temperature constant at T_w. The current density (amps/cm²) is J and the resistivity of the solution is ρ_R (ohms-cm). If the fluid properties are constant and the flow is laminar and fully developed (with respect to <u>both</u> velocity and temperature), solve for the temperature distribution in the fluid.

8-9 (C*)

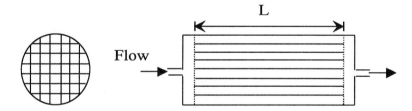

Catalytic combustion is often used to remove trace quantities of CO or hydrocarbons from flue gas streams. In order to minimize pressure drop, the catalyst is often in the form of a "monolith"; i.e., a series of parallel channels (see sketch). If it is assumed that the width/height ratio of the channels is large (actually, the channels are typically square), then any one channel is in a configuration similar to that described in Problem 8-7. However, the reaction is typically first order with respect to the CO concentration and the rate constant is usually a strong function

of temperature ($k_s = A_o \exp\left[-\dfrac{E}{R_g T}\right]$). In addition, a more realistic assumption for the energy

transport boundary condition at the wall is to assume that the wall is adiabatic. Assuming steady-state, laminar flow with "flat" profiles, and reactor inlet conditions of C_{A_0} and T_0,

(a) Derive the differential equations necessary to solve for the concentration of CO as a function of reactor length.

(b) Place the equations in dimensionless form and show that the solution will depend on the following three dimensionless parameters

$$\Phi_1 = \frac{A_0 L}{\bar{v}H} \ , \quad \Phi_2 = \frac{E}{R_g T_0} \ , \quad \Phi_3 = \frac{|\Delta \tilde{H}_r| C_{A0}}{\rho \hat{c}_P T_0}$$

(c) Use MATLAB to generate plots of dimensionless CO concentration versus dimensionless length as a function of the pertinent dimensionless parameters. The suggested range of parameters are:

$$\Phi_1 : \ (1 - 5) 10^{-5} \ , \quad \Phi_2 : \ 15 - 30 \ , \quad \Phi_3 : \ 5 - 20$$

8-10   Use the generalized Equations of Change presented in Tables A-1 - A-4 in Appendix A to derive Equations (8.22) and (8.26). Be sure to list the REASONS for eliminating the various terms in the Equations of Change.

8-11   Referring to Problem 5-21, simplify the generalized Equations of Change in Appendix A to derive the equation which describes the variation of v_θ with radius.

CHAPTER 9

Turbulent Transport

Up to this point, we have encountered the subject of turbulence at various points throughout the text, but without delving into the subject in much detail. For example, in Chapter 2 it was pointed out that turbulence results in vigorous mixing, leading to convective fluxes which are highly efficient transport mechanisms. In this chapter, turbulence is discussed in a more formal manner. First, a basic qualitative description is given and these observations are then used in an attempt to describe turbulent momentum transport from a fundamental, differential balance point of view. This is followed by a description of semi-empirical methods which attempt to predict velocity, temperature and concentration profiles in turbulent flow. Finally, the theory of <u>surface renewal</u> is described in order to relate molecular transport mechanisms to the sporadic convective nature of turbulent flows.

9-1 THE NATURE OF TURBULENT FLOW

Any introductory discussion of turbulent flow should start with a description of a rather famous set of experiments carried out by Osborne Reynolds in 1883. Figure 9-1 illustrates the essence of these experiments. Reynolds immersed a clear glass tube in a glass tank full of liquid, allowing the liquid to flow into one end of the tube at various flow rates. In order to visually observe the flow patterns within the tube, he had provisions for injecting a small stream of fluid containing a dark-colored dye into the center of the tube. At the lower flow rates (laminar flow), the dye stream kept its identity and no mixing took place (Figure 9-1b). At higher flow rates (turbulent flow), the dye stream no longer appeared as a thread, but mixed with the main body

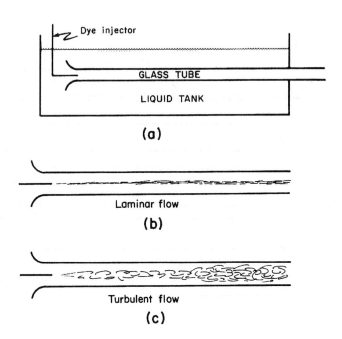

Figure 9-1 Reynolds' Experiment

of the fluid as a series of erratic eddies and cross currents. As can be seen from Figure 9-1c, the character of turbulent flow is sporadic and three dimensional; in fact, there are velocity components in all directions. In addition the magnitude and directions of the velocity components change with time even though the flow system, on the average, is time independent. In a word, turbulence is a synonym for <u>mixing</u>. If we were able to keep track of all velocity components in both direction and time we could attempt to formulate a mathematical model of the system and solve for the momentum flux imparted to the wall. However, to date this has only been marginally successful, and then only at a cost of large units of computer time.

In Reynolds' experimentation, the point at which the "transition" from laminar to turbulent flow took place was found to be dependent on the <u>Reynolds number</u>, a dimensionless parameter which was named after him as a result of these classic experiments. As a rough "rule of thumb," the critical value of the Reynolds number in circular ducts is 2,100. That is, at lower Reynolds numbers we can expect to find laminar flow, and turbulent flow is likely to prevail at higher Reynolds numbers. Actually this is somewhat simplistic. A better description of the transition Reynolds number is to state that at values below 2,100, the fluid is capable of dampening out any instabilities which might occur (obstructions in the pipe, vibrations, etc.) but at higher Reynolds numbers the instabilities will propagate. For example, laminar flow conditions in pipes have been reported at Reynolds numbers as high as 10^4 but as soon as the pipe experiences a slight vibration, the flow rapidly becomes turbulent. The fact that the Reynolds number is the determining parameter here can be better appreciated by realizing that it is a ratio of inertial (convective) to viscous forces; i.e.,

$$Re = \frac{\rho \, \bar{v}}{\dfrac{\mu}{D}}$$

Thus the viscous forces serve to dampen the convective forces and are overwhelming at lower Reynolds numbers.

Now let's look in more detail at how local velocity measurements at various points in a pipe might vary with time. Figure 9-2b shows the output of two velocity probes located at two different radial positions and sufficiently far from the entrance section to ensure fully developed conditions. A number of observations can be made here. First, the probe at the center of the pipe exhibits larger velocity fluctuations than the probe very near the wall, a natural consequence of the more vigorous mixing state near the center of the pipe[*]. Second, it should be noted that although the fluctuations are rather random, with a range of frequencies and amplitudes, there is an equal variation (given a sufficient sampling time) about some mean velocity (Figure 9-2c). If a sufficient sampling time is employed, this mean velocity will be time invariant and is called the <u>time-averaged velocity</u>, [v]. Formally, it can be computed by

$$[v] \equiv \frac{\displaystyle\int_{t_1}^{t_2} v \, dt}{\displaystyle\int_{t_1}^{t_2} dt} \tag{9.1}$$

Referring again to Figure 9-2c, at any instant in time, the instantaneous velocity, v, can be related to [v] by "time averaging" the instantaneous velocity, or

$$v = [v] + v' \tag{9.2}$$

where v' is the <u>fluctuating velocity</u>. Using Equations (9.1) and (9.2) we can show that the time-average of the fluctuating velocity (or <u>any</u> fluctuating variable, for that matter) is equal to zero. For example, v from Equation (9.2) can be substituted into Equation (9.1) and integrated

---

[*] Recall that eventually, molecular transport will dominate right at the wall.

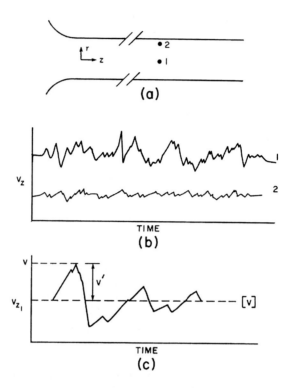

Figure 9-2 Time-Varying Turbulent Velocities

$$[v] = \frac{\int_{t_1}^{t_2} ([v] + v')\, dt}{(t_2 - t_1)} = \frac{[v]\,(t_2 - t_1)}{(t_2 - t_1)} + \frac{\int_{t_1}^{t_2} v'\, dt}{(t_2 - t_1)}$$

$$[v] = [v] + \frac{\int_{t_1}^{t_2} v'\, dt}{(t_2 - t_1)}$$

and therefore,

$$\int_{t_1}^{t_2} v'\, dt = 0$$

To generalize, the basic definition of Equation (9.1) can be used to formulate the following rules for time averaging[#]

<u>Time-Averaging Rules</u>

$$[v'] = 0$$

$$[[v]] = [v]$$

$$[[v]v'] = [v][v'] = 0 \tag{9.3}$$

$$[v^2] = [v]^2 + [v'^2]$$

$$[v_i v_j] = [v_i][v_j] + [v_{i'} v_{j'}]$$

9-2 TIME-AVERAGED MOMENTUM EQUATION

In turbulent flow, there will be velocity components in all directions and there will be instantaneous variations of velocities with time. As we shall see, even though the time variations can be averaged over some suitable time scale, nonlinear terms (such as convective momentum fluxes) will remain. Thus, if differential momentum balances are to be rigorously applied to turbulent flow, they will be exceedingly complex. One approach is to first construct a differential momentum balance, allowing velocities to vary with time and in all three dimensions and to then time-average the entire momentum balance. To illustrate this approach, consider fully developed turbulent pipe flow and only the z-component of momentum. The DVE and applicable fluxes are as shown in Figure 9-3. For "steady-state" conditions, it results in[##]

[#] Although the velocity is used to illustrate these rules, they apply equally to <u>any</u> fluctuating variable (T and C_A, for example).

[##] Here, steady-state is taken to mean no changes with time in a time-averaged sense. Also, this equation is considerably simplified due to order-of-magnitude considerations. See Problem 9-3 for a more thorough analysis.

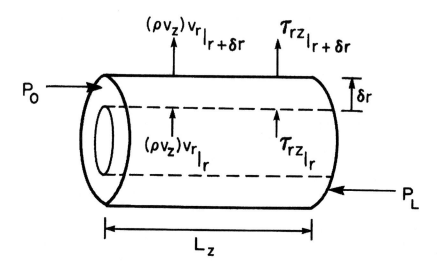

Figure 9-3 DVE for Turbulent Pipe Flow

$$-\frac{1}{r}\frac{d}{dr}(r\tau_{rz}) - \frac{1}{r}\frac{d}{dr}(r\rho v_z v_r) + \frac{(P_0 - P_L)}{L} = 0 \tag{9.4}$$

Using Newton's Law of viscosity and assuming constant fluid properties we can also write

$$\frac{\mu}{r}\frac{d}{dr}(r\frac{d v_z}{dr}) - \frac{\rho}{r}\frac{d}{dr}(r v_z v_r) + \frac{(P_0 - P_L)}{L} = 0 \tag{9.5}$$

The next step in the time-averaging procedure is to express the instantaneous velocities (in addition to the pressure) as a summation of their time-averaged values and fluctuating components and then time-average the entire equation, term-by-term.

As pointed out above, the time-averaging of linear terms results in the instantaneous value being replaced by the time-averaged value. For example, time-averaging the first term in Equation (9.5) gives

$$\left[\frac{\mu}{r}\frac{d}{dr}\left(r\frac{d}{dr}([v_z] + v_{z'})\right)\right] = \frac{\mu}{r}\frac{d}{dr}(r\frac{d[v_z]}{dr})$$

On the other hand, time averaging the convective transport term [the second term on the left hand side of Equation (9.5)] results in

$$\left[\frac{\rho}{r}\frac{d}{dr}(r v_z v_r)\right] = \frac{\rho}{r}\left(\frac{d}{dr}(r[v_z][v_r]) + \frac{d}{dr}(r[v_{z'} v_{r'}])\right)$$

With these results and the fact that $[v_r] = 0$,[#] the time-averaged momentum equation becomes

$$\frac{\mu}{r}\frac{d}{dr}(r\frac{d[v_z]}{dr}) + \frac{([P_0]-[P_L])}{L} - \frac{\rho}{r}\frac{d}{dr}(r[v_{z'} v_{r'}]) = 0 \qquad (9.6)$$

If Equation (9.6) is compared to Equation (5.17), which describes fully developed pipe flow under laminar flow conditions, we can see that the first two terms of (9.6) are identical. The velocity and pressures are simply replaced by their time-averaged values. However, in turbulent flow, there is one additional term resulting from the time averaging of the product of the z- and r-components of the fluctuating velocity vector. Note that, not only is this additional term non-linear, but it arises from the convective momentum flux term in the momentum balance. Unfortunately, the introduction of this term has taken place without an additional equation or constitutive relation. In other words we now have more unknowns than equations. Over the years, elaborate statistical analyses have been employed in an effort to circumvent this closure problem, but success has been limited to highly constrained flow problems.

Whereas the particular time-averaged nonlinear term shown in Equation (9.6) is dominant for fully developed pipe flow, it is actually only one element of the nine-component tensor which is known as the Reynolds stress tensor

$$\begin{vmatrix} \rho[v_{z'} v_{z'}] & \rho[v_{z'} v_{r'}] & \rho[v_{z'} v_{\theta'}] \\ \rho[v_{r'} v_{z'}] & \rho[v_{r'} v_{r'}] & \rho[v_{r'} v_{\theta'}] \\ \rho[v_{\theta'} v_{z'}] & \rho[v_{\theta'} v_{r'}] & \rho[v_{\theta'} v_{\theta'}] \end{vmatrix}$$

Figures 9-4 and 9-5 show experimental turbulent pipe flow data for some of the elements of the Reynolds stress tensor. These data were collected by Laufer [1] using *hot-wire anemometry*[##] Note that the velocity data in these figures are normalized with respect to the friction velocity, v_*, which is defined by

---

[#] This can be formally proven by time-averaging the continuity equation (see problem 9.1).

[##] This is a velocity probe consisting of a heated, fine wire which is cooled by fluctuations in the local velocity. Instantaneous velocities are measured by calibrating the wire's resistance with velocity.

$$v_* \equiv \sqrt{\frac{\tau_w}{\rho}} \tag{9.7}$$

The following can be gleaned from these two figures: the fluctuating velocity in the flow direction has the highest magnitude, all of the fluctuating components have their maximum values very close to the wall, and all decay to about the same value at the center of the pipe. Figure 9-4 also shows the radial variation of the most important element of the Reynolds stress tensor, $[v_{z'} v_{r'}]$. Its importance lies in the fact that, in fully developed turbulent flow, it is the only element of the stress tensor to appear in the differential momentum balance [Equation (9.6)] and it has the physical significance of representing the convective transport of z-momentum in the r-direction; a term which arises only in turbulent flow. Figure 9-5 shows how the Reynolds number affects the fluctuating components very near the wall; in this case, the radial variation of $v_{z'}$.

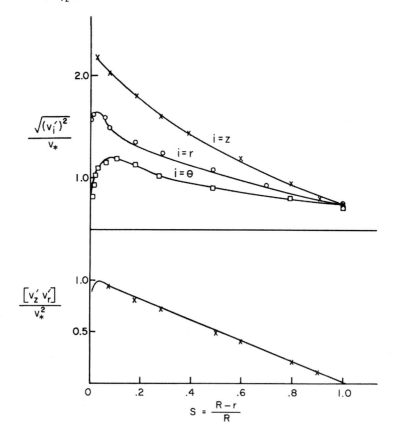

Figure 9-4 Reynolds Stress Tensor in Pipe Flow

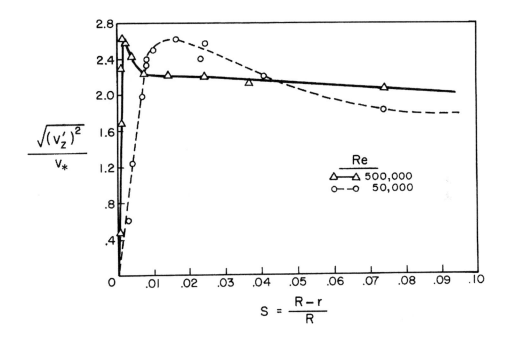

Figure 9-5 Reynolds number Effect near Pipe Walls

9-3 SEMI-EMPIRICAL METHODS

While the treatment of turbulent flow presented in the first two sections of this chapter has certain fundamentally appealing features, as pointed out earlier, there are always more unknowns than there are equations; the so-called *closure problem*. Thus it is not surprising that engineers and scientists have resorted to semi-empirical methods in order to quantify turbulent transport. Some of these methods take the form of transfer coefficient correlations which are only intended to calculate overall transfer rates of energy, momentum, and mass. However, other semi-empirical methods have also been adopted in an attempt to predict velocity, temperature, and concentration distributions in turbulent flow. It is these latter approaches which will be discussed here, since they can also be used to calculate overall transfer rates (by differentiation at the wall/interface).

9-3.a Eddy Viscosity and the "Universal Velocity Profile"

One of the earliest approaches to the prediction of turbulent velocity profiles was to draw a direct analogy with molecular momentum transport and to assume that the turbulent contribution to momentum transport could also be described by a relationship similar to Newton's Law of Viscosity. Thus the momentum flux was split into two components, a laminar transport flux, τ_{rz}^l, and a *turbulent flux*, τ_{rz}^t. The total transport flux is the sum of the two, which can be related to the time-averaged velocity gradient by

$$\left(\frac{\tau_{rz}^l + \tau_{rz}^t}{\rho}\right) = \frac{\tau_{rz}^{tot}}{\rho} = -(\nu + \varepsilon)\frac{d[v_z]}{dr} \tag{9.8}$$

where ε is the <u>eddy viscosity</u>. Figure 9-6 shows the expected radial distribution of the turbulent momentum flux as a function of Reynolds number. Note that the contribution of the turbulent momentum flux to the total momentum flux is low near the wall but dominates near the center of the pipe. It is not too surprising that the radial variation of τ_{rz}^t is similar to the radial variation of the elements of the Reynolds stress tensor as shown in Figure 9-4, since the turbulent momentum flux is replacing these elements. Thus, if the eddy viscosity concept is employed, the comparable equation to Equation (9.4) is

$$-\frac{1}{r}\frac{d}{dr}(r\tau_{rz}^{tot}) - \frac{dP}{dz} = 0 \tag{9.9}$$

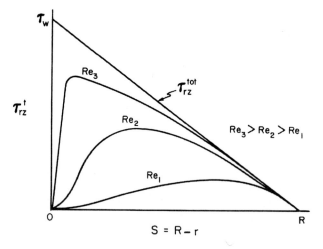

Figure 9-6 Radial Variation of τ_{rz}^t

which, from Equation (9.8), can be written in terms of the velocity gradient as

$$\frac{1}{r}\frac{d}{dr}(r\,\tau'_{rz}) - \frac{\mu}{r}\frac{d}{dr}(r\,\frac{d[v_z]}{dr}) + \frac{d[P]}{dz} = 0 \qquad (9.10)$$

Comparing this equation with Equation (9.6) it is easy to see that

$$\frac{\tau'_{rz}}{\rho} = [v_{z'}\,v_{r'}] \qquad (9.11)$$

The same sort of approach can also be taken in turbulent energy and mass transport to define turbulent values of the thermal and mass diffusivities. So, in summary we have

$$\frac{\tau'_{rz}}{\rho} = [v_{r'}\,v_{z'}] = -\varepsilon\frac{d[v_z]}{dr}$$

$$\frac{q^t_r}{\rho\,\hat{c}_P} = [v_{r'}\,T'] = -\alpha^t\frac{d[T]}{dr} \qquad (9.12)$$

$$J^t_{A_r} = [v_{r'}\,C_{A'}] = -D^t\frac{d[C_A]}{dr}$$

Note, that since the time-averaged products vary with position, the turbulent diffusivities (ε, α', D') will all be functions of radial position (as well as the degree of turbulence). Thus, in order to use this approach, a certain amount of data fitting will be required. Its advantage over the transfer-coefficient approach is that this approach results in predictions of the velocity, temperature, and concentration profiles as well as the fluxes at the walls/interfaces. Recall that the latter is all that transfer coefficients can give.

In order to use this concept to solve for the time-averaged velocity profile in turbulent pipe flow, we start with Equation (9.9) which can be integrated once to give

$$\tau^{tot}_{rz} = -\frac{dP}{dz}\frac{r}{2} + \frac{C_1}{r} \qquad (9.13)$$

C_1 is zero by virtue of the symmetry boundary condition at the center of the pipe and substituting for τ^{tot}_{rz} in terms of its laminar and turbulent components, we have

$$(\mu + \mu') \frac{d[v_z]}{dr} = \frac{dP}{dz} \frac{r}{2}$$

where μ' is the turbulent viscosity. Integrating this expression, we obtain

$$[v_z] = \frac{dP/dz}{2\mu} \int_R^r \frac{r + dr}{\left(1 + \dfrac{\mu'}{\mu}\right)} \tag{9.14}$$

Thus we need to have a knowledge of how μ' varies with r before we can predict the velocity profile.

The first step in the derivation of a generalized velocity profile is to segment the profile into the three distinct regions shown in Figure 9-7. The region closest to the wall is assumed to be a region where turbulent eddies cannot penetrate and thus molecular momentum transport is prevalent in that region (the laminar sublayer). The region in the center of the pipe is considered to be in a high state of turbulence and is called the turbulent core. The region(s)[#] between these two extremes is called the buffer layer. The second step in the development is to "scale" the velocity so that its dependence on the degree of turbulence will be assured. This is most conveniently done by defining appropriate characteristic values for velocity and radial position and then casting the applicable differential equation into dimensionless form. Equation (9.14) can be placed in dimensionless form by the following choice of dimensionless variables

$$v_z^+ = \frac{[v_z]}{v_*}, \quad s^+ = \frac{s v_*}{\nu}$$

where v_*, the friction velocity has already been defined in Equation (9.7) and $s = R - r$. With these definitions and realizing that $\dfrac{dP}{dz}$ can be expressed in terms of the momentum flux at the wall, τ_w [evaluate Equation (9.13) at $r = R$], the dimensionless form of Equation (9.14) becomes[##]

[#]   Sometimes more than one region is used to describe this "transition" region.

[##] See Problem 9-5.

$$v^+ = \int_0^{s^+} \frac{\left(1 + \frac{s^+}{s_R^+}\right) ds^+}{\left(1 + \frac{\varepsilon}{v}\right)} \tag{9.15}$$

where s_R^+ corresponds to the center of the pipe; i.e., at $s = R$.

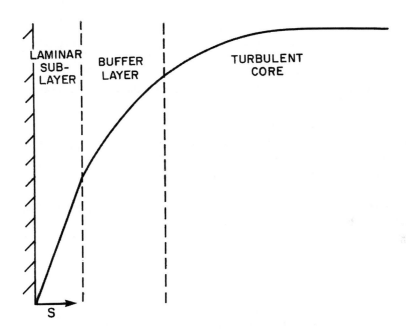

Figure 9-7 Segmented Regions in Turbulent Pipe Flow

Now lets consider the laminar sublayer region. In this region, the eddy viscosity is zero and it is reasonable to assume that $1 - \frac{s^+}{s_R^+} \sim 1$ so that the integration of Equation (9.15) is simply

$$v^+ = s^+ \tag{9.16}$$

which is generally found to be valid for $0 \le s^+ \le 5$.

The expression for the velocity profile in the turbulent core is typically based on a theoretical development by Prandtl, known as the <u>Prandtl mixing length theory</u>. Basically it was assumed that the turbulent momentum flux comes about by the convective transport of time-averaged momentum by the root mean square of the fluctuating velocity component. The time-averaged velocity was then assumed to vary linearly over a "mixing length," l, and furthermore, the root mean square velocity was assumed to be equal to the product of l and the velocity gradient. With these assumptions, the turbulent momentum flux is given by

$$\tau'_{rz} = -\rho \, l^2 \left| \frac{d[v_z]}{ds} \right| \frac{d[v_z]}{ds} \qquad (9.17)$$

Two more assumptions were then made: that the flux in the core was equal to the wall flux[#] and that the mixing length was proportional to the distance from the wall ($l = \kappa s$). With these last two assumptions, Equation (9.17) can be placed in dimensionless form in terms of s^+ and v^+ so that

$$\left| \frac{d\,v^+}{ds^+} \right|^2 = \frac{1}{\kappa^2 s^{+2}}$$

which can be integrated from the edge of the turbulent core region, s_I^+, to any point, s^+, within the region, to yield

$$v^+ = \left(v_I^+\right) + \frac{1}{\kappa} \ln s^+$$

Empirically, κ is found to be approximately 0.4 and the bracketed term to be equal to 5.5 so that the final expression for the dimensionless velocity profile in the turbulent core is

$$v^+ = 2.5 \ln s^+ + 5.5 \qquad (9.18)$$

which is generally valid for $s^+ \geq 26$.

The velocity profile in the buffer layer is due to an empirical fit between these two extremes; specifically,

$$v^+ = 5.0 \ln s^+ - 3.05 \qquad (9.19)$$

[#]   In view of Figure 9-5, this might seem like a rather outrageous assumption but even the turbulent core can be very close to the wall (see Problem 9-3).

Figure 9-8 shows a comparison of these predictions with actual velocity profile data. The fit to this so-called <u>universal velocity profile</u> is quite remarkable given some of the drastic assumptions which were made in the derivation.

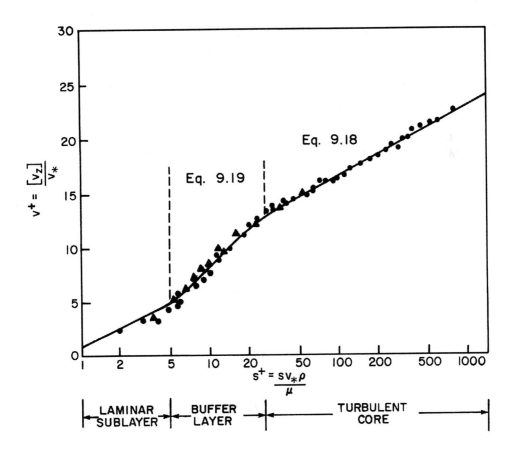

Figure 9-8 The Universal Velocity Profile (● from [2], ▲ from [1])

9-3.b Surface Renewal Theory

Another method to describe turbulent transport with molecular transport mechanisms is the <u>surface renewal theory</u>. The approach consists of formulating the mathematics associated with molecular transport between an "eddy," recently convected from the bulk fluid, and the interface (or surface) at which it has arrived. The eddy only remains at the interface for some finite period of time, after which it is replenished by a "fresh" eddy. The interface can then be thought of as being periodically "renewed" at a rate which is dependent on the degree of turbulence. Because of the time dependence of the transport rates, the basic mathematics corresponding to unsteady-state transport must be employed in order to describe the transport rate during the residence time of the eddy at the interface. Once again we use energy transport to illustrate the concept. The model is represented pictorially in Figure 9-9 where a turbulent eddy is imagined to come to rest on a heat transfer surface[*] at time = zero. Since it arrives with a temperature equivalent to the bulk temperature of the fluid, T_b, which is lower than the wall temperature, T_w, energy will be transported from the surface into the eddy. Furthermore, since the eddy is at rest, the energy transport will take place by molecular mechanisms (i.e., by conduction). Thus we can see that the mathematical model is simply one of unsteady energy transport in one dimension. In addition, if the eddy only remains at the surface for a short period of time, then the energy does not penetrate to its far boundary and the eddy can be considered to be of "infinite" extent. In this case, the correct model is one-dimensional unsteady-state conduction in an infinite medium.

The convective nature of the problem is represented by the time an average eddy remains at the surface; that is, the rate at which the surface is renewed by fresh eddies (the "i+1" eddy in Figure 9-9) arriving from the bulk fluid.

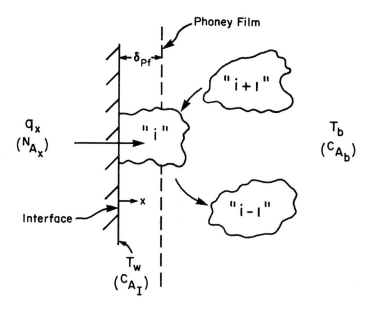

Figure 9-9 Eddy Replacement at a Surface

---

[*]   The model is equally representative of mass transport, as indicated by the parenthetical symbols in Figure 9-9.

If the wall temperature remains constant while the eddy remains at the surface, the temperature distribution will vary with time in the manner described by Equation (7.23); i.e.,

$$T - T_b = (T_w - T_b) \left[1 - \frac{2}{\sqrt{\pi}} \int_0^{x/\sqrt{4\alpha t}} \exp(-\eta^2)\, d\eta \right] \tag{9.20}$$

However, we are interested in the <u>energy</u> transported into the eddy from the wall, which can be obtained from Fourier's Law by

$$(q_w)_i = q_{x|_{x=0}} = -k \frac{\partial T}{\partial x}\bigg|_{x=0} \tag{9.21}$$

where the subscript, i, is used to denote that this is for the i <u>th</u> eddy. If the temperature distribution given by Equation (9.20) is differentiated with respect to x, evaluated at x = 0 and substituted into (9.21), we obtain

$$(q_w)_i = -k(T_w - T_b) \frac{\partial}{\partial x} \left[1 - \frac{2}{\sqrt{\pi}} \int_0^{x/\sqrt{4\alpha t}} \exp(\eta^2)\, d\eta \right]_{x=0}$$

or

$$(q_w)_i = \frac{k(T_w - T_b)}{\sqrt{\pi \alpha t}} \tag{9.22}$$

As it stands, Equation (9.22) is not very useful from a calculational point of view since we have no idea how long a particular eddy will reside on the surface and thus we do not have a value for t. However, we can use this theory together with some statistical analyses to determine the dependence of turbulent heat fluxes on fluid properties (see Chapter 10). For now it is a nice mathematical illustration of the discussion in Section 1-2 where it was pointed out that transport eventually is achieved on a molecular scale [$k(T_w - T_b)$ in Equation (9.22)] and the role of convection is to shorten the distance (reduce the time) over which the transport occurs.

REFERENCES

[1] Laufer, J., "The Structure of Turbulence in Fully Developed Pipe Flow", NACA Tech. Rept. No. 1174 (1953).

[2] Deissler, R.G., "Analysis of Turbulent Heat Transfer, Mass Transfer and Friction in Smooth Tubes at High Prandtl and Schmidt Numbers", NACA Tech. Note No. 3016 (1953).

PROBLEMS

9-1   Show, by time-averaging the continuity equation, that $[v_r] = 0$ for steady-state, fully developed, incompressible turbulent flow in a pipe.

9-2   Following the development given in Section 9-2, derive the time-averaged energy equation which describes steady-state turbulent energy transport in a pipe. Assume that the velocity profile is fully developed and that molecular transport in the flow direction can be neglected.

9-3   Starting with the generalized momentum equation in Appendix A, use time averaging to obtain Equation (9.6) for fully developed, steady-state turbulent flow in a pipe. State the reason(s) for eliminating the various terms in the 3-dimensional momentum equation. HINT: Use the continuity equation and an order of magnitude assessment of the remaining terms

9-4   Starting with the expression for the <u>total</u> momentum flux, τ^{tot}, given in Equation (9.13) for steady-state, fully developed flow in a tube and the empirical expression for the velocity profile,

$$v_z = \bar{v} \left[\frac{R-r}{R} \right]^{\frac{1}{7}},$$ obtain an expression for the "turbulent" viscosity as a function of r. It can be assumed that $\tau^{tot} \sim \tau^t$

9-5   Starting with Equation (9.14), derive Equation (9.15).

9-6   If the eddy viscosity varies linearly with distance from the wall (with a proportionality constant, k_ε, solve for the velocity distribution in the steady-state, fully developed turbulent flow of a newtonian fluid in a pipe with inside radius, R.

9-7   (a) Derive an expression for the eddy viscosity as a function of s^+ in the "buffer" layer.

(b) If $S^+ << S_R^+$, determine the ratio of the eddy viscosity to the kinematic viscosity at both ends of the buffer layer.

9-8   The universal velocity profile is described by two semi-empirical equations of the form
$v^* = a \ln s^* + b$. Using this relationship together with the assumption that, in Equation (9.8),
$\tau_{rz}^{tot} \sim \tau_{rz}^{t}$, show that the eddy viscosity in a pipe varies with r according to

$$\varepsilon = \frac{r}{R} \frac{(R-r)}{a} \left(\frac{\tau_w}{\rho} \right)^{1/2}$$

HINT: Remember that, in fully developed flow, $\tau_{rz}^{tot} = \dfrac{r}{R} \tau_w$

9-9   For water flowing in turbulent flow in a 1" diameter pipe at a Reynolds number of 10^5 :

(a) Calculate the distance from the pipe wall corresponding to the edge of the laminar sublayer and
the edge of the buffer layer. The shear stress at the wall can be calculated from the "*friction
factor*", $f = \dfrac{2\tau_w}{\rho \overline{v}^2}$, which is numerically equal to 0.0045 under these conditions.

(b) During the development of the Prandtl Mixing Length analysis it was assumed that the shear
stress in the turbulent core was equal to the wall stress. How bad is this assumption at
$s^+ = 500$?

9-10

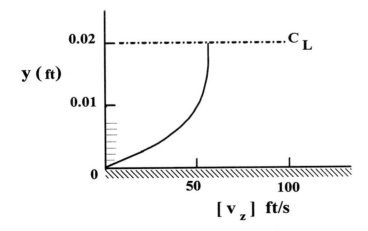

For the turbulent velocity profile of water over a flat wall, as shown above,

(a) What is the thickness of the laminar sublayer?

(b) What is the boundary layer thickness?

(c) What is the value of the eddy viscosity at a distance of 0.008 ft from the wall ($T = 20\ {}^{0}C$)?

9-11   Calculate the average eddy residence time as predicted by Surface Renewal Theory for turbulent heat transfer to water in a 1"-tube at $Re = 10^{5}$.

CHAPTER 10

Transfer Coefficients

It has already been pointed out a number of times, that even laminar convection problems are difficult to solve and the necessity of dealing with time variations in turbulent flows introduces added complexity. Consequently the concept of <u>transfer coefficients</u> was developed so that engineers could get on with the business of the practical calculations necessary to size heat exchangers, separation equipment and piping systems. The detailed use of transfer coefficients and the many empirical correlations which have been developed over the years are left to Chapters 11through14. The emphasis in this chapter is placed on the concept of transfer coefficients and how it relates to the differential approach of transport phenomena.

10-1 GENERAL DEFINITIONS

First of all it should be appreciated that the transfer coefficient concept is not totally empirical but is based on scientific observations. In that sense, it is really a semi-empirical approach to rate processes. Given the fact that the transport of energy, mass, and momentum only occurs when there are driving force gradients, a generalized transfer coefficient can be defined as

$$Trnsfr.Coeff. = \frac{Flux}{DrivingForce} \tag{10.1}$$

While this is a definition, the utility of defining transfer coefficients lies in the fact that if a value of the transfer coefficient is available, Equation (10.1) can be solved for the transport flux. Thus, once transfer coefficients are defined, experimental data can be correlated so that the transfer coefficients can be predicted (within experimental accuracy) for a variety of convection problems, and these predictions can be utilized to calculate fluxes and, ultimately, to size equipment.

Equation (10.1) can be generally extended to each of the transport processes if we define the driving forces in terms of the volumetric concentrations of energy, mass, and momentum. Furthermore, since transfer coefficients represent a macroscopic approach to transport problems, the fluxes referred to in Equation (10.1) are those that pass between phase interfaces. For our purposes we will assume that the interface in question is a wall and, as pointed out earlier, the fluxes right at the wall will be molecular. Consequently, the applicable transfer coefficients can be defined by

$$(C_{tf})_E = \frac{q_w}{\Delta(\rho\,\hat{H})}$$

$$(C_{tf})_m = \frac{N_{Aw}}{\Delta C_A} \tag{10.2}$$

$$(C_{tf})_{mom} = \frac{\tau_w}{\Delta(\rho v)}$$

While this is a generalized representation, previous practice and convenience have resulted in the specific transfer coefficients being defined in slightly different forms than is shown in Equation (10.2). Consequently, the transfer coefficients for each will be separately addressed below.

10-1.a Energy: Heat Transfer Coefficients

Although the driving force in molecular energy transport is a temperature gradient, the macroscopic approach represented by the transfer coefficient approach requires a macroscopic difference. While the driving force shown in Equation (10.2) is macroscopic, temperature differences rather than differences in enthalpy concentration are normally employed. In other words, since $\Delta\rho\hat{H} = \Delta\rho\,\hat{c}_P T$, the fluid properties are "lumped" in with the transfer coefficient.

Moreover, there is usually a choice of which temperature difference to use. For example, in a heat-exchanger tube, we could use the difference between the centerline (maximum) fluid temperature and the wall temperature or the difference between the average tube-side fluid temperature and the average temperature of the fluid on the outside of the tube, or, any combination of these temperature differences. Since the choice is arbitrary, we must be careful to learn just how the driving force is defined so that the proper values of the transfer coefficients are matched to the driving force upon which they were originally defined.

Of course practical considerations as to which temperatures to use usually dictate those which are easy to measure, and thus average fluid temperatures are most often used.[#] With these considerations, the *heat transfer coefficient* can be defined from Equation (10.2) as

$$h = \frac{q_w}{\overline{T} - T_w} \tag{10.3}$$

When written explicitly in terms of the energy flux

$$q_w = h\left[\overline{T} - T_w\right]$$

this relationship is known as *Newton's Law of Cooling*. Recall that this relationship was used in a number of the examples discussed in Chapter 3. Note also, that the units of h are E/l^2-t-deg (eg., J/m^2-sec-K).

10-1.b Mass: Mass Transfer Coefficients

When Equation (10-2) is applied to mass transport, the molecular flux at the "wall" (or interface) is the flux relative to fixed coordinates, N_{Aw}, and <u>not</u> the diffusion flux, J_{Aw}. Since most mass transfer processes occur across interfaces, the appropriate analogy to the wall heat flux is the interfacial molar flux, N_{Al}. Furthermore, we are not restricted to just volumetric concentration driving forces, but must decide on which type of concentrations to use in addition to where they should be measured. As a result, there are many different forms of mass transfer coefficients, and it can become quite confusing.

Using the form of Equation (10.2), a generalized mass transfer coefficient can be defined as

$$k_{m(x,c,p)} = \frac{N_{A_l}}{\Delta(conc_A)_{x,c,p}} \tag{10.4}$$

[#] It is also relatively easy to place a <u>thermocouple</u> in the center of a pipe. However, in highly turbulent flow, the centerline temperature is very nearly equal to the average temperature.

where x, C, and p represent mole fraction, molar concentration, and partial pressure driving forces, respectively. Because all of these driving forces are different representations of concentration, they are interchangeable. To illustrate this, assume we have a gas and rewrite Equation (10.4) explicitly in terms of flux. Thus, equivalently,

$$N_{A_I} = k_{mx}[\overline{X}_A - X_{A_I}] = k_{mc}[\overline{C}_A - C_{A_I}] = k_{mp}[\overline{P}_A - P_{A_I}] \tag{10.5}$$

Note that the units of the various mass transfer coefficients are very different:

$$k_{mx} \sim \left[\frac{moles}{l^2 \cdot t}\right], \quad k_{mc} \sim \left[\frac{l}{t}\right], \text{ and } k_{mp} \sim \left[\frac{moles}{l^2 \cdot t \cdot press}\right].$$

Starting with the defining equation for k_{mc} and since, $C_A = C X_A$ and using $C = \dfrac{P}{R_g T}$

$$N_{A_I} = k_{mc}[\overline{C}_A - C_{A_I}] = k_{mc}C[\overline{X}_A - X_{A_I}] = \frac{k_{mc}P}{R_g T}[\overline{X}_A - X_{A_I}] \tag{10.6}$$

Similarly, starting with the defining equation for k_{mp} and using <u>Dalton's Law</u>, $P_A = P X_A$, then

$$N_{A_I} = k_{mp}[\overline{P}_A - P_{A_I}] = k_{mp} P[\overline{X}_A - X_{A_I}] \tag{10.7}$$

Equating all of the terms with $\overline{X}_A - X_{A_w}$ driving forces in Equations (10-5) through (10-7), it follows that

$$k_{mx} = k_{mc} C = \frac{k_{mc} P}{R_g T} = k_{mp} P \tag{10.8}$$

10-1.c Momentum: The Friction Factor

Referring to Equation (10.2) for momentum transport, both the numerator (wall flux) and the denominator (velocity) are vectors. Thus, in order to simplify matters, the momentum transfer coefficient has been historically defined in a somewhat different manner. The vectorial nature of the denominator can be removed by forming the *dot product* of the velocity vector and this still retains a driving force related to the fluid velocity. Furthermore, if there is no slip at the wall, then the denominator becomes, $\Delta\rho v^2 = [\rho \overline{v}^2 - 0] = \rho \overline{v}^2$. Since the denominator now has units of energy/volume, it is customary to consider it as kinetic energy/volume and to arbitrarily multiply the denominator by 1/2.[#] With these modifications the units of both the numerator and denominator in the defining equation for the momentum transfer coefficient will be identical and thus the refined momentum transfer coefficient is dimensionless. This relationship for the momentum transfer coefficient is known as the <u>friction factor</u> and is defined by

$$f = \frac{|\tau_w|}{\frac{\rho \bar{v}^2}{2}} \qquad (10.9)$$

where the absolute value of the wall shear stress is used since the friction factor is not a vector quantity.

Because transfer coefficients are intended for use in practical calculations, the momentum flux at the wall is not a convenient parameter which can be directly measured. However, f can be expressed in terms of the pressure drop in a pipe (which <u>is</u> easily measured) by recalling that, in fully developed flow, there is a static force balance between the accelerating pressure difference and the retarding forces at the wall.[##] While we used a differential approach in Chapter 5, we can apply the same principle here, but in a macroscopic manner. Thus if we apply a force balance over a length of pipe, L, which has a radius, R, then

$$(P_o - P_L)\pi R^2 = \tau_w \, 2\pi \, RL \qquad (10.10)$$

Solving for τ_w in Equation (10.10) and substituting into Equation (10.9), we obtain a related definition for the friction factor

$$f = \frac{D}{2\rho\bar{v}^2}\frac{\Delta P}{L} \qquad (10.11)$$

where f is written in terms of the pipe diameter (which is more typical) and $\Delta P = P_o - P_L$

Keep in mind that Equation (10.11) only applies to a straight, cylindrical pipe. Momentum transfer coefficients can be applied more generally to regular (i.e., constant cross-sectional area) but noncircular or unfilled ducts as well. For example, consider the triangular duct in Figure 10-1. The pressure acts on the triangular cross sectional area, A_x, and the shear stress acts on the *wetted area*, A_w; that is, the surface area of the duct which is in contact with the fluid. In this case, a static force balance results in

---

[#] $\frac{1}{2}\rho v^2$ is often referred to as the <u>dynamic pressure</u>, since it has units of pressure.

[##]   See the discussion on the "Alternative Derivation of Equation (5.17)" in Chapter 5.

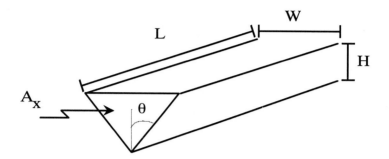

Figure 10-1 Flow in an Irregular Duct

$$\tau_w = \frac{(P_o - P_L)\,A_x}{A_w} \tag{10.12}$$

The duct surface can be related to a straight length of the duct, L, by defining the wetted area in terms of a *wetted perimeter*, $\wp_w$,

$$A_w = \wp_w L \tag{10.13}$$

Substituting Equations (10.12) and (10.13) into Equation (10.9), the friction factor can be expressed as

$$f = \frac{\Delta P}{\frac{1}{2}\rho\bar{v}^2 L}\,\frac{A_x}{\wp_w}$$

The parameter, $\dfrac{A_x}{\wp_w}$, is known as the *hydraulic radius,* R_H, and it is common practice[#] to define an *equivalent pipe diameter,* D_{eq}, as

$$D_{eq} = 4\,R_H \tag{10.14}$$

In which case, the friction factor expression becomes

[#]   This is done so that, in a straight cylindrical pipe of diameter, D, $D_{eq} = 4R_H = 4\dfrac{A_x}{\wp} = \dfrac{4\pi R^2}{2\pi R} = D$

$$f = \frac{\Delta P}{2\rho\bar{v}^2}\frac{D_{eq}}{L}$$

For the specific case of the triangular duct in Figure 10-1,

$$D_{eq} = \frac{2\,HW}{W + \dfrac{2H}{\cos\theta}}$$

Because the transfer coefficient definitions have so many arbitrary factors associated with them, care should be exercised relative to their definition before using transfer coefficient correlations. For example, another well known friction factor definition is the *Blasius friction factor, f*, which is defined as *4f*. In other words,

$$f' = 4f = \frac{2\Delta P}{\rho\bar{v}^2}\frac{D_{eq}}{L} \tag{10.15}$$

10-2 TRANSFER COEFFICIENT PREDICTIONS IN LAMINAR FLOW

In order to illustrate how mathematical solutions can be utilized to calculate transfer coefficients, we now consider some simple convective systems which are in laminar flow and thus have analytical solutions.

The simplest system to analyze is fully developed laminar flow in a straight pipe. The solution for the velocity profile has already been derived in Section 5-4, namely

$$v_z = 2\bar{v}\left[1 - \left(\frac{r}{R}\right)^2\right] \tag{5.26}$$

According to Equation (10.9), we need to evaluate τ_w in order to obtain an expression for the friction factor. Therefore, using Newton's Law of viscosity

$$\tau_w = -\mu\left(\frac{dv_z}{dr}\right)\bigg|_R$$

we can differentiate Equation (5.26) and substitute into Equation (10.9) to obtain

$$f = \frac{8\mu}{R\bar{v}\rho} = \frac{16}{\text{Re}}$$

which is the accepted experimental relationship for smooth pipes [1].

Turning now to energy transport, recall that the problem of "fully developed" energy transport in tubular laminar flow was analyzed in Section 8-2. For a constant temperature gradient in the flow direction, the solution for the temperature profile was given by

$$T - T_{w0} = \frac{\Delta T}{\Delta z}\left[z + \frac{\bar{v}R^2}{8\alpha}\left(4\eta^2 - \eta^4 - 3 \right) \right] \qquad (8.25)$$

The heat transfer coefficient is defined by Equation (10.3) and both q_w and $\bar{T} - T_w$ can be obtained from Equation (8.25) by using Fourier's Law and the definition of the average temperature in the cross section of the tube. That is

$$q_w = -k\frac{\partial T}{\partial r}\Big|_R$$

$$\bar{T} - T_w = \frac{\int_0^R (T - T_w)\, 2\pi\, r dr}{\int_0^R 2\pi\, r dr}$$

The Nusselt number, Nu, which is a *dimensionless* heat transfer coefficient , can be calculated from the definition of the heat transfer coefficient; i.e.,

$$Nu = \frac{h}{\dfrac{k}{D}} = \frac{hD}{k} = \frac{D}{k}\ \frac{q_w}{\bar{T} - T_w}$$

When the solution for the temperature profile is differentiated and integrated according to the above equations, the Nusselt number calculates to be equal to 6.0, a constant.[#]

[#] If a mixing cup average temperature is used in place of the bulk temperature, the Nusselt number is calculated to be 4.36 (see Problem 10-12).

The fact that the Nusselt number is a constant is a direct consequence of the assumption of a constant axial temperature gradient in the derivation of Equation (8.25), which is an idealization. Nevertheless, for water and a Reynolds number of 1000, experimental correlations predict Nusselt numbers ranging between 5 and 7 for a 1"-pipe between 10 and 20 foot long; not terribly different from the predicted value of 6.0

Finally, let's turn to mass transport in a rectangular geometry. For a rectangular duct with a large *aspect ratio* ($\frac{W}{H} >> 1$, Figure 10-2), the situation can be approximated as flow between two parallel plates. If we have mass transport between the fluid and the wall and we again employ the concept of "fully developed" conditions; in this case, a constant molar flux at the wall which leads to a constant axial concentration gradient, it can be shown (see Problem 10-15) that the solution for the concentration profile is

$$c_A^* = -\frac{3}{2}(4y^{*2}-1) \tag{10.16}$$

where $c_A^* = \dfrac{C_A - C_{Aw}}{\overline{C}_A - C_{Aw}}$ and $y^* = \dfrac{y}{H}$. Defining the mass transfer coefficient in terms of a volumetric concentration driving force

$$k_{mc} = \frac{N_{Aw}}{\overline{C}_A - C_{Aw}}$$

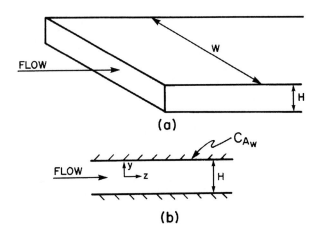

(a)

(b)

Figure 10-2 Mass Transport in a Rectangular Duct

$$k_{mc} = \frac{-D_{AB}\left(\dfrac{\partial C_A}{\partial y}\right)_{\Big|H/2}}{\overline{C}_A - C_{A_w}}$$

which, after differentiation of Equation (10.22) gives

$$k_{mc} = \frac{6\,D_{AB}}{H}$$

Analogous to the Nusselt number, a dimensionless mass transfer coefficient can also be defined and is commonly referred to as the <u>Sherwood number</u>, Sh, which in this case is

$$Sh = \frac{k_{mc}H}{D_{AB}} = 6 \tag{10.17}$$

Again, the assumption of a constant axial gradient produces a constant value of the dimensionless transfer coefficient.

10-3 THE "PHONEY FILM" THEORY

There is still another way in which to view the concept of transfer coefficients; one that has both the advantage of simplicity and of relating the concept to molecular transport. This method, the *Phoney Film* theory, assumes that the actual transport taking place at an interface (wall) can be equivalently calculated by evaluating the molecular transport taking place across a thin film adjacent to the interface. This concept has its roots in reality since we know that only molecular transport occurs very near the interface and the role of turbulent convection is to force this region progressively closer to the interface as the degree of turbulence increases. The difficulty with this approach is not in its descriptive concept but in its lack of quantitative capability since the thickness of this film is not known *a priori*. As we shall see, experimental data are still needed in order to relate transfer coefficients to Phoney Film theory and thus provide a more fundamental basis for their use.

The goal of Phoney Film theory is to invent a film in which only molecular transport takes place and with a thickness which will yield accurate calculations of the actual transport rate. To illustrate the approach, consider energy transport between a flowing fluid and a pipe wall. In this example the Phoney Film theory is best visualized by referring to Figure 10-3 which shows an actual temperature profile in a region close to the pipe wall. Although not necessary to the theory, the profile in Figure 10-3 is assumed to be of a turbulent nature; that is, steep near the wall and flat throughout most of the pipe cross section. Because we only have molecular transport in the phoney film and, in laminar flow, only molecular transport mechanisms prevail in the y-direction, the Phoney Film theory is often called <u>Laminar Film theory</u>. However, due to the inventive nature of this concept, calling it a "phoney" film is probably more descriptive.

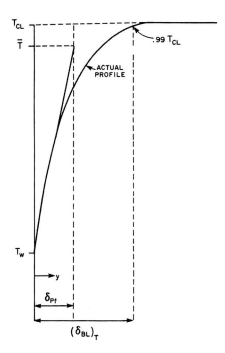

Figure 10-3 Temperature Profiles near a Wall

10-3.a Applications to Energy and Momentum Transport

With these concepts in place, we can now proceed to analyze for the energy transport taking place in the phoney film. Assuming steady-state conditions and that the thickness of the film, δ_{pf}, is sufficiently thin that rectangular geometry can be assumed, a differential energy balance within the film yields

$$-\frac{d\,q_y}{dy} = 0 \qquad (10.18)$$

The boundary condition at $y = 0$ is clear and is simply the stipulation of the wall temperature. However, we have a choice as to where to place the outer edge of the film. Since the goal is strictly utilitarian, it is more convenient to use the average temperature so that the overall driving force will be the difference between the average temperature and the wall. Thus, substituting Fourier's Law into Equation (10.18), we have

$$\frac{d}{dy}\left[-k\frac{dT}{dy}\right] = 0$$

$$\text{at } y = 0 \text{ , } T = T_w$$

$$\text{at } y = \delta_{pf} \text{ , } T = \overline{T}$$

The solution for the temperature profile through the film is linear and equal to

$$T = T_w + (\overline{T} - T_w)\frac{y}{\delta_{pf}} \tag{10.19}$$

Although Equation (10.19) is not very useful in a quantitative sense (since we do not have a value for δ_{pf}), it does give a graphical illustration of the presence of the film as is shown in Figure 10-3. Notice that, even though the actual temperature profile is not really linear, we can still calculate the correct value of the energy flux at the wall. That is, the linear temperature profile in Figure 10-3 is drawn so that it is tangent to the actual profile at the wall. Notice too, that the phoney film thickness is <u>not</u> the same as the <u>thermal boundary layer thickness</u>, $(\delta_{bL})_T$, which is defined as the distance from the wall where the temperature reaches 99% of the free stream (centerline) value.

While we were careful to draw the linear temperature profile so that it was tangent to the actual profile at the wall, the slope of the linear profile will depend on the value of δ_{pf}. In other words, we effectively chose δ_{pf} when we drew the tangent line. As pointed out earlier, we need experimental data in order to be able to predict the value of δ_{pf} for a specific flow situation. This is most easily done by relating it to the heat transfer coefficient. To achieve this, the energy flux can be calculated by differentiating Equation (10.19) to obtain

$$q_{y|_0} \equiv q_w = -k\left(\frac{dT}{dy}\right)|_w = -\frac{k(\overline{T} - T_w)}{\delta_{pf}} \tag{10.20}$$

Since the heat transfer coefficient can be defined by

$$|q_w| = h(\overline{T} - T_w) \tag{10.21}$$

a comparison of Equations (10.20) and (10.21) results in[#]

$$h = \frac{k}{\delta_{pf}} \tag{10.22}$$

As can be seen, either h or δ_{pf} could be correlated with experimental data and they are inversely related. Historically, it is the transfer coefficients which have been experimentally correlated and the phoney film concept is primarily used as a qualitative tool.

Similar approaches can also be taken in mass and momentum transport and since the phoney film approach finds more utility in the former (due to bulk diffusion and the possible presence of chemical reactions), it will be discussed last. For momentum transport, the temperature profile in Figure 10-3 can be replaced with a velocity profile, and the analogy is direct except for the fact that the velocity at the wall is zero. The solution to the molecular momentum balance is easily derived to yield

$$v_z = \frac{\overline{v}\,y}{\delta_{pf}}$$

and since

$$|\tau_w| = \left| -\mu \frac{\overline{v}}{\delta_{pf}} \right| = f\frac{1}{2}\rho\overline{v}^2$$

then it follows that

$$f = \frac{2\mu}{\rho\overline{v}\,\delta_{pf}} = \frac{2}{Re_{pf}} \tag{10.23}$$

where Re_{pf} is the Reynolds number based on the phoney film thickness.

[#]   The negative sign in Equation (10.20) merely indicates that q_w is in the negative-y direction, whereas h is defined in terms of the absolute value of q_w.

10-3.b Mass Transport in Gas-Liquid Films

Although we were not dealing specifically with the Phoney Film theory, the subject of molecular mass transport through a film adjacent to an interface was previously discussed in Chapter 4-2. To illustrate the utility of film theory in dealing with the complexities introduced to convective mass transport by the presence of chemical reactions, bulk diffusion and a two-phase system, consider the films adjacent to either side of the gas-liquid interface as shown in Figure 10-4. This type of a model in gas-liquid transport is sometimes referred to as the Two Film theory. First lets consider the gas side of the interface which only includes species A and B. If there are no chemical reactions within the film and if it is assumed that we have equimolar counter diffusion (i.e., $N_{A_z} = -N_{B_z}$), then a steady-state mass balance for species A within the film yields

$$-\frac{dN_{A_z}}{dz} = 0$$

$$N_{A_z} = J_{A_z} = -C\, D_{AB}^G\, \frac{dX_A}{dz} = constant$$

which, when combined with the boundary conditions

$$\text{at } z=0; \; X_A = X_{AI}$$

$$\text{at } z=\delta_{pf}^G; \; X_A = \overline{X}_A$$

gives

$$N_{AI} = -\frac{C\, D_{AB}^G}{\delta_{pf}^G}(\overline{X}_A - X_{AI}) \tag{10.24}$$

Thus, with equimolar counter diffusion, the concentration profile through the film is linear and a comparison with Equation (10-4) gives a prediction for the mass transfer coefficient which is analogous to Equation (10.22) for heat transfer; viz.,

$$k_{mx} = \frac{C\, D_{AB}^G}{\delta_{pf}^G} \tag{10.25}$$

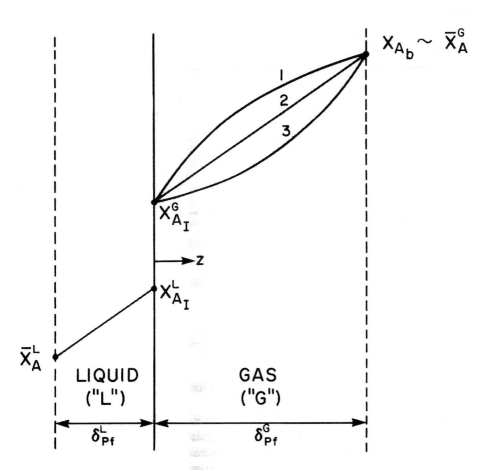

Figure 10-4 Gas-Liquid Mass Transport ("1" = Stagnant Gas Diffusion, "2" = Equimolar Counter Diffusion, "3" = Diffusion with Chemical Reaction)

Since it would indeed be fortuitous if $N_{A_z} = -N_{B_z}$, let's now consider the opposite extreme of stagnant film diffusion; that is, when $N_{B_z} = 0$. In this case, N_{A_z} is still constant but is equal to $\dfrac{J_{A_z}}{(1 - X_A)}$, and we have

$$N_{A_I} = -\frac{C D_{AB}^G}{(1 - X_A)} \frac{dX_A}{dz}$$

$$N_{A_I} \int_0^{\delta_{pf}^G} dz = -C D_{AB}^G \int_{X_{A_I}^G}^{\overline{X}_A^G} \frac{dX_A}{(1 - X_A)}$$

$$N_{A_I} = \frac{C D_{AB}^G}{\delta_{pf}^G} \ln \frac{1 - \overline{X}_A^G}{1 - X_{A_I}^G}$$

Using the definition of the mass transfer coefficient

$$k_{mx}^G (\overline{X}_A^G - X_{A_I}^G) = \frac{C D_{AB}^G}{\delta_{pf}^G} \ln \left[\frac{1 - \overline{X}_A^G}{1 - X_{A_I}^G} \right]$$

$$k_{mx}^G = \frac{C D_{AB}^G}{\delta_{pf}^G} \frac{1}{\dfrac{\overline{X}_A^G - X_{A_I}^G}{\ln \left[\dfrac{1 - \overline{X}_A^G}{1 - X_{A_I}^G} \right]}}$$

The functionality of X_A^G in the above equation can be rearranged to give

$$\frac{\overline{X}_A^G - X_{A_I}^G}{\ln \dfrac{1 - X_{A_I}^G}{1 - \overline{X}_A^G}} = \frac{-\left[(1 - \overline{X}_A^G) - (1 - X_{A_I}^G) \right]}{\ln \dfrac{1 - \overline{X}_A^G}{1 - X_{A_I}^G}} \equiv -[1 - X_A^G]_{\overline{LM}}$$

where $[1 - X_A^G]_{\overline{LM}}$ is known as the <u>log mean</u> of $1 - X_A^G$ which, for a binary system, is also $[X_B^G]_{\overline{LM}}$. As a result, the mass transfer coefficient corresponding to the gas side of the film becomes[#]

[#]   Quite often the log-mean correction is not included in the defining equation for the mass transfer coefficient (see Problem 10-6, for example).

$$k_{mx}^G = \left| \frac{-C D_{AB}^G}{\delta_{pf}^G \ [X_B^G]_{LM}} \right| \tag{10.26}$$

One of the difficulties in expressing the mass transfer coefficient as in Equation (10.26) is that the interfacial concentration, X_{A_I}, cannot be readily measured. Consequently, the mass transfer coefficient in gas-liquid systems is often defined in terms of the driving force between the bulk gas and liquid concentrations. To do this, let's now concentrate on the mass transport on the liquid side of the interface. Once again, at steady-state and with no chemical reactions, the molar flux of A in the liquid is constant and equal to N_{A_I}[*]. Since the solubilities of gases in liquids are usually quite small, it is not unreasonable to assume that $N_A^L \sim J_A^L$ and, integrating across the liquid side phoney film, we have

$$\left| N_{A_I} \right| = \left| \frac{-(C D_{AB})^L}{\delta_{pf}^L} (X_{A_I}^L - \overline{X}_A^L) \right| = k_{mx}^L (X_{A_I}^L - \overline{X}_A^L)$$

If the molar flux of A is equated in terms of the gas and liquid side mass transfer coefficients, then

$$k_{mx}^L (\overline{X}_A^L - X_{A_I}^L) = -k_{mx}^G (\overline{X}_A^G - X_{A_I}^G) \tag{10.27}$$

It is reasonable to assume gas-liquid equilibrium right at the interface and for many systems, $X_{A_I}^G = m X_{A_I}^L$, where m is a parameter, dependent on temperature[**]. Using this relationship to express, $X_{A_I}^G$ in terms of $X_{A_I}^L$ in Equation (10.27), $X_{A_I}^L$ can be written in terms of the bulk gas and liquid concentrations as

$$X_{A_I}^L = \frac{k_{mx}^G \overline{X}_A^G + k_{mx}^L \overline{X}_A^L}{k_{mx}^L + m k_{mx}^G}$$

which can be substituted into the left-hand side of Equation (10.27) to give

---

[*]   At steady-state, this is identical to the molar flux through the gas side phoney film.

[**] When the gas side concentration is expressed in terms of partial pressures, this type of equilibrium relationship is known as <u>Henry's Law</u> and is generally valid at low solubilities of A

$$N_{A_I} = - \left[\frac{1}{\dfrac{m}{k_{mx}^L} + \dfrac{1}{k_{mx}^G}} \right] [\overline{X}_A^G - m \overline{X}_A^L]$$

Notice that, in view of the assumed equilibrium relationship, $m \overline{X}_A^L$ is really the composition of the bulk gas which <u>would</u> be in equilibrium (it really isn't) with the bulk liquid concentration. If this value is designated as $[\overline{X}_A^G]^*$, we can now define an <u>overall mass transfer coefficient</u> based on gas composition driving forces, namely,

$$|N_{A_I}| = K_{OG}[(\overline{X}_A^G - [\overline{X}_A^G]^*)$$

so that

$$K_{OG} = \left[\frac{1}{\dfrac{m}{k_{mx}^L} + \dfrac{1}{k_{mx}^G}} \right] \tag{10.28}$$

Thus, the overall mass transfer coefficient is not only a function of the degree of convective mixing but also of the particular equilibrium relationship which prevails for a particular system. While this kind of definition is convenient since it is based on measurable concentrations, K_{OG} is not directly correlated with parameters such as Reynolds and Schmidt numbers as was the case with k_{mx} values. In this sense it is comparable to <u>overall heat transfer coefficients</u> which typically include conduction through pipe walls and which will be discussed in Chapter 12. Other definitions of overall mass transfer coefficients are also discussed in detail in Chapter 13.

EXAMPLE 10-1: Phoney Film – Mass Transport and Chemical Reaction

Consider the liquid-side film in Figure 10-4 but with a first-order homogeneous chemical reaction, A → B. This is a technique that is often used commercially (amine scrubbing of acid gases, for example) and thus we wish to derive an expression for the prevailing mass transfer coefficient, based on the concentration driving force between the bulk liquid and the interface. For convenience, the coordinate system will be reversed so that positive z proceeds from the interface into the bulk liquid.

Solution

First of all, a mass balance for species 'A' within the film yields

$$-\frac{dN_{A_z}}{dz} - k_r C_A = 0 \tag{10.29}$$

Because of the reaction stoichiometry, $N_{A_z} = -N_{B_z}$, and thus $N_{A_z} = J_{A_z}$. For a constant overall molar concentration, Equation (10.29) can then be written

$$\frac{d^2 C_A}{dz^2} - \frac{k_r}{D_{AB}} C_A = 0$$

This differential equation can be placed in dimensionless form by defining $z^* = z/\delta_{pf}$, $C_A^* = C_A/\overline{C}_A$ which when solved with the boundary conditions

$$\text{at } z^* = 0; \ C_A^* = C_{A_I}^*$$

$$\text{at } z^* = 1; \ C_A^* = 1$$

gives a solution for C_A^* [#]

$$C_A^* = \frac{1 - C_{A_I}^* \ \sinh[\beta(z^* - 1)]}{\sinh \beta} \tag{10.30}$$

where the dimensionless parameter, β, is defined as $\beta = \left(\frac{k_r \delta_{pf}^2}{D_{AB}}\right)^{\frac{1}{2}}$ and is often called the Hatta number, named after S. Hatta who first considered this problem in 1932 [2]. The molar flux at the interface can be computed by differentiating the concentration profile; i.e.,

$$N_{A_I} = -\frac{D_{AB} \overline{C}_A}{\delta_{pf}} \left(\frac{d C_A^*}{dz^*}\right)_{|z^* = 0} \tag{10.31}$$

[#] Note that in deriving this particular form of the solution, we have used the identity
$$\sinh(u - v) = \cosh v \ \sinh u - \cosh u \ \sinh v$$

When Equation (10.29) is differentiated according to (10.31) and N_{AI} is equated to the defining equation for the mass transfer coefficient, k_{mc}, based on the driving force between the liquid interfacial and bulk concentrations, we obtain

$$k_{mc} = (k_r D_{AB})^{\frac{1}{2}} \frac{C_{AI}}{C_{AI} - \overline{C}_A} \coth \beta \qquad (10.32)$$

If we now compare the mass transfer coefficients with and without chemical reaction, we can define an <u>Enhancement Factor</u>, ξ,

$$\xi \equiv \frac{k_{mc}}{(k_{mc})_o}$$

where $(k_{mc})_o$ is the mass transfer coefficient *without* chemical reaction, which we have already seen [Equation (10.25)] is simply $(k_{mc})_o = \dfrac{D_{AB}}{\delta_{pf}}$. Therefore, using Equation (10.32), the enhancement factor can be expressed as

$$\xi = \left(\frac{\delta_{pf}^2 \, k_r}{D_{AB}} \right)^{\frac{1}{2}} \frac{C_{AI}}{C_{AI} - \overline{C}_A} \coth \beta = \beta \frac{C_{AI}}{C_{AI} - \overline{C}_A} \coth \beta$$

In situations where the reaction rate is fast relative to the diffusion rate;[#] that is $\dfrac{k_r}{D_{AB}} >>> 1$, then $\beta \rightarrow \infty$ and $\coth \beta \rightarrow 1$ so that

$$\xi = \frac{C_{AI}}{\overline{C}_A - C_{AI}} \beta$$

Typically in a liquid, solubilities are low and, if the reaction is reasonably effective, a good assumption is that $\overline{C}_A \approx 0$, in this situation, then

$$\xi = \beta$$

or, in other words, the Enhancement Factor is equal to the Hatta number.

_____

[#]   This is referred to as a <u>diffusion-limited</u> reaction.

10-4 TRANSFER COEFFICIENTS AND DIFFERENTIAL BALANCES

Now that the transfer coefficients have been defined, let us look at how they relate to the differential balance approach which has been described in the first seven chapters of the text. Looking first at energy transport, we know that if the differential equations which arise from differential balances can be solved, solutions will be obtained for temperature as a function of spatial coordinates and time. With those solutions and the knowledge that all transport will eventually occur via molecular mechanisms, Fourier's Law applies, and the energy flux at a pipe wall, for example, is simply $q_w = -k\dfrac{\partial T}{\partial r}\Big|_R$. While this applies at every point along the wall, we wish to calculate the _total_ energy transported over a length of the pipe, or

$$\bar{q}\, A_w = \pi D \int_0^L q_{r_{|R}}\, dz = \pi D \int_0^L \left(-k\frac{\partial T}{\partial r}\right)\Big|_R dz \tag{10.33}$$

where $\bar{q}$ is the average energy flux over the area of the pipe being considered. The total energy can also be calculated using the definition of the heat transfer coefficient in Equation (10.3) and thus we have[#]

$$h(\bar{T}-T_w)\,\pi\, DL = \pi D \int_0^L q_{r_{|R}}\, dz = \pi D \int_0^L \left(-k\frac{\partial T}{\partial r}\right)\Big|_R dz \tag{10.34}$$

It is important to realize that if we were able to obtain a solution for the temperature as a function of r and z, then we could analytically evaluate the integral in Equation (10.33) and obtain a value for h. Of course, if a solution for T were available, then there would be no need to define a heat transfer coefficient in the first place. Another way to view heat transfer coefficients in the context of Equation (10.33) is to realize that, in the usual situation where we do not have an analytical solution for T, we must resort to experimental correlations and the experimental data really provide the numerical values for the integral in Equation (10.33).

Even though analytical solutions for T(r,z) are not available in complex or turbulent convection situations, Equation (10.34) can provide useful information on how to correlate the experimental data necessary to determine h. Using the similarity techniques of Chapter 7

[#] Note that $\dfrac{\partial T}{\partial r}\Big|_R$ is independent of r.

Equation (10.34) can be placed in dimensionless form by using the following dimensionless dependent and independent variables:

$$q_r^* = \frac{q_r}{\rho\,c_P\bar{v}(T-\bar{T})}, \quad T^* = \frac{T-T_w}{\bar{T}-T_w}$$

$$r^* = \frac{r}{D}, \quad z^* = \frac{z}{D}$$

Note that the characteristic value of the molecular energy flux has been arbitrarily chosen to be the convective energy flux with the reference enthalpy based on the average temperature of the fluid. The reason for this particular choice will become apparent and is based on traditional definitions of specific dimensionless parameters.

Using these definitions together with Equation (10.3), the integrals in Equation (10.34) can be placed in dimensionless form to give

$$h(\bar{T} - T_w\,)\pi DL = \pi D^2 \rho\, c_P\bar{v}(\bar{T} - T_w\,)\int_0^{L/D} q_r^* \big|_{IR}\, dz^* =$$

$$\hspace{8cm} (10.35)$$

$$\pi Dk(\bar{T} - T_w)\int_0^{L/D} -(\frac{\partial T^*}{\partial r^*})|_{\frac{1}{2}}\, dz^*$$

Simplifying and multiplying through by $\dfrac{D}{kL}$, the dimensionless form of Equation (10.35) is

$$\frac{hD}{k} = \frac{D^2\rho\,\hat{c}_P\bar{v}}{kL}\int_0^{L/D} q_r^* \big|_{r_R}\, dz^* = \frac{1}{L/D}\int_0^{L/D} -(\frac{\partial T^*}{\partial r^*})|_{\frac{1}{2}}\, dz^* \qquad (10.36)$$

The first term in Equation (10.36) is the Nusselt number, which was introduced in 10-2. The dimensionless parameter in front of the integral in the second term can also be re-arranged by multiplying and dividing by the viscosity, μ, so that

$$\frac{D^2\rho\,\hat{c}_P\bar{v}}{kL}\frac{\mu}{\mu} = \frac{\hat{c}_P\mu}{k}\frac{D\bar{v}\rho}{\mu}\frac{1}{L/D}$$

The first dimensionless group on the right hand-side is the Prandtl number and the second, $\dfrac{D\bar{v}\rho}{\mu}$, is the <u>Reynolds number</u>, Re. Equation (10.36) now becomes

$$Nu = \frac{Re\ Pr}{L/D} \int_0^{L/D} q^*_{r_{|R}}\ dz^* = \frac{1}{L/D} \int_0^{L/D} -\left(\frac{\partial T^*}{\partial r^*}\right)_{|\frac{1}{2}}\ dz^* \tag{10.37}$$

Note that, for a pipe, $Re\ Pr\dfrac{D}{L} = \dfrac{4}{\pi}Gz$, where Gz is known as the Graetz number and can

be defined by $Gz = \dfrac{\dot{m}c_P}{kL} = \dfrac{\pi}{4}\dfrac{\rho \bar{v}c_P D^2}{kL}$. Mathematical solutions for $T^*(r^*, z^*)$ are available in

laminar flow, and therefore the integrals in Equation (10.37) can be computed to give

$$Nu = 1.75\,(Gz)^{1/3} \tag{10.38}$$

Since the mathematical model does not fully account for all of the physical phenomena, experimental correlations in laminar flow give a slightly higher value for the Nusselt number; viz.,

$$Nu = 2.0\,(Gz)^{1/3} \tag{10.39}$$

However in situations where mathematical solutions are <u>not</u> available, the development given in Equations (10.33) through (10.37) only serves to <u>identify</u> the pertinent dimensionless parameters with which experimental data should be correlated. The calculation of the total energy transferred to the walls will then depend on either the ability to solve the appropriate differential equations (in order to compute the integrals) or on the availability of experimental data. In turbulent flow we must still rely on the latter and, in these cases, the data really provides an experimental evaluation of the integrals in Equation (10.37).

Up to this point, the entire discussion relative to dimensionless heat transfer coefficients has centered on cylindrical pipe geometry. In fact, the Nusselt number can be defined more generally as

$$Nu \equiv \frac{h}{k}\,L_c \tag{10.40}$$

where L_c is a <u>characteristic dimension</u> appropriate to the system being analyzed. Of course in pipe flow, the characteristic dimension is typically chosen to be the pipe diameter. For flow in irregular conduits, D_{eq} would be a more appropriate choice and, indeed, many of the experimental correlations present in the literature use the equivalent diameter in place of pipe diameter in the definitions of Nusselt and Reynolds numbers. For transport over packing materials (catalysts, for example), the particle diameter is usually chosen as the characteristic dimension.

Analogous expressions can also be obtained for the mass transfer coefficient depending on the particular driving force being employed. Thus, for mole fraction driving forces, the analogous expression to Equation (10.33) is

$$\overline{N}_A A_w = \pi DL\, k_{mx}(\overline{X}_A - X_{Aw}) = \pi D \int_0^L N_{A\,r|_R}\, dz =$$

(10.41)

$$\pi D \int_0^L \frac{J_{A\,r|_R}}{(1 - \Phi X_{Aw})}\, dz = \pi D \int_0^L \frac{-cD_{AB}\left(\dfrac{\partial X_A}{\partial r}\right)_{|_R}}{(1 - \Phi X_{Aw})}\, dz$$

Where, to be general, allows for situations with a wide range of inter-relationships between the species' fluxes (see discussion in 4-2).

With the choice of the dimensionless variables

$$J^*_{A_r} = \frac{J_{A_r}}{C(\overline{X}_A - X_{Aw})\overline{v}}, \quad X_A{}^* = \frac{(X_A - X_{Aw})}{(\overline{X}_A - X_{Aw})}$$

together with the previously defined dimensionless independent variables, Equation (10.41) becomes

$$Sh = \frac{Re\,Sc}{L/D}\int_0^{L/D} J^*_{A_r|_{\frac{1}{2}}}\, dz^* = \frac{1}{L/D}\int_0^{L/D}\left(-\frac{\partial X_A{}^*}{\partial r^*}\right)_{|_{\frac{1}{2}}}\, dz^*$$

(10.42)

where the *Sherwood number* , Sh, is defined by

$$Sh = \frac{k_{mx}(1 - \Phi X_{Aw})D}{C D_{AB}}$$

(10.43)

and Sc is the <u>Schmidt number</u>, which has been previously defined in Chapter 7. Equation (10.43) also holds for other driving forces with k_{mx} replaced by $k_{mc}\,C$ and by $k_{mp}\,P$ for concentration and pressure driving forces, respectively.

In momentum transport the transfer coefficient, f , is already dimensionless and the analogous expression to Equations (10.33) and (10.41) is

$$\bar{\tau}_w A_w = f\left(\frac{1}{2}\rho\bar{v}^2\right) A_w = \int_{A_s} \tau_{rz}\big|_R \, dA_s = \int_{A_s} -\mu\left(\frac{\partial v_z}{\partial r}\right)\big|_R \, dA_s$$

If τ_{rz}^* and v_z^* are chosen as

$$\tau_{rz}^* = \frac{\tau_{rz}}{\rho\bar{v}^2} \, , \quad v^* = \frac{v}{\bar{v}}$$

then it is easily shown that

$$f = \frac{2}{L/D} \int_0^{L/D} \tau_{rz}^* \big|_{\frac{1}{2}} \, dz^* = \frac{2}{Re\dfrac{L}{D}} \int_0^{L/D} \left(-\frac{\partial v_z^*}{\partial r^*}\right)_{\big|\frac{1}{2}} \, dz^* \qquad (10.44)$$

10-5 TRANSFER COEFFICIENTS AND TURBULENCE

Since transfer coefficients are an empirical means to account for convective mixing, they can also be related to the various semi-empirical models for turbulence. While there are any number of turbulent transport models, only the two introduced earlier, Surface Renewal theory and Phoney Film theory, will be discussed here.

10-5.a Surface Renewal Theory

The surface renewal concept which was introduced in Chapter 9 to describe turbulent heat transfer at a surface can also be formulated in terms of the heat transfer coefficient. For a particular eddy, "i", the heat transfer coefficient based on the difference between the wall temperature and the bulk temperature, is

$$(h)_i = \frac{(q_w)_i}{(T_w - T_b)}$$

and, upon substituting for $(q_w)_i$ from Equation (9.22), we obtain

$$(h)_i = (t)^{-1/2} \sqrt{\frac{k\rho \, \hat{c}_p}{\pi}} \qquad (10.45)$$

Note that at time zero, Equation (10.45) predicts an infinite heat transfer coefficient which subsequently decreases with the square root of time. Of course, the occurrence of an infinite heat transfer coefficient is only a consequence of the failure of the model to describe nature right at time = zero. Nevertheless, it is safe to say that the heat transfer coefficient will be high if the eddy remains at the surface for only a short time. Since this will occur at high surface renewal rates (high degree of mixing) and is consistent with empirical knowledge, the model is qualitatively encouraging.

Let us assume that the I'th eddy remains at the surface for a "residence time", τ_i. Over this time period the average heat transfer coefficient for this eddy may then be calculated by

$$(\bar{h})_i = \frac{1}{\tau_i} \int_0^{\tau_i} h_i \, dt \qquad (10.46)$$

substituting for h_i from Equation (10.45), Equation(10.46) becomes

$$(\bar{h})_i = \frac{1}{\tau_i} \sqrt{\frac{k\rho \, \hat{c}_p}{\pi}} \int_0^{\tau_i} t^{-1/2} \, dt$$

$$(10.47)$$

$$= 2 \sqrt{\frac{k\rho \, \hat{c}_p}{\pi}} \frac{1}{\tau_i^{1/2}}$$

However, this is only for the I'th eddy. The heat transfer coefficient for the system should be calculated by averaging Equation (10.47) over all eddies which arrive at the surface during some repeatable time interval, t_o. Mathematically this can be expressed as

$$h = \frac{\Sigma (\bar{h})_i \tau_i}{\Sigma \tau_i}$$

If Equation (10.47) is substituted for $(\bar{h})_i$, then the heat transfer coefficient is given by

$$h = 2 \sqrt{\frac{k\rho \, \hat{c}_p}{\pi}} \frac{1}{\bar{\tau}^{\frac{1}{2}}} \qquad (10.48)$$

where the average eddy residence time, $\bar{\tau}$, is

$$\bar{\tau} = \left(\frac{\Sigma \tau_i}{\Sigma (\tau_i)^{\frac{1}{2}}} \right)^2$$

We can also use the surface renewal theory to estimate the dependence of the Nusselt number on the Reynolds and Prandtl number. Thus, if Equation (10.48) is multiplied by $\dfrac{D}{k}$,

$$Nu = \frac{2D}{k} \left(\frac{k\rho \, \hat{c}_p}{\pi} \right)^{1/2} \frac{1}{\bar{\tau}^{\frac{1}{2}}} = \frac{2}{\sqrt{\pi}} \left(\frac{D \dfrac{D}{\bar{\tau}}}{v} \frac{v}{\alpha} \right)^{1/2}$$

$$(10.49)$$

$$\propto \left[Re \, Pr \right]^{1/2}$$

where we have assumed that $\dfrac{D}{\bar{\tau}}$, an "eddy velocity," is proportional to the average fluid velocity in the pipe.

While this has been an interesting exercise, we are not able to achieve the goal of predicting h, or the dimensionless heat transfer coefficient, Nu , unless we can predict the average eddy residence time, $\bar{\tau}$. On the other hand, it is now possible to estimate the dependence of h or Nu on the thermal properties of the fluid. Equation (10.48) predicts that the only dependence of the heat transfer coefficient on thermal fluid properties is to the 0.5 power of k and $\hat{c}_p$ (since $\bar{\tau}$ should only be dependent on the degree of mixing) and Equation (10.49) predicts that the Nusselt number should vary with the square root of the Reynolds and Prandtl numbers. It is interesting to compare this dependency with that predicted by the Dittus-Boelter equation, which correlates experimental data for heat transfer coefficients in turbulent pipe flow [4]; viz.,

$$Nu = .023 \, Re^{.8} \, Pr^{1/3}$$

so that

$$h \propto [\, \hat{c}_p^{.33} k^{.67} \,] \left[\frac{\rho^{.8} \, v^{.8}}{D^2 \mu^{.47}} \right] \tag{10.50}$$

If we compare this dependency with that predicted by the surface renewal theory (Equations (10.48) and (10.49)), we can see that the prediction of the model is not entirely consistent with experimental evidence; that is, the Dittus-Boelter equation predicts that h will depend on $k^{2/3}$ and $\hat{c}_p^{1/3}$ and that the Nusselt number is more strongly dependent on the Reynolds number and less dependent on the Prandtl number. Whereas the model does not exactly match the Dittus-Boelter equation, it is not unreasonable. Of course these discrepancies should not be unexpected since this rather idealized model is not a perfect fit to physical reality and the true situation is probably only partially described by such a representation.

10-5.b Surface Renewal vs. Phoney Film theory

In Section 10-3 we introduced the concept of the "phoney" film to give us a conceptual grasp of transfer coefficients in convective transport. In this chapter, we also added a different view of convective phenomena by describing surface renewal theory as it applies to turbulent energy transport. As might be expected, the surface renewal theory is also applicable to mass transport. In fact, it was first developed in an attempt to obtain a more realistic picture of convective mass transport.[#]

Rather than just repeat the development of 10-5.a for mass transport, instead let us compare the surface renewal theory with the more classical Phoney Film theory. For simplicity let us assume that conditions are such that bulk diffusion effects are negligible (equimolar counter diffusion for example) and the total molar concentration is constant. Therefore the situation at any time in the i'th eddy is described by

$$\frac{\partial C_A}{\partial t} = D_{AB} \frac{\partial^2 C_A}{\partial x^2} \tag{10.51}$$

which is completely analogous to Equation (7.12) in energy transport

Following the identical approach described in 10-5.a, we can define the instantaneous mass transfer coefficient in the "i"th eddy as

---

[#]   Actually, in the original work [3] it was first called *penetration theory* and the eddies were assumed to arrive at the surface in a random manner.

$$(k_{mc})_i = \frac{N_{Ax_{|x=0}}}{(C_{AI} - C_{Ab})} = -\frac{D_{AB}\frac{\partial C_A}{\partial x}|_{x=0}}{(C_{AI} - C_{Ab})} \tag{10.52}$$

Solving Equation (10.51) as was done in Section 7-3, together with the initial and boundary conditions

$$C_A = C_{Ab} \ at \ t = 0, \ all \ x$$

$$C_A = C_{AI} \ at \ x = 0, \ all \ t \geq 0$$

$$C_A = C_{Ab} \ at \ x = \infty, \ all \ t$$

we have

$$C_A - C_{Ab} = (C_{AI} - C_{Ab})\left[1 - \frac{\sqrt{\pi}}{2}\int_0^{x/\sqrt{4D_{AB}t}} \exp(-\eta^2)\, d\eta\right] \tag{10.53}$$

The instantaneous mass transfer coefficient can then be calculated from Equations (10.52) and (10.53) to yield

$$(k_{mc})_i = \sqrt{\frac{D_{AB}}{\pi t}} \tag{10.54}$$

which, when averaged over the eddy residence time, τ_i, can be expressed in terms of the average eddy residence time, $\bar{\tau}$, to give

$$k_{mc} = 2\sqrt{\frac{D_{AB}}{\pi \bar{\tau}}}$$

and thus

$$N_A = 2\sqrt{\frac{D_{AB}}{\pi \bar{\tau}}}\left[C_{AI} - C_{Ab}\right] \tag{10.55}$$

This result can be compared to the Phoney Film theory where Equation (10.24) predicts that the molar flux of 'A' at the interface is given by

$$N_A = \frac{D_{AB}}{\delta_{pf}}(C_{AI} - C_{Ab})$$ (10.24)

Comparing Equations (10.24) and (10.55), notice that Phoney Film theory predicts a linear dependency of N_A on D_{AB} whereas the surface renewal theory predicts a square root dependency on D_{AB}. These two equations can be compared further to show that

$$\delta_{pf} = \frac{1}{2}\sqrt{\pi D_{AB}\bar{\tau}}$$ (10.56)

Now let us compare the phoney film thickness δ_{pf} with the <u>penetration thickness</u>, δ_p, of a given eddy which we shall define as that thickness at which the concentration of 'A' in the eddy is within 99% of the bulk concentration. From the solution presented in Equation (10.53) it can be shown that $(C_A - C_{Ab}) \geq .99(C_{AI} - C_{Ab})$ whenever $\eta > 2$, or $x \geq 4\sqrt{D_{AB}t} \equiv \delta_p$. Thus the penetration thickness, δ_p, for mass transport in an eddy with a residence time, τ_i would be

$$\delta_p = 4\sqrt{D_{AB}\bar{\tau}}$$ (10.57)

Comparing Equations (10.56) and (10.57) for any given eddy $(\bar{\tau} = \tau_i)$, we can see that

$$\delta_{pf} = \frac{\sqrt{\pi}}{8}\delta_p = .22\delta_p$$ (10.58)

Thus, since δ_p is the distance into the eddy corresponding to a concentration within 99% of the bulk concentration, δ_{pf} corresponds to that point in an eddy which is within 22% (approximately) of the bulk concentration. Not too surprisingly this is roughly equivalent to that portion of the eddy over which the concentration gradient is reasonably linear. In other words, the surface renewal theory is related to the Phoney Film theory in the sense that it would provide a statistical average of the distance over which the concentration profile is approximately linear.

Before abandoning this discussion it should also be pointed out that both these theories are at odds with experimental data relative to the dependence of the mass transfer coefficient on the diffusivity. Whereas the surface renewal theory predicts a square root dependency (Equation (10.54), and Phoney Film theory predicts a linear dependency, experimental correlations indicate that

$$k_{mc} \propto (D_{AB})^n$$

where $0.3 < n < 0.7$. It is somewhat encouraging that the surface renewal theory predicts a value halfway between these two extremes.

REFERENCES

[1] McCabe, W.L. and J.C. Smith, *Unit Operations of Chemical Engineering*, 3rd ed., McGraw-Hill Co., N.Y., 1976

[2] Hatta, S., Tohoku Imp. Univ. Tech. Rept., <u>10</u>, p. 119 (1932)

[3] Higbie, R., Trans. Am. Inst. Chem. Engrs., <u>31</u>, p.365 (1935)

PROBLEMS

10-1 Referring to the situation shown in FIGURE 4-1, Derive an expression for the mass transfer coefficient, k_{mx}, based on the driving force between the bulk gas and the interface, for the case of equimolar counter diffusion.

10-2 Mass transfer coefficients between gases and packed bed catalyst particles can be correlated (Smith, J.M., *Chemical Engineering Kinetics*, 3rd ed., McGraw-Hill, 1981, p. 94) by:

$$\frac{k_m \rho}{G} = 1.01 (Re_p)^{-.407}$$

where G is the superficial mass flux in the bed and Re_p is the Reynolds number based on the diameter of the catalyst particles. For catalytic combustion of trace hydrocarbons in air, calculate all three mass transfer coefficients, $k_{m(x,c,p)}$, for a case where $Re_p = 100$, $G = .3$ lb$_m$/s-ft^2, at 20 atm pressure and a temperature of 600 K. Assume ideal gas behavior.

10-3 Use the solution for the temperature profile obtained in Problem 8-4,

$$T - T_{wo} = \frac{\Delta T}{L} z + \frac{5}{12} \frac{\Lambda H^2}{k} - \frac{\Lambda H^2}{12 k} \left[6 \left(\frac{y}{H} \right)^2 - \left(\frac{y}{H} \right)^4 \right]$$

where $\quad \Lambda = \dfrac{\rho \hat{c}_P \Delta P\, H^2}{2\mu l}\, \dfrac{\Delta T}{L}$

to obtain a numerical value for the Nusselt number if the latter is based on a characteristic dimension of H and a driving force of $T_w - \overline{T}$.

10-4   Starting with the definition of the friction factor given in Section 10-4, derive Equation (10.44).

10-5

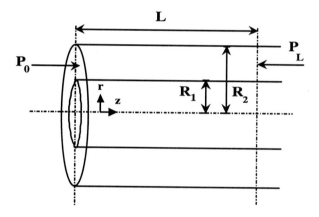

(a) For fully developed laminar flow of a newtonian fluid in the horizontal annulus shown above, derive an expression for the velocity profile.

(b) Derive an expression for the average velocity in the annulus

(c) Defining the friction factor as

$$f = \dfrac{(F_f)_{ToT}\, / \, A_w)}{1/2\, \rho\, \bar{v}^{\,2}}$$

where $(F_t)_{ToT}$ is the total friction force at the walls, take an overall force balance on the annulus to show that f can also be expressed in terms of the pressure drop in the annulus as

$$f = \frac{(\Delta P / L) \; R_H}{1/2 \, \rho \bar{v}^2}$$

where R_H is the hydraulic radius.

(d) Use the results of (b) and (c) to derive an expression for the friction factor as a function of Reynolds number and κ, where $Re = \dfrac{D_{eq} \bar{v} \rho}{\mu}$ and $\kappa = \dfrac{R_1}{R_2}$.

10-6   Based on the *universal velocity profile*, it can be shown (see Problem 9-8) that the eddy viscosity varies with r, according to

$$\varepsilon = \frac{r}{R} \frac{(R-r)}{a} \left(\frac{\tau_w}{\rho} \right)^{1/2}$$

Calculate the value of the eddy viscosity at a point 2 mm from the wall in a smooth, 10-cm diameter pipe which has water (at 25 °C) flowing at a Reynolds number of 70,000 and compare this value with the value for the kinematic viscosity of water at 25 °C. For a smooth pipe, the friction factor can be calculated from [see Chapter 11-1, Equation (11.13)].

$$f = 0.0014 + \frac{0.125}{Re^{0.32}}$$

10-7   The "gas-side" mass transfer coefficient correlation given by Gilliland and Sherwood (Ind. Eng. Che., 26 p. 516, 1934) for a wetted wall column is,

$$Sh = .023 \, Re^{.83} \, Sc^{.44}$$

where the authors defined the Sherwood number as $\; Sh = \dfrac{k_g D}{D_{AB}}$

Use this correlation to determine the thickness of the phoney film for CCl_4 absorption from air at a Reynolds number of 10^5 and at atmospheric pressure and 325 K in a 1.5 in-diameter column. Assume that the gas properties correspond to those for pure air.

10-8

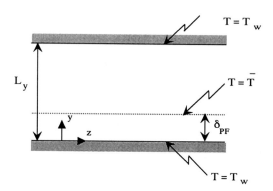

An exothermic chemical reaction takes place while a fluid is flowing between parallel plates, as shown in the above sketch. Using Phoney Film theory, solve for the temperature profile through the phoney film and then derive an expression for the Nusselt number (defined as hL_y/k) corresponding to this situation. The heat of reaction is $\Delta \tilde{H}_r$ and the reaction is zero order with a rate constant, k_r (moles reacting/vol-time).

10-9

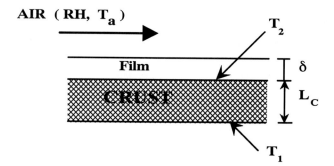

Referring to the nuclear waste tank described in Problem 3-14, the sketch shows the crust as well as a "film" which represents the convective transport between the crust surface and the purge air. The energy of radioactive decay is primarily removed by water evaporation and, due to the presence of the crust, the water must first move through the crust (probably by capillary action), evaporate at the top surface of the crust, and then be transported into the purge air stream (which is at temperature, T_a, and relative humidity, RH). That is, energy is transported through the crust by both molecular and convective mechanisms. Referring only to the crust and assuming that the molecular transport through the solid portion of the crust can be described by an "effective thermal conductivity" of the crust, k_{eff}, use the experimental measurements given below, to obtain a value for k_{eff}.

$T_1 = 60$ C, $T_2 = 45$ C, $\dot{Q} = 40,000$ BTU/hr, $L_c = 1.2$ m

10-10 In Problem 10-9, values of the temperature at the top of the crust (T_2) and the total heat loss were both given. In reality these are only estimates since the exact location of the top of the crust is not known with certainty and there are also heat losses through the tank walls etc. Derive the equations necessary to solve for the temperature distribution through the crust <u>without</u> knowing either T_2 or $\dot{Q}$. The following assumptions apply:

- The mass transport resistance in the crust is negligible compared to the convective transport resistance between the top of the crust and the purge air.

- The mass and heat transfer coefficients are known.

- The effective thermal conductivity and liquid velocity through the crust are known and equal to the values determined in 10-9.

- The relationship between the concentration of water vapor and the temperature at the top of the crust is of the form $P^* = K_1 \, \text{EXP}\left[-\dfrac{K_2}{T}\right]$ and is quantitatively known.

Without solving the equations, describe how you would go about obtaining the solution.

10-11

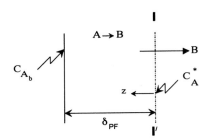

(a) For the situation shown in the sketch, derive an expression for the Sherwood number (in terms of δ_{PF}) for the situation shown in the sketch where species A is being transported through the interface, I-I', and is undergoing a first order reaction ($A \rightarrow B$) within the gaseous phoney film. Assume that species B can cross the interface.

(b) What is the expression for the Sherwood number if the diffusivity is much less than the reaction rate constant ($\dfrac{D_{AB}}{k_r} \gg 1$)?

10-12 For fully developed turbulent flow in tubes, the Dittus-Boelter relationship

$$Nu = .023 \, Re^{.8} \, Pr^{1/3}$$

is used to correlate experimental heat transfer coefficients.

(a) Use this relationship to derive an expression for $\left(\dfrac{\partial T^*}{\partial r^*} \right)_{\left|\frac{1}{2}\right.}$ in Equation (10.37) [assume that

$\left(\dfrac{\partial T^*}{\partial r^*} \right)_{\left|\frac{1}{2}\right.}$ is not a function of $z^{\cdot}$].

(b) If in a given situation, water is the fluid at 60 °F, and $Re = 10^5$, determine the thickness of the phoney film in a 1 in schedule-40 pipe.

(c) Empirical expressions for temperature profiles in turbulent pipe flow often take the form (see Problem 2-2c)

$$\frac{T - T_w}{\overline{T} - T_w} = 1 - \left(\frac{r}{R} \right)^n$$

where n will typically have values on the order of 7-10. Determine the value of n which is compatible with the Dittus-Boelter correlation under the conditions given in (b). What do you conclude relative to the usefulness of such temperature distributions for calculating heat transfer rates to the pipe wall?

10-13 Repeat the calculation for the "limiting" Nusselt number" in Section 10-2 if the average temperature is defined as the "mixing cup temperature." That is,

$$\overline{T}_{MC} = \frac{\int_0^R v_z\, T\ 2\pi\, r dr}{\int_0^R v_z 2\pi\, r dr}$$

10-14 For the laminar flow in a rectangular duct as described in Problem 5-1, derive an expression for the friction factor in terms of the Reynolds number based on the duct height (dimension, $2B$, in Problem 5-1.

10-15 For ducts with very high aspect ratios ($\dfrac{W}{B} >> 1$, as shown in Problem 5-11), derive an expression for the laminar flow friction factor as a function of Reynolds number (defined in terms of B).

10-16 Derive Equation (10.22) and then use the "mixing cup" average concentration

$$\overline{C}_{A_{mc}} = \frac{\int_0^R v_z\, C_A\ 2\pi\, r dr}{\int_0^R v_z 2\pi\, r dr}$$

to calculate a limiting Sherwood number.

10-17

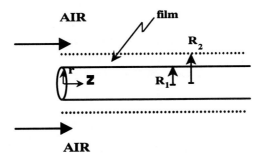

A very long cylinder of naphthalene is exposed to air at steady-state and at a constant temperature where it sublimes at the air-solid surface. If the mole fraction of naphthalene in the bulk air is X_{Ab}, and the sublimation (i.e., equilibrium) mole fraction of A is X_A^*, use Phoney Film theory to obtain an expression for the molar flux of A at the surface, $(N_{A_r})_{R_1}$. It may be assumed that the air is insoluble in the naphthalene.

10-18 The temperature distribution in the turbulent flow of water on the tube-side of a shell-and-tube heat exchanger is given by

$$T - T_w = 238 \left[1 - \left(\frac{r}{R} \right)^{10} \right]$$

where the temperature is given in degrees Fahrenheit and the tube is 1-1/4 in BWG-14.

(a) Calculate a numerical value of the tube-side heat transfer coefficient (based on $T_{CL} - T_w$)

(b) Calculate the thickness of the phoney film.

10-19 Nitrogen is flowing in a smooth horizontal tube (D = .5 cm, L = 35 m) at an inlet pressure of 300 kPa. The volumetric flow rate is .05 cfm and the temperature is 60 °C. Calculate the pressure drop over the length of the pipe. The dependence of the friction factor on Reynolds number for turbulent flow can be calculated by $f = .046 \, Re^{-.20}$.

10-20

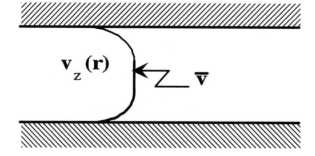

The sketch shows a turbulent velocity profile superimposed on the cross section of a cylindrical tube of radius, R, in steady-state, fully developed flow. We wish to use Phoney Film theory to analyze the momentum transport to the wall.

(a) Use a differential momentum balance and solve for the velocity profile <u>within the film</u>.

(b) Using the conventional definition of the friction factor (Equation (10.9)), use the solution in (a) to determine a relationship for the friction factor as a function of the Reynolds number (based on the pipe diameter) and $\dfrac{\delta_{PF}}{R}$.

10-21 In fixed-bed adsorbers, a fluid containing species i moves through the bed with a constant velocity, $\bar{v}$, and species, i is transported to the bed solids where it adsorbs to the solid surface. In analyzing such systems it is generally assumed that the concentration of i on the solid surface, C_{i_s} , is approximately zero. Following the approach of Cooney (AIChE J., <u>37</u>, p. 1270, 1991), take a differential mass balance on the fluid assuming that molecular transport in the flow direction is negligible and that the mass transfer rate between the fluid and solid is given by $K'(C_i - C_{is})$ moles/vol-t, the concentration of i at the bed entrance is C_{i_0} and

(a) Solve for C_i as a function of z , assuming that $C_{i_s} = 0$

(b) Assume the existence of a *Langmuir adsorption isotherm*; i.e., an equilibrium relationship between C_i and C_{i_s} of the form

$$C^* = 2C_s^* + \frac{11C_s^*}{1 + 10C_s^*}$$

where the asterisk values indicate that the concentrations are normalized with respect to the inlet value; eg., $C^* = \dfrac{C_i}{C_{i_0}}$. Use numerical methods (MATLAB) to generate a plot of C^* vs. z for

$K' = 2\,min^{-1}$, $\bar{v} = 100$ cm/min .

(c) Compare the solutions in (a) and (b) for the same values of K' and $\bar{v}$.

10-22 Solve the differential equation and use the associated boundary conditions to obtain the solution for the concentration profile given as Equation (10.30) in Example 10-1.

10-23 Use a development similar to that presented in section 10-5b to obtain a relationship between the phoney film thickness and the average eddy residence time, $\bar{\tau}$, for turbulent energy transport.

PART III

Macroscopic Calculations

Illustrative methods to size the equipment necessary to achieve desired levels of momentum, energy, and mass transport are covered in Chapters 11through13. In principle, this can be achieved by applying the methods described in Chapters 3 through 5 of the text. That is, the velocity, temperature, and concentration profiles are calculated and then differentiated at the system boundaries to calculate the rates of momentum, energy, and mass transport. Because the mixing state in most large, commercial process units is usually complex, they rarely lend themselves to these type of calculations. Consequently the conventional approach is to abandon the goal of predicting the profiles and to focus, instead, on the overall rates by using transfer coefficients and empiricism. To distinguish this macroscopic approach from that used in molecular transport, we will refer to this approach as momentum, energy and mass <u>transfer</u>. Nevertheless, the methodology and concepts introduced in the first ten chapters of the text are still useful for understanding and improving the macroscopic approaches. Consequently, differential analysis is employed whenever possible, and then simplified so that conventional macroscopic calculations can be conducted. The coverage of these topics is not intended to be exhaustive. If additional details and applications are desired, they are amply described in the traditional unit operations textbooks [1,2].

REFERENCES

[1] McCabe, W.L., Smith, J.C. and Harriot, P., *Unit Operations of Chemical Engineering*, McGraw-Hill, N.Y., 1976

[2] Coulson, J.M. and Richardson, J.F., *Chemical Engineering*, 4th ed., Pergamon Press, Oxford, UK, 1990

CHAPTER 11

Macroscopic Calculations: Momentum Transport

11-1 APPLICATIONS OF BERNOULLI'S EQUATION

11-1.a Piping Systems

The starting point for macroscopic calculations to size piping systems is the "point" form of the Bernoulli Equation, which was derived in Chapter 8

$$-\frac{d}{dz}(\rho v_z v_z) - \frac{dP}{dz} - \rho g_z = 0 \tag{8.17}$$

Recall that this equation resulted from a "frictionless" differential momentum balance which, strictly speaking, is only applicable far from the walls of a pipe (e.g., the center of a pipe in developing flow). For a system with constant density (incompressible flow), Equation (8.17) can be written as

$$-\frac{d}{dz}(\frac{\rho v_z^2}{2})-\frac{dP}{dz}-\rho g_z = 0 \tag{11.1}$$

Equation (11.1) can be integrated between *any* two points, 1 and 2, in a piping system to give

$$\Delta(\frac{\rho \bar{v}^2}{2})+\Delta P+\rho g_z \Delta z = 0 \tag{11.2}$$

where Δ is taken to indicate the difference in the terms between points 2 and 1 and it is assumed that the velocity profile in the pipe is essentially "flat," so that the point velocity, v_z, is replaced by the average velocity, $\bar{v}$.

In order to use this equation for practical piping calculations, a correction term must be added to account for the friction at the walls. Consequently, when Equation (11.2) is applied between two points in a real piping system, a term, ρh_f, is added. This is an empirical correction for the energy/volume generated by friction at the walls,[#] and Equation (11.1) is then written as

$$\Delta(\frac{\rho \bar{v}^2}{2})+\Delta P+\rho g_z \Delta z+\rho | h_f | = 0 \tag{11.3}$$

The integration of Equation (11.2) is really the conversion of the differential momentum balance at any point within the fluid to a macroscopic momentum balance over the entire system. In terms of units, the integration turns the equation from a force/volume balance to an overall force balance acting on the boundaries of the system (force/area units). The original differential equation describing the momentum balance at any point in a pipe was derived in Chapter 8 as Equation (8.5)

$$-\frac{\partial}{\partial z}(\rho v_z v_z)-\frac{\partial \tau_{zz}}{\partial z}-\frac{1}{r}\frac{\partial}{\partial r}(r \tau_{rz})-\frac{\partial P}{\partial z} = 0 \tag{8.5}$$

If $\tau_{zz} \sim 0$, this equation can be compared to Equation (11.2), with the result

$$\rho | h_f | = \int_1^2 \left[\frac{1}{r}\frac{\partial}{\partial r}(r \tau_{rz}) \right]_{|R} dz \tag{11.4}$$

[#]   The term, h_f, is actually the conversion of mechanical energy into heat and is sometimes referred to as the <u>friction head</u>, or the friction energy loss per unit mass of fluid.

Note that since the integration is applied over the entire system, the derivative of the shear force is evaluated at the wall (boundaries) of the pipe. The integrand in Equation (11.4) is just the friction force per unit volume, so that when it is applied over the volume of a cylindrical pipe of diameter, D, and length, L

$$\int_1^2 \left[\frac{1}{r} \frac{\partial}{\partial r} (r \tau_{rz}) \right]_{|R} dz = \frac{\tau_w \pi \, DL}{\frac{\pi \, D^2 L}{4}} \int_0^L dz = \frac{4 \tau_w L}{D}$$

and, from Equation (11.4), h_f is then

$$h_f = \frac{1}{\rho} \frac{4 \tau_w L}{D} = \frac{4L}{D} \left[-\nu \frac{\partial v_z}{\partial r} \right]_{|R} \tag{11.5}$$

Thus h_f can be calculated from the velocity distribution, if it is known. As discussed in Chapter 5, this is easily done in fully developed laminar flow, but empirical correlations are needed in complex flows involving turbulence.

Note that each term in Equation (11.3) also has units of energy/volume. Therefore, it can also be derived by taking an energy balance over the system. To illustrate this approach, the first law of thermodynamics for an open system can be written between points 1 and 2 of the piping system shown in Figure 11-1. This system consists of three separate lengths of pipe with different diameters, a pump and a valve. Point 2 is at a vertical distance, ΔH, above point 1. At steady state the energy input at point 1 must equal the energy output at point 2 and the mathematical description of the energy balance per unit mass of fluid is

$$\hat{U}_1 + P_1 \hat{V}_1 + (\hat{PE})_1 + (\hat{KE})_1 + \eta \, W_P = \hat{U}_2 + P_2 \hat{V}_2 + (\hat{PE})_2 + (\hat{KE})_2 + h_f$$

where U, PE, and KE are the internal, potential, and kinetic energies, respectively, W_p is the energy input to the pump per mass of fluid, η is the efficiency of the pump (i.e., the fraction of the pump energy that is transferred to the fluid) and h_f is the energy loss per mass of fluid due to friction in the piping system. If the system is isothermal and incompressible, then $\hat{U}_1 = \hat{U}_2$. Recognizing that $\hat{V} = \frac{1}{\rho}$, $\hat{PE} = g_z$ and $\hat{KE} = \frac{\overline{v}^2}{2}$, the energy balance can be written as

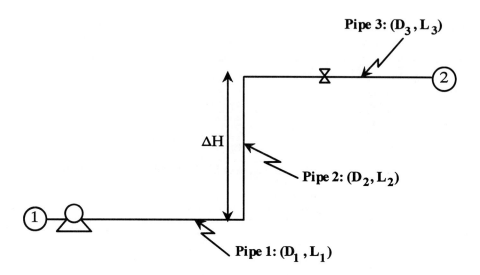

Figure 11-1 Piping System

$$\Delta \frac{\bar{v}^2}{2} + \frac{1}{\rho}\Delta P + g_z \Delta H + h_f - \eta W_P = 0 \qquad (11.6)$$

which is actually Equation (11.3) with the addition of a pump (a SOURCE) and $\Delta z = \Delta H$.

The next issue to be dealt with is the velocity in the kinetic energy term. Since the derivation assumed a flat profile, it needs to be corrected for deviations from that assumption. The deviation becomes more significant as the Reynolds number decreases and is a maximum in laminar flow where the velocity in the center of the pipe is twice the average velocity. To correct for the profile, an average kinetic energy can be calculated so that the first term in Equation (11.6) is written as $\Delta(\beta \bar{v}^2)$ where $\bar{v}$ is the average velocity in the pipe and β is defined by

$$\beta \bar{v}^2 = \frac{\displaystyle\int_{A_x} v^2 \, v \, dA_x}{\displaystyle\int_{A_x} v \, dA_x} \qquad (11.7)$$

It is readily shown that $\beta = 2$ in laminar flow (see Problem 11.2).

EXAMPLE 11.1: Using Equation (11.6) To Size A Pump

Determine the horsepower of a pump (68% efficient) necessary to pump 380,000 lb_m /h of a diesel oil from the middle of a fractionator (10 m above ground) through 350 m of 4"-schedule 40 pipe to the bottom of an atmospheric storage tank which is 80 m above ground level. The depth of the diesel in the tank is 20 m and the fractionator is at 100 psig pressure. The specific gravity of the diesel is 0.95 and the energy loss due to friction at the walls of the pipe can be taken to be 85 ft-lb_f /lb_m.

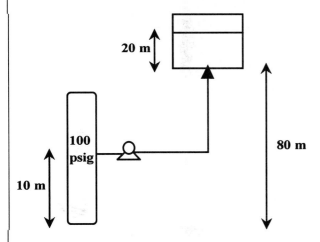

Solution

In this case, point 2 is at the top of the fluid in the storage tank where $\bar{v}$ is zero and P is equal to atmospheric pressure, and point 1 is at the fractionator where $\bar{v}$ is also zero but the pressure is at 100 psig. A consistent unit system must be used and, to illustrate the difference between lb_f and lb_m, the AES of units is chosen.

From Equation (11.6)

$$\eta W_P = \Delta \bar{v}^2 + \frac{1}{\rho}\Delta P + g_z \Delta z + h_f$$

The kinetic energy term is zero and

$$\frac{1}{\rho}\Delta P=\frac{(14.7-114.7)\frac{lb_f}{in^2}\,144\frac{in^2}{ft^2}}{59.28\frac{lb_m}{ft^3}}=-242.9\,\frac{ft\text{-}lb_f}{lb_m}$$

The potential energy term is

$$g_z\Delta z=\frac{32.17\frac{ft}{s^2}(80+20-10)\,m\,3.28\frac{ft}{m}}{32.17\frac{lb_m\text{-}ft}{lb_f\text{-}s^2}}=295.2\,\frac{ft\text{-}lb_f}{lb_m}$$

Therefore, the energy delivered to the fluid by the pump is

$$\eta W_P=(-242.9+295.2+85)=137.3\,\frac{ft\text{-}lb_f}{lb_m}$$

and the energy input to the pump is

$$W_P=\frac{137.3}{.68}=201.9\,\frac{ft\text{-}lb_f}{lb_m}$$

The required pump horsepower is then $PW=(W_P)\,\dot{m}$, or

$$PW=201.9\,\frac{ft\text{-}lb_f}{lb_m}\,3.8\times10^5\,\frac{lb_m}{hr}=7.67\times10^7\,\frac{ft\text{-}lb_f}{hr}$$

$$PW=6.51\times10^7\,\frac{ft\text{-}lb_f}{hr}\,5.0505\times10^{-7}\,\frac{hp\text{-}hr}{ft\text{-}lb_f}=38.75\ hp$$

11-1.b Pressure Losses in Pipes and Fittings

Because the friction loss energy was given in Example 11-1, the diameter of the pipe did not figure into the calculations. In order to calculate h_f, empirical correlations are needed to account for the energy losses due to "skin" friction at the pipe walls (h_{fs}) as well as losses due to flow through valves and fittings (h_{ff}) and changes in fluid velocity as the fluid passes through regions of different cross-sectional areas (expansion and contraction, h_{fe}, h_{fc}). The total energy loss, h_f, can be calculated from the sum total of all these contributions

$$h_f = h_{fs} + h_{ff} + h_{fe} + h_{fc} \qquad (11.8)$$

The loss due to friction at the pipe walls has already been discussed in terms of its relation to the momentum balance in a pipe but since all of the energy losses will result in pressure "losses" (drops), h_f can also be expressed in terms of the total pressure drop in the system

$$h_f = \frac{\Delta P}{\rho} \qquad (11.9)$$

For skin friction at the pipe walls, it was shown in Chapter 10 that a dimensionless pressure gradient, the friction factor, can be defined as

$$f = \frac{\tau_w}{\dfrac{\rho \bar{v}^2}{2}} = \frac{\Delta P D}{2\rho \bar{v}^2 L} \qquad (11.10)$$

Thus, if Equations (11.9) and (11.10) are combined, h_{fs} can be expressed in terms of the friction factor

$$h_{fs} = \frac{2\bar{v}^2 L f}{D} \qquad (11.11)$$

In general, the friction factor will be dependent on the Reynolds number and $\dfrac{L}{D}$, as was shown in Chapter 8. For example, it was shown in Chapter 10 that for fully developed laminar flow in a pipe,

$$f = \frac{16}{Re}$$

However, for complex flows such as turbulent flow, experimental data are measured and then correlated by relating the friction factor to the Reynolds number. For example, for fully developed flow of a newtonian fluid in highly turbulent flow (Re > 50,000), one such correlation for "hydraulically smooth" pipes is

$$f = \frac{0.046}{Re^{.20}} \qquad (11.12)$$

The reason for the emphasis on smooth pipes is that the roughness on commercial pipe walls can contribute to the degree of turbulence and increase the energy loss. Consequently, the friction factor is also correlated with a dimensionless "roughness factor," $\frac{k}{D}$, and usually presented in the form of friction factor plots such as illustrated in Figure 11-2. Roughness factors are highly dependent on the nature of the pipe and typical values are given in Perry's Handbook [1]. Note that the Reynolds number range between laminar and turbulent flow is designated as the "transition" region. Data in this region is highly scattered and more difficult to correlate. An experimental correlation which applies to this region as well as to turbulent flow is given in Equation (11.13)

$$f = 0.0014 + \frac{0.125}{Re^{.32}} \qquad (11.13)$$

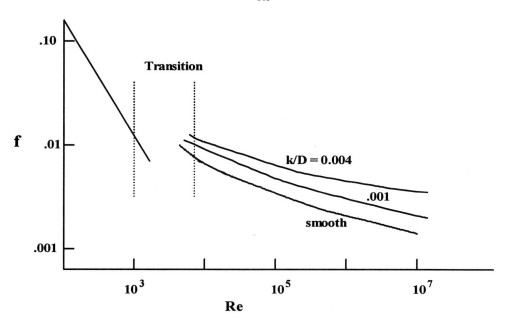

Figure 11-2 Friction Factor Plot

Figure 11-2 can also be used for flow in conduits other than pipes as long as the diameter is replaced by the equivalent diameter as defined by Equation (10.14)

$$D_{eq} = 4 R_H \qquad (10.14)$$

It should be kept in mind that the friction factor plot in Figure 11-2 applies only to friction at the walls of pipes. Energy losses due to flow through valves and fittings or due to abrupt changes in fluid direction (a 90° elbow, for example) can completely dominate the total energy loss in short piping systems with numerous fittings. The energy losses in these fittings are due to inertial effects rather than skin friction. As a result, the energy loss per volume is expressed in terms of the kinetic energy per volume, or

$$\rho\, h_{ff} = K_f \frac{\rho\, \bar{v}^2}{2}$$

$$\qquad (11.14)$$

$$h_{ff} = K_f \frac{\bar{v}^2}{2}$$

where K_f is an empirically determined parameter which is dependent on the type of fitting. A representative list of these parameters is given in Table 11-1 for different fittings.[#]

Figure 11-3 shows typical flow patterns for a fluid undergoing expansion and contraction while moving between two different sized pipes and for a fluid passing through an orifice. As can be seen from Figure 11-3a, when the fluid expands, it is unable to conform to the abrupt change in the pipe diameter and eddy currents form near the outer edge of the flange as a result of boundary layer separation. This causes a locally low pressure in this region which results in an energy loss. Similar to the expressions for fitting losses, the energy loss per unit mass of fluid due to expansion can be expressed as

$$h_{fe} = K_e \frac{\bar{v}_A^{\;2}}{2} \qquad (11.15)$$

[#]   Pressure losses in fittings can also be expressed in terms of "equivalent" pipe lengths.

Table 11-1 Energy Losses in Fittings

FITTING	K_f	FITTING	K_f
45 degree std elbow	0.35	Valve, Gate (open)	0.17
90 degree std elbow	0.75	Valve, Gate (1/2 open)	4.5
180 degree bend, close return	1.5	Valve, diaphragm (open)	2.3
Tee, std.	0.40	Valve, diaphragm (1/2 open)	4.3
Coupling	0.04	Valve, globe (open)	6.0
		Valve, globe (1/2 open)	9.5
		Valve, check (swing)	2.0
		Valve, check (disk)	10.0

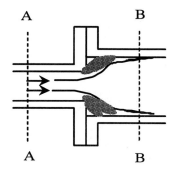

(a) Expansion

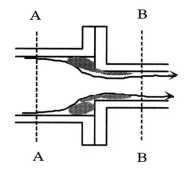

(b) Contraction

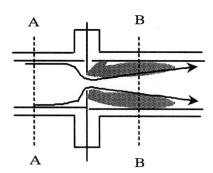

(c) Orifice: Contraction & Expansion

Figure 11-3 Expansion and Contraction Flow Patterns

For expansion, K_e can actually be derived from a combined momentum and energy balance with the result

$$K_e = \left[1 - \frac{(A_x)_A}{(A_x)_B} \right]^2 \tag{11.16}$$

For contraction losses,

$$h_{fc} = K_C \cdot \frac{\bar{v}_B^{\ 2}}{2} \tag{11.17}$$

In general, the parameter, K_C, is obtained from experimental data but it is usually less than 0.1 in laminar flow and, at higher Reynolds numbers it is approximately equal to

$$K_C = 0.4 \left[1 - \frac{(A_x)_B}{(A_x)_A} \right] \tag{11.18}$$

Expansion and contraction losses can also be exploited to measure volumetric flow rate. For example, an "orifice meter" consists of a plate with a hole in the center that is placed between flanges of two adjoining sections of pipe. The fluid contracts and then expands as it moves through the orifice (Figure 11-3c) and this results in a pressure drop across the orifice which can be measured. The pressure taps are located on either side of the orifice at upstream and downstream distances, points A and B, that are sufficiently removed from the eddy regions of the flow. The magnitude of the pressure drop can be related to the volumetric flow rate through the orifice by means of Equation (11.19)

$$\dot{V} = C_o \frac{\pi D_o^{\ 2}}{4\sqrt{1 - \gamma^4}} \left(\frac{2\Delta P}{\rho} \right)^{1/2} \tag{11.19}$$

where D_o is the orifice diameter, γ is the ratio of the orifice diameter to the pipe diameter, and ΔP is the pressure drop across the orifice. The coefficient, C_o, is usually determined by calibration although, with a carefully manufactured orifice plate and for Reynolds numbers greater than 30,000 (based on the orifice diameter), C_o is approximately equal to 0.61.

One disadvantage of orifice meters is the large, irreversible pressure losses across the orifice. However, the same principle can be exploited with only minimal pressure loss with the use of a "venturi meter." In this case, the meter consists of a section of pipe with both a smooth contraction and a smooth expansion. Because of the smoothness of the contraction and expansion, the irreversible pressure loss is much less with a venturi meter. However, in order to obtain a significant measurable pressure drop, the downstream pressure tap is placed at the "throat" of the meter; i.e., at the point of smallest diameter. The equation relating flow rate to pressure drop is the same as Equation (11.19) except that the coefficient, C_v, is usually about 0.98 for a well designed venturi.

11-1.c Optimum Pipe Diameter

Choosing a pipe size (diameter) is a tradeoff between pumping costs and the "fixed" costs of purchasing, installing, and maintaining the pipe. For example, the purchase cost of a pipe increases with pipe diameter but, because friction losses are greater for a smaller diameter pipe, pumping costs are also larger. An optimum pipe diameter can be arrived at by relating the pumping and fixed costs to pipe diameter and then minimizing the cost with respect to the pipe diameter. To do this, let's look at the energy loss per mass of flowing fluid due to skin friction losses in a pipe. From the Bernoulli equation, we have

$$\eta W_P = h_{fs} = \frac{2 f \bar{v}^2 L}{D} \tag{11.20}$$

In a smooth tube, the friction factor in turbulent flow is related to Reynolds number by Equation (11.12), which, when substituted into Equation (11-20), results in

$$\eta W_P = \frac{0.092 \bar{v}^2 L}{D \left(\dfrac{D \bar{v} \rho}{\mu} \right)^{0.20}}$$

This equation can be expressed in terms of pipe diameter by realizing that $\bar{v} = \dfrac{4 \dot{V}}{\pi D^2}$, so that

$$W_P \propto \frac{\mu^{.20} \dot{V}^{1.8} L}{\eta \rho^{.20} D^{4.8}} \tag{11.21}$$

The total pumping cost is obtained by multiplying Equation (11.21) by the mass flow rate of the fluid ($\rho \dot{V}$) and by the specific cost of power, K_{power}

$$Pumping\,Cost \propto W_P \propto K_{Power} \frac{\mu^{.20} \dot{V}^{2.8} \rho^{.80} L}{\eta D^{4.8}} \tag{11.22}$$

The fixed cost can be represented by (Ref [2])

$$Fixed\ Cost \propto K_{pipe}\ K_{install}\ D^n \qquad (11.23)$$

Where K_{pipe} is the cost per length of pipe and $K_{install}$ is the installation cost. The total cost is the sum of the pumping and fixed costs (Equations 11.22 and 23), or

$$Total\ Cost = A_1\ K_{Power}\ \frac{\mu^{.20}\ \dot{V}^{2.8}\ \rho^{.80}\ L}{\eta\ D^{4.8}} + A_2\ K_{pipe}\ K_{install}\ D^n \qquad (11.24)$$

where A_1 and A_2 are the numerical constants arising from Equations (11.22 and 23).

To minimize the total cost with respect to pipe diameter, we take the derivative of Equation (11.24) with respect to D, set it equal to zero and solve for D

$$D = \left[\frac{1}{\eta}\ \frac{A_1}{A_2}\ \frac{K_{Power}}{K_{pipe}\ K_{install}}\ \mu^{.20}\ \rho^{.80}\ \dot{V}^{2.8}\ \frac{L}{n} \right]^{\frac{1}{n + 4.8}} \qquad (11.25)$$

The value of n depends on pipe size; it is equal to 1.5 for $D \geq 1"$ and equal to 1.0 for $D < 1"$.

EXAMPLE 11-2: Pressure Drop in A Piping System

Referring to the piping system shown in Figure 11-1, calculate the pressure at a point in pipe # 3 which is located 2000 ft from its connection to pipe # 2 if the pipe is delivering 70 gpm of water at 20 °C. The pressure at the outlet of the pump is 250 psig, the valve is a globe valve (half-open), and the pipe diameters and lengths are as follows: #1 = 2"-schedule 40 with length 500 ft, #2 = 1-1/2"-schedule 40 with length 350 ft, #3 = 4"-schedule 40 with length 2000 ft.

Solution

With the pipe diameters expressed in inches, the velocity and Reynolds number in each pipe can be calculated from

$$\bar{v} = \frac{4\ \dot{V}}{\pi\ D^2} = \frac{4\,(70)\ \text{gal/min}\,(60)\ \text{min/hr}}{(3.14)(7.48)\ \text{gal/ft}^3} \frac{144\ \text{in}^2/\text{ft}^2}{D^2} = \frac{1.03 \times 10^5}{D^2}\ \text{ft/hr}$$

$$Re = \frac{D\,\bar{v}\,\rho}{\mu} = \frac{2.21 \times 10^5}{D}$$

The results of these calculations are given in the table, below.

Pipe	$\bar{v} \times 10^{-4}$ ft/hr	$Re \times 10^{-5}$	$h_{fs} \times 10^{-10}$ ft²/hr²	$h_{ff} \times 10^{-10}$ ft²/hr²	$h_{fc} \times 10^{-10}$ ft²/hr²	$h_{fe} \times 10^{-10}$ ft²/hr²
1	2.41	1.07	1.53	-	0.013	-
2	4.02	1.38	3.66	0.06	-	0.013
3	0.62	0.54	2.38	.06 + .09	-	-

The friction energy losses in the pipe are due to skin friction in each of the three pipes, two 90° elbows, a valve in pipe #3 and contraction between pipes #1 and 2 and expansion between pipes #2 and #3. The skin friction losses are computed from Equation (11.11) and since the Reynolds numbers are all greater than 50,000, (see Table), the friction factor is calculated from Equation (11.12)

$$h_{fs} = \frac{2\,\bar{v}^2\,L\,f}{D} = 1.17 \times 10^{10}\, \frac{L}{D^5 Re^{.20}}$$

The fitting losses are calculated from Equation (11.14) and the coefficients listed in Table 11-1. The losses in each elbow are computed from the velocities in the smaller tubes and are, thus, identical

$$h_{ff} = 2\,(0.75)\,\frac{\bar{v}_2^{\,2}}{2} + (9.5)\,\frac{\bar{v}_3^{\,2}}{2} = 0.75\,(4.02 \times 10^4)^2 + 4.75\,(6.2 \times 10^6)^2$$

$$h_{ff} = 1.39 \times 10^9 \text{ ft}^2/\text{hr}^2$$

The expansion and contraction losses are calculated from Equations (11.15) and (11.17) and the two coefficients are calculated from (11.16) and (11.18)

$$K_e = \left[1 - \left(\frac{1.6}{2.067} \right)^2 \right]^2 = 0.16$$

$$K_C = 0.4 \left[1 - \left(\frac{1.6}{2.067} \right)^2 \right] = 0.16$$

and so, the losses are also identical.

$$h_{fc} = h_{fe} = 0.16 \frac{(4.02 \times 10^4)^2}{2} = 1.29 \times 10^8 \ ft^2/hr^2$$

Summing all these losses, $h_f = 5.6 \times 10^{10} \ ft^2/hr^2$

The pressure drop between the pump outlet and the desired point in pipe #3 can be computed by applying Equation (11.3)

$$\Delta P = -62.4 \left[\frac{(6.2 \times 10^3)^2 - (2.41 \times 10^4)^2}{2} + (32.2)(3600)^2 (350) + 5.6 \times 10^{10} \right]$$

$$= -1.26 \times 10^{13} \ lb_m/ft\text{-}hr^2$$

and the pressure in pipe #3 is

$$P = 250 - 1.26 \times 10^{13} \ lb_m/ft\text{-}hr^2 \ \frac{hr^2}{(3600)^2 \ s^2} \ 2.1584 \times 10^{-4} \ \frac{psia}{lb_m/ft\text{-}s^2}$$

$$= 41 \ psig$$

11-2 FLOW IN PACKED BEDS

A subject of considerable interest to engineers is flow "external" to solid objects. An understanding of momentum transport when solids move through fluids is important for analyzing flow over air foils, the pneumatic transport and settling of solids and for chemical processing in packed bed catalytic reactors and separators. To illustrate macroscopic momentum calculations in external flows, only flow in packed and fluidized beds will be discussed here. Figure 11-4 is a sketch of flow in a packed bed. As the fluid moves between the individual packed particles, it experiences friction at the surface of the packing as well as numerous changes in direction. Both phenomena involve momentum transport; friction at the solid surfaces due to molecular transport and convective momentum transport resulting from changes in flow direction. Both momentum transport mechanisms result in a "drag" force which manifests itself as a pressure drop across the bed. If the particles are the appropriate size for the diameter of the bed and are packed properly, the fluid, on the average, will flow in a uniform manner through the bed.

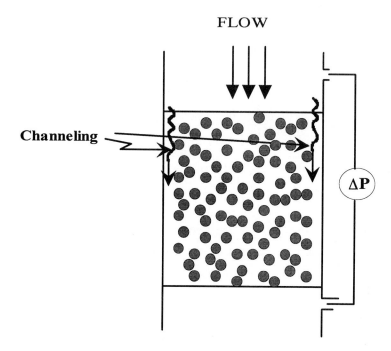

Figure 11-4 Flow in a Packed Bed

However, if the particle diameter is too large for the bed ($d_p / D > 10$) then the void fraction near the walls will be higher than in the center of the bed and the fluid will flow preferentially through these "channels" near the wall ("channeling") and fluid-solid contacting will be less efficient.

11-2.a Pressure Losses: Single-Phase Flow

The pressure drop across the bed is an important parameter in packed bed operations; not only because it is an energy loss but also because it is an indirect measure of the fluid–solid contacting. To predict the pressure drop in such a system, both momentum transport mechanisms must be taken into account and then correlated with experimental pressure drop measurements. The form of the correlation equation is derived as follows. First of all, it is assumed that the bed is packed uniformly; that is, the voids are distributed evenly in all directions. If the packing consists of particles where one dimension is significantly larger than another (cylinders with $L/d > 1$, for example), it is possible that the void cross section could be smaller than the void volume. For this reason, care is usually taken to "dump" the packing into the vessel in a careful, prescribed manner. Next, it is assumed that the flow through the bed can be described as flow through parallel channels with a radius equivalent to the hydraulic radius,

R_H. With these assumptions, the molecular transport contribution to the momentum flux is taken to be

$$\tau \propto \frac{dv}{dr} = C_1 \frac{\mu \bar{v}}{R_H}$$

The total drag force per unit surface area of the packing is assumed to be a linear combination of the molecular momentum flux and the convective flux, $(\rho v) v$, or

$$\frac{F_D}{A_{SP}} = C_1 \frac{\mu \bar{v}}{R_H} + C_2 \rho \bar{v}^2 \qquad (11.26)$$

where F_D is the drag force and A_{SP} is the total surface area of the packing. In a vessel of volume, V, the void fraction of the bed is defined as

$$\varepsilon = \frac{V - V_{packing}}{V_{Bed}}$$

In a uniform bed, the cross-sectional area available for flow is the product of the vessel cross-sectional area and the void fraction. Therefore the fluid velocity in the voids (channels) is equal to the superficial velocity divided by the void fraction, or

$$\bar{v} = \frac{v_0}{\varepsilon} \qquad (11.27)$$

The hydraulic radius is defined as the fluid-filled cross section divided by the wetted perimeter of the channel and if the numerator and denominator are multiplied by the length of the bed, L,

$$R_H = \frac{A_x \varepsilon}{\wp_w} \frac{L}{L} = \frac{V \varepsilon}{A_{Sp}} \qquad (11.28)$$

If the bed contains N_P particles with a surface area, A_P, then

$$A_{Sp} = N_P \, A_P$$

and

$$N_P = \frac{Volume \; Particles}{Volume \; per \; Particle} = \frac{V \, (1 - \varepsilon)}{V_P}$$

Therefore

$$A_{SP} = \frac{V(1-\varepsilon)}{V_P} A_P = \frac{A_x L(1-\varepsilon)}{V_P} A_P \tag{11.29}$$

Substituting Equation (11.29) for A_{SP} into (11.28), the hydraulic radius is

$$R_H = \frac{\varepsilon}{(1-\varepsilon)} \frac{V_P}{A_P} \tag{11.30}$$

The total drag force is just the pressure drop multiplied by the void cross-sectional area, so that

$$\frac{F_D}{A_{SP}} = \frac{\Delta P (A_x \varepsilon)}{A_{SP}} \tag{11.31}$$

Using Equations (11.27), (11.30), and (11.31), Equation (11.26) becomes

$$\frac{\Delta P (A_x \varepsilon)}{A_x L (1-\varepsilon)} \frac{V_P}{A_P} = C_1 \frac{\mu v_0 (1-\varepsilon)}{\varepsilon^2} \frac{A_P}{V_P} + C_2 \frac{\rho v_0^2}{\varepsilon^2} \tag{11.32}$$

For a sphere, $\dfrac{V_P}{A_P} = \dfrac{d_p}{6}$, and Equation (11.32) can be placed in dimensionless form

$$\frac{\Delta P}{\rho v_0^2} \frac{d_p}{L} = 36 C_1 \frac{(1-\varepsilon)^2}{\varepsilon^3 Re_P} + 6 C_2 \frac{(1-\varepsilon)}{\varepsilon^3}$$

where Re_P is defined in terms of the particle diameter and the bed superficial velocity. With the experimentally determined coefficients, $36 C_1 = 150$ and $6 C_2 = 1.75$, this equation is known as the <u>Ergun</u> equation

$$\frac{\Delta P}{\rho v_0^2} \frac{d_p}{L} = 150 \frac{(1-\varepsilon)^2}{\varepsilon^3 Re_P} + 1.75 \frac{(1-\varepsilon)}{\varepsilon^3} \tag{11.33}$$

This equation can also be used for other shaped particles by defining an equivalent spherical diameter,

$$d_{eq} = \frac{6 V_P}{A_P} \tag{11.34}$$

The basis for Equation (11.34) is that the effective particle diameter be a sphere which has the same ratio of the volume of the particle to its surface area. Since that ratio for a sphere is $6/d_p$, Equation (11.34) results. This is sometimes referred to as the <u>sphericity</u> of the particle.

Two properties of the Ergun equation should be noted. First of all, when the Reynolds number is very high, the last term on the right-hand side of the equation is dominant and the first term is dominant at low Reynolds numbers. Second, the pressure drop is a sensitive inverse function of the bed void fraction, as pointed out earlier. A 20% error in the void fraction can result in a 50% or more error in the pressure drop prediction.

EXAMPLE 11-3: Pressure Drop in a Packed Bed Reactor

A "methanator" is a catalytic reactor designed to convert residual CO concentrations in a hydrogen plant to methane. The catalytic reactor is to be a packed bed (void fraction = 0.45) consisting of catalyst "tablets" (1/4" by 1/4") with a specific gravity of 1.55. In order to achieve the desired CO conversion, the reactor operates at a space velocity (volumetric flow rate at standard conditions / volume of catalyst) of 10,000 hr^{-1}. If the reactor diameter is chosen to be 3', what will be the pressure drop across the bed if the feed gas is fed to the reactor at 300 $^\circ$C and 0.4 MPa at a flow rate 10,000 SCFM. For calculational purposes, the gas can be assumed to have the properties of CO_2.

Solution

Length of bed:

$$10^4 \text{ hr}^{-1} = \frac{10^4 \text{ scfm}}{V_{bed}} = \frac{10^4}{\dfrac{\pi(3^2) \text{ ft}^2}{4} L}$$

$$L = 8.5 \text{ ft}$$

Equivalent diameter of tablet: $d_{eq} = \dfrac{6[\frac{\pi}{4}(1/2)^2]1/4}{\pi(1/2)(1/4) + 2\frac{\pi}{4}(1/2)(1/2)} = 0.375"(0.0313 \text{ ft})$

Superficial velocity: $v_0 = \dfrac{10^4 \text{ scfm}(0.1/0.4)(573/273)}{\frac{\pi}{4}(3^2) \text{ ft}^2 \, 60 \text{ s/min}} = 12.4 \text{ ft/s}$

viscosity (CO_2 at 300 C) $= 0.026$ cp $(6.72 \times 10^{-4} \text{ lb}_m /\text{ft-s}^2 / \text{cp}) = 1.75 \times 10^{-5} \text{ lb}_m /\text{ft-s}^2$

density $= \dfrac{(4)(44)}{(0.73)(573)(1.8)} = 0.234 \; lb_m / ft^3$

$$Re_p = \frac{(0.375/12)(12.4)(0.234)}{1.75 \text{ x } 10^{-5}} = 5163$$

pressure drop:

$$\frac{\Delta P}{\rho v_0^2} \frac{d_{eq}}{L} = 150 \frac{(1-\varepsilon)^2}{\varepsilon^3 Re_p} + 1.75 \frac{(1-\varepsilon)}{\varepsilon^3}$$

$$\frac{\Delta P}{\rho v_0^2} \frac{d_{eq}}{L} = \frac{(150)(1-.45)^2}{(.45)^3 (5163)} + \frac{(1.75)(1-.45)}{(.45)^3}$$

$$= 0.096 + 10.56 = 10.66$$

$$\Delta P = (10.66)\,(0.23\ \text{lb}_m/\text{ft}^3)\,(12.4)^2\ \text{ft}^2/s^2\, \frac{8.5}{0.375/12} = 2.56 \text{ x } 10^4\ \frac{\text{lb}_m\text{-ft}}{s^2}$$

or, using the conversion factors from Appendix E

$$\Delta P = (2.56 \text{ x } 10^4)\,(2.1584 \text{ x } 10^{-4}) = 5.52\ \text{psi}$$

11-2.b Pressure Losses: Countercurrent Two-Phase Flow

There are many applications of countercurrent gas-liquid flows in packed beds, usually involving mass transfer operations such as distillation or gas absorption/stripping. Here we will emphasize gas absorption since it is specifically covered in Chapter 13 as a practical mass transfer application of transport phenomena. The best way to approach this subject is to start with a fixed gas flow over a packed bed and visualize what happens as liquid is applied at increasing flow rates in a direction opposite to the gas flow. If the liquid is well distributed (this may require periodic redistribution in large columns with low height-to-diameter ratios), it will efficiently wet the packing in the form of a thin film. This is desirable in gas absorption because it maximizes the interfacial area between gas and liquid, resulting in higher mass transport rates. The consequence here is that the void cross section for gas flow is reduced, causing an increase in pressure loss. Figure 11-5 illustrates this in terms of the pressure loss per unit length of packing as we move from point a to point b on the figure. If we now hold the liquid flow steady and increase the gas flow (b → c), the pressure drop increases linearly with approximately the same dependence on gas flow rate as exists in a dry packing; i.e., with a slope on the log-log scale of 1.8 - 2.0[#]. As the gas flow rate is further increased, point c is reached where the liquid becomes locally entrained within the void cross section, causing the liquid to completely fill

---

[#] This is the predicted dependence of the Ergun Equation at high flow rates, $\dfrac{\Delta P}{L} \propto v^2$.

the cross section at those locations. When this occurs, the pressure drop dependence on gas flow is much higher, quickly approaching a vertical line on the logarithmic plot. At this inflection point, the column is said to be "flooded." Note that the same phenomena will occur if the gas flow is maintained constant and the liquid flow is increased (b → d in Figure 11-5).

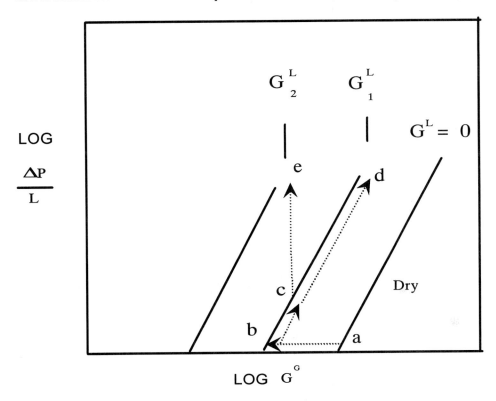

Figure 11-5 Pressure Drop in Gas-Liquid Packed Beds

Predictions of the pressure drop in these flows is rather limited and thus it is necessary to resort to empirical correlations of the available data. Much of the data has been collected and correlated by commercial vendors of column packings (e.g., the Norton Co.) but recently, Robbins [3] published a generalized correlation of pressure drop in these flows by extending equations used in simpler gas flow over packings ("dry" beds). Criticizing the fact that most correlations contain corrections for liquid flow even when it is not justified, he started with Leva's [4] modification of the effect of liquid flow on the predicted pressure drop over dry packings

$$\Delta P = C_0 \, \rho^G (v^G)^2 \; [10^{C_1 v^L}] \tag{11.35}$$

where C_0 and C_1 are empirically determined constants. This particular equation was chosen because it gives the correct dependency for dry packings when v^L is zero.

Robbins found that the parameters C_0 and C_1 could be correlated with the "packing factor," F_P, which had been traditionally employed to characterize pressure losses over packed beds and he developed the generalized pressure drop correlation given in Figure 11-6. To use this correlation, he defined the gas and liquid loading factors as

$$G_{G_f} = G_G \left(\frac{.075}{\rho_G} \right)^{\frac{1}{2}} \left(\frac{F_P}{20} \right)^{\frac{1}{2}} \tag{11.36}$$

and

$$G_{L_f} = G_L \left(\frac{62.4}{\rho_L} \right) \left(\frac{F_P}{20} \right)^{\frac{1}{2}} (\mu_L)^{.10} \tag{11.37}$$

where G_L and G_G have units of lbm/ft²-hr, the densities have units of lbm/ft³, and the liquid viscosity is in centipoise. In addition to these loading factors, it is also necessary to have values for F_{PD}. Robbins presented data for more than 90 commercial packings, and representative values are given in Table 11-3.

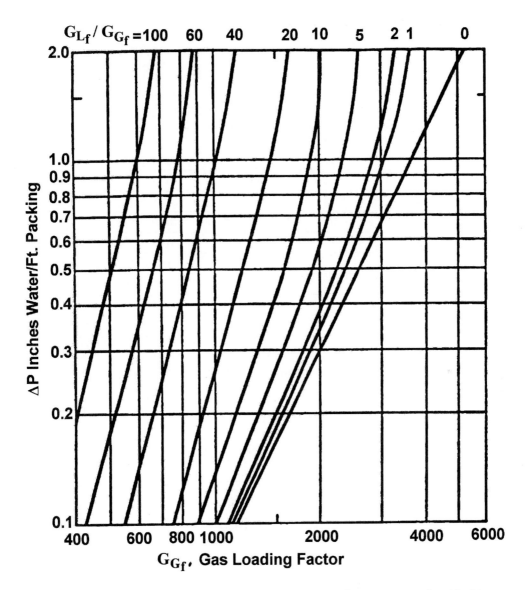

Figure 11-6 Generalized Pressure Drop Correlation — Countercurrent Gas-Liquid
Flow in Packed Columns, from Robbins, L.A.,Chem. Engr. Prog., <u>87</u> (5),
p. 87, 1991.6. Reproduced with permission of the American Institute of
Chemical Engineers. Copyright © 1991 AIChE. All rights reserved.

Table 11-3 Characteristics for Some Typical Packings

Packing	Size (d_p - in)	Wall Thick (in)	Sp. Surf Area (a_p - ft^{-1})	Bed Void Fraction (ε)	Pack Factor[*] (F_p)
Raschig	1/2	3/32	112	0.62	580 (c)
Rings	3/4	3/32	74	0.72	255 (c)
	1	1/8	58	0.74	155 (c)
	1 1/2	3/16	37	0.73	95 (c)
	2	1/4	28	0.74	65 (c)
	3	3/8	19	0.75	37 (c)
	4	3/8	14	0.80	
Pall Rings	5/8		104	0.87	97(p), 70(m)
	1		63	0.94	52(p), 48(m)
	1 1/2		39	0.95	32(p), 28(m)
	2		31	0.96	25(p), 20(m)
Berl	1/2		142	0.62	240 (c)
Saddles	3/4		87	0.66	170 (c)
	1		76	0.68	110 (c)
	2		32	0.72	45 (c)
Intalox	1/2		190	0.78	200 (c)
Saddles	3/4		102	0.77	145 (c)
	1		78	0.77	98(c), 33(p)
	2		36	0.79	40(c), 21(p)

[*] (c) = ceramic,   (m) = metal,   (p) = plastic

EXAMPLE 11-4: Pressure Drop in A Gas Absorber

An air stream containing 12% SO_2 is to be delivered at 800 scfm to a 4'-diameter column packed with 3/4"-ceramic Berl Saddles where it is to be contacted with 350 gpm of water. If the air stream is at 30 C, and atmospheric pressure, determine the pressure drop in the column (inches H_2O /ft of packing).

<u>Solution</u>

The gas and liquid mass velocities in the column are:

$$G_G = \frac{(800)\,\dfrac{\text{scf}}{\text{min}}\,(1/359)\,\dfrac{\text{lb-mole}}{\text{scf}}\,(29)\,\dfrac{\text{lb}_m}{\text{lb-mole}}\,(60)\,\dfrac{\text{min}}{\text{hr}}}{\dfrac{\pi(4)^2}{4}\,\text{ft}^2} = 309\,\dfrac{\text{lb}_m}{\text{ft}^2\text{-hr}}$$

$$G_L = \frac{(350)\,\dfrac{\text{gal}}{\text{min}}\,(8.34)\,\dfrac{\text{lb}_m}{\text{gal}}\,(60)\,\dfrac{\text{min}}{\text{hr}}}{\dfrac{\pi(4)^2}{4}\,\text{ft}^2} = 13{,}944\,\dfrac{\text{lb}_m}{\text{ft}^2\text{-hr}}$$

From Table 11-3, the packing factor, F_p, is 170 and the density of the gas is

$$\rho_G = \frac{PM}{R_g T} = \frac{(1)(29)}{(0.73)(303)(1.8)} = 0.073\,\frac{\text{lb}_m}{\text{ft}^3}$$

The loading factors to be used with Figure 11-6 are:

$$G_{G_f} = 309\left(\frac{.075}{.073}\right)^{1/2}\left(\frac{170}{20}\right)^{1/2} = 913\ \text{lb}_m/\text{ft}^2\text{-hr}$$

$$G_{L_f} = 13{,}955\left(\frac{62.4}{62.4}\right)\left(\frac{170}{20}\right)^{1/2}(1.0)^{0.20} = 40{,}685\ \text{lb}_m/\text{ft}^2\text{-hr}$$

Thus, $\dfrac{G_{L_f}}{G_{G_f}} = 44.5$ and, from Figure 11-6, $\Delta P = 0.7$ inches water / ft of packing.

11-3 FLOW IN FLUIDIZED BEDS

Still another application of flows external to solid objects is "fluidization." In these situations, the fluid is passed upwards over the solids at a high-enough velocity to actually suspend the solid particles and the two-phase mixture actually behaves very much like a liquid. Because these "fluidized beds" have many industrial applications (catalytic reactors, solids drying, coal gasification/combustion), it is appropriate to introduce at least some of the more elementary aspects of fluidization. For a more complete discussion of the subject, see the text by Kunii and Levenspiel [5].

A good way to visualize fluidization phenomena is to start with a packed bed and record the pressure drop over the bed as the fluid velocity is slowly raised. This is shown in Figure 11-7. At low superficial velocities, the fluid passes upwards through the voids in the bed and the Ergun equation provides a good description of the pressure drop across the bed. This continues until the superficial velocity reaches the point where the product of the vessel cross sectional area and the pressure drop is exactly equal to the gravitational force exerted on the mass of particles in the bed. At this point the incoming fluid is able to "lift" the solids, resulting in a marked expansion in the bed volume and the solid particles become individually suspended. This point is characterized by a local maximum in the pressure drop and is termed the point of *minimum fluidization*. The superficial velocity of the fluid at this point is called the minimum fluidizing velocity, v_{mf} , and this value is often used to characterize fluidized beds. As the fluid velocity is increased further beyond this point the pressure drop has only a slight dependency on velocity. Visual studies of a fluidized bed in this region of operation show that much of the gas travels through the bed in the form of "bubbles," as shown in Figure 11-8.

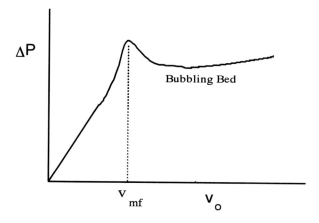

Figure 11-7 Minimum Fluidization

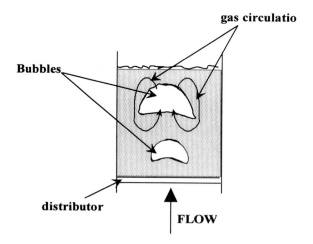

Figure 11-8 Bubbling Fluidized Bed

These are not true bubbles as one would observe in a gas-liquid system, although they do have very similar shapes. Rather, they are local void volumes which have a lower concentration of solid particles and this results in high gas re-circulation rates through the voids. It is these circulation patterns which provide the high degree of mixing associated with fluidized beds. While efficient mixing is desirable, these bubbles grow as they move through the bed and the bubble velocity is dependent on the size of the bubbles. This means that much of the gas spends less time in the bed and if the bed is a reactor, this means that a large fraction of the gas has a lower residence time in the bed, leading to lower conversions. This region of fluidized bed operations is often referred to as a "bubbling bed." It should be pointed out that this is not usually a problem in a liquid phase fluidized bed. If the fluid velocity is increased still further (not shown in Figure 11-8), the bed can become unstable, characterized by large and rapid fluctuations in the pressure drop. This region of operation is called the "slugging bed" region. At still higher velocities, the fluid is able to physically remove individual particles, and this is the region where "pneumatic transport" of solids takes place. Since the solid concentrations are usually low in this region, methods have also been developed to feed solids at a high rate, resulting in higher concentrations. When this is done, the operation is termed a "fast" fluidized bed [6].

Because fluidized beds are characterized by the minimum fluidizing velocity, it is useful to derive predictive equations for v_{mf} . At minimum fluidizing conditions the total drag force on the particles will be balanced by the total weight of the particles, or

$$\Delta P \, A_x = (\rho_s - \rho_f) \, A_x \, L_{mf} \, (1 - \varepsilon_{mf}) g \qquad (11.38)$$

where L_{mf} is the length of the bed at minimum fluidizing conditions. The methodology is to utilize a suitable prediction of the pressure drop as a function of velocity in a packed bed and then extrapolate it to minimum fluidizing conditions. Substitution of that pressure drop into Equation (11.38) then allows for the prediction of the minimum fluidizing velocity. A suitable vehicle here is the Ergun equation evaluated at minimum fluidizing conditions. Therefore, Equation (11.33) is written as

$$\frac{\Delta P}{\rho_f \, v_{mf}^2} \frac{d_p}{L_{mf}} = 150 \, \frac{(1 - \varepsilon_{mf})^2}{\varepsilon_{mf}^3 \, Re_{P_{mf}}} + 1.75 \, \frac{(1 - \varepsilon_{mf})}{\varepsilon_{mf}^3} \qquad (11.39)$$

If Equation (11.39) is solved for ΔP and substituted into Equation (11.38), the minimum fluidizing velocity must be solved in a quadratic equation. Alternatively, explicit equations for v_{mf} can be obtained by simplifying (11.39) at low and high Reynolds numbers. At low Reynolds numbers ($Re_{P_{mf}} < 20$), the first term on the right-hand side of (11.39) is dominant and when it is combined with Equation (11.38) the minimum fluidizing velocity can be calculated from

$$v_{mf} = \frac{d_P^2}{150} \frac{(\rho_s - \rho_f)}{\mu} \frac{\varepsilon_{mf}^3}{(1 - \varepsilon_{mf})} g \tag{11.40}$$

When the Reynolds number is high ($Re_{P_{mf}} > 100$), the first term on the right hand side of Equation (11.39) is negligible and the minimum fluidizing velocity can be calculated from

$$v_{mf} = \left[\frac{d_P(\rho_s - \rho_f)\varepsilon_{mf}^3 g}{1.75 \rho_f} \right]^{1/2} \tag{11.41}$$

Equations (11.40) and (11.41) are for solid beds with a uniform particle size. In many applications, there is a particle size distribution and the particles may not be spherical. Non-spherical particles can be handled by using an equivalent diameter as defined in Equation (11.34). To apply these equations to mixtures of particles, an average of the particle sizes must be employed. There are a number of methods for calculating average diameters in mixtures but for packed and fluidized beds, a "surface area weighted average" is most appropriate.[#] Kunii and Levenspiel (Ref [5]) discuss this in detail.

EXAMPLE 11-5: Calculation of Minimum Fluidizing Velocity

Calculate the diameter of a fluidized bed reactor which is to operate at a superficial velocity which is 3 times the minimum. The bed consists of 120 μ m spherical zeolite particles ($\rho = 2.1$ g/cm^3) and the hydrocarbon feed to the reactor is 500 lb-moles/hr and has the properties of gaseous cyclohexane at 150 C and 1 atm pressure. The bed void fraction at minimum fluidizing conditions is 0.65.

Solution

Since the particles are small, assume that the Reynolds number will be small and Equation (11.40) can be used. The molecular weight of cyclohexane is 84 and its viscosity at 150C is 0.0097 cp [7]). The density of the gas can be calculated from the ideal gas law and is 0.0024 g/cm^3. Therefore the minimum fluidizing velocity is predicted to be

$$v_{min} = \frac{(0.012)^2}{150} \text{ cm}^2 \left[\frac{2.1 - 0.0024 \text{ g/ cm}^3}{9.7 \times 10^{-5} \text{ g/cm-s}} \right] \frac{(0.65)^3}{(1 - 0.65)} 980 = 16 \text{ cm/s}$$

[#]   The size and shape of the "effective" particle is defined so that the surface area/mass of the particle is unchanged.

At this point a check of the Reynolds number, $\dfrac{d_P \, v_{mf} \, \rho}{\mu}$, yields a value of 3, well below the criteria of 20 needed to use equation (11.40).

The reactor diameter can now be calculated from

$$3v_{mf} \frac{\pi D^2}{4} = \dot{V} = (500) \text{ lb-moles/hr} \, (359) \text{ scf} \; / \; \text{lb-mole} \left(\frac{423}{273} \right) \frac{(1) \text{ hr}}{3600 \text{ s}}$$

which results in a diameter of 8 ft.

REFERENCES

[1] *Chemical Engineers Handbook,*, R.H. Perry and C.H. Chilton, 5th ed., Mcgraw-Hill, NY, 1973, p. 5-21

[2] Peters. M.S. and K.D. Timmerhaus, "Plant Design and Economics for Chemical Engineers", McGraw-Hill Co., N.Y., 1968

[3] Robbins, L.A., "Chem. Engr. Progr.", p. 87, May, 1991

[4] Leva, M., "CEP Symp. Ser.", 50, p. 51 (1954)

[5] Kunii, D. and O. Levenspiel, *Fluidization Engineering*, R.E. Krieger, N.Y., 1977

[6] Yerushalmi, J. and A.M. Squires, "AIChE Symp. Ser., #161, 73, p. 44, (1977)

[7] *Chemical Engineers Handbook,*, R.H. Perry and C.H. Chilton, 5th ed., Mcgraw-Hill Co., N.Y., 1973, p. 3-211

PROBLEMS

11-1   Use the continuity equation, Equation (8.2), to show how Equation (8.17) leads to Equation (11.1) for fully developed, steady state flow.

11-2   Show that the parameter, β , as defined by Equation (11.7) is equal to 2 for fully developed laminar flow in a tube.

11-3 Water at 20 °C is flowing in a horizontal run of smooth, 2"-schedule 40 steel pipe. The pipe has two pressure taps located 30-ft apart and the differential pressure between the two taps is 0.5 inches of H₂O. What is the average velocity in the pipe?

11-4 A large <u>elevated</u> tank (open to the atmosphere) is supplying water to a spray chamber. The pressure at the inlet of the spray nozzle must be 40 psig and the flow rate is to be 150 gpm. The system consists of a 2"-schedule 40 galvanized pipe with a 10-ft horizontal run, four 90°-elbows and an open gate valve. If the level in the tank is maintained at 10 ft, what should be the vertical distance between the bottom of the tank (the tank outlet) and the nozzle?

11-5

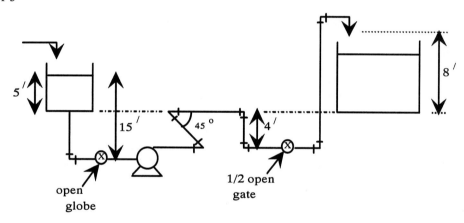

The piping system shown above is delivering a newtonian liquid at 30,000 lb/hr. Calculate the additional pump capacity needed to double the flow rate. The piping system has the following characteristics:

Liquid: μ = 2.3 cp, ρ = 1.37 g/cm³,   Pipe: 1"-schedule 40 steel, actual length = 47 ft

Pump: 0.5 hp motor (motor-pump efficiency = 60%)

11-6

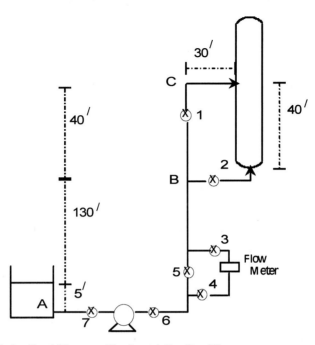

Pipe Length A – B = 150'   Pipe Length B – C = 40'

Equivalent Lengths: Tees = 1.6', Elbows = 2.4', Globe valves = 12', Open gate valves = 0.6', Flow meter = 20'

A mixture of 10% toluene-in-benzene (ρ = 56 lb$_m$/ft^3) is to be pumped from a vented tank to a distillation column which is at 3 atm pressure. A pump which is 80% efficient is equipped with a 1.5 hp motor. The flow rate is to be 6600 lb$_m$/hr and the mixture can be assumed to have the properties of benzene. Valves 1–4 are globe valves, Valve 5 is closed and Valves 6 & 7 are open gate valves. The pressure losses through fittings are expressed in terms of equivalent lengths of straight pipe as shown above. The temperature in the tank is 60 ^{0}F and kinetic energy contributions to the energy balance can be neglected.

(a) If all the fluid is delivered to the middle of the column (Valve 2 closed), will 1"-schedule 40 pipe be adequate?

(b) Would this pipe size and pump be adequate if the flow rate were doubled and 50% of the flow entered the bottom of the column? If not, stipulate the hp of the motor required.

11-7 Taylor (*Int. J. H.t & Mass Trnsf.r*, 10, p. 1123–1128, 1967) claim that the typical friction factor – Reynolds number correlation for flow in pipes should be modified for <u>gases</u> whenever the pipe wall temperatures are much higher than the bulk fluid temperatures. Specifically, they recommend the following relationship for Re > 3000

$$\frac{f}{2} = \left[0.007 + \frac{0.0625}{Re_s^{0.32}} \right] \left(\frac{T_b}{T_s} \right)^{0.5}$$

where Re_s refers to the Reynolds number evaluated at the wall surface temperature.

Use this expression to calculate the friction factor for air, at a temperature of 200°C, flowing at atmospheric pressure and 400 SCFM in a heated 4" schedule 40 pipe which is at 1000 °C. Then compare this calculation with the predictions of Equations (11.12) and (11.13) with the Reynolds number evaluated at a temperature equal to the average temperature of the bulk gas and the wall (the so called <u>mean film temperature</u>). What is the percent difference in the pressure drop per foot of pipe length between the "Taylor" method and the other two methods?

11-8 Adams (*Chem. Engr. Progress*, December 1997, p. 55-58) has proposed a "quick" method for initial pipe sizing for turbulent liquid flows, which he calls "Jack's Cube" (as you might suspect, his name is Jack). His equations are:

$$\dot{V} = 1.2\,(D+2)^3 \quad for \ \ D > 2"$$

$$\dot{V} = \frac{D}{2}\,(D+2)^2 \quad for \ \ D < 2"$$

where $\dot{V}$ is in gal/min and the pipe diameter, D, is in inches. On the other hand, Peter's & Timmerhaus (*Plant Design and Economics for Chemical Engineers*, McGraw-Hill, N.Y., 1968) state that a good rule of thumb for <u>reasonable</u> velocities is 3 – 10 ft/s. Compare the pressure drops per length of pipe that are predicted by these methods to deliver water at 25 °C and 500 gal/min.

11-9 An orifice meter with a diameter of 1.75" is installed in a 4"-schedule 40 pipe which delivers a distillate product to storage. The pressure drop across the orifice is 7 psi. Assuming the orifice coefficient is 0.61, what is the flow rate of the distillate. The specific gravity of the fluid is 0.88.

11-10 A packed bed catalytic reactor utilizes 3/8" spherical catalyst particles to convert a 2:1 (molar) mixture of H_2 and CO to methanol. Because of catalyst pore diffusion limitations (see Chapter 4), it is proposed to reduce the catalyst size to 4 mm. Downstream operations require that the pressure drop over the bed be no greater than twice its current value. The superficial velocity is 0.2 ft/s at the reactor conditions of 200 atm and 275 °C, and it can be assumed that the packing void fraction will be the same. For calculation purposes, assume that the properties of the gas are those of carbon monoxide.

(a) Will the proposed catalyst size reduction meet the pressure drop restrictions?

(b) Another option is to load less catalyst in the reactor and reduce the flow rate so as to keep the residence time the same. What will be the percent reduction in the catalyst loading in order to give a pressure drop twice its current value?

11-11 A fixed-bed catalytic reactor is being used to produce methane from CO and H_2. Because of the exothermic nature of the reaction, a 7:1 recycle ratio (volume recycle/volume fresh feed) is being used. The recycle gas is essentially methane, the catalyst size is 1/8"-cylinders ($L/d_p = 5$), the methane production rate is 50×10^6 SCF/Day and the reactor has a diameter of 8' and a height of 16'.

(a) If the reactor is isothermal (400 C) and the inlet pressure is 1000 psia, what would be the annual operating cost of the recycle compressor. The operating cost can be estimated at 2.5×10^{-8} $ per ft-lb$_f$ / hr of the delivered gas.

(b) What would be the annual operating cost savings if the recycle ratio were cut to 4:1?

11-12 Calculate the pressure drop in a counter current ammonia absorber operating at 10 psig, 50 °C, with liquid and gas mass velocities of 15,000 and 500 lb$_m$/ft^2-hr, respectively. The packing consists of 3/4"-intalox saddles.

11-13 What is the minimum fluidizing velocity for 60-μm zeolite particles ($\rho = 1.2$ g/cm^3) in a hydrocarbon mixture which has the properties of gaseous cumene? The bed-void fraction at minimum fluidizing conditions is 0.65.

11-14 In many fluidized bed applications, the efficiency of gas-solid contacting is related to the so called "bubble diameter." As pointed out by Kunii and Levenspiel [*Fluidization Engineering*, R.E. Krueger Pub. Co., 1977, p.130], the bubble velocity can be calculated from

$$u_B = v_0 - v_{mf} + u_{Br}$$

where u_{Br} is the bubble rise velocity and is related to the bubble diameter by

$$u_{Br} = .711(g_z d_B)^{1/2}$$

It is proposed to adsorb CS_2 from an air stream by contacting the stream with activated carbon particles ($\rho = 55$ lbm/ft^3) in a fluidized bed unit. The air flow rate is 420,000 ft^3/min at atmospheric pressure and 70 F and an existing unit is available with a vessel diameter of 48'. For efficient contact it is necessary that $\dfrac{u_B}{v_{mf}} < 10$ and that $d_B < 13$ cm. Calculate the activated carbon particle size which will meet this criterion. Assume that the bed-void fraction at minimum fluidizing conditions is 0.45.

CHAPTER 12

Macroscopic Calculations: Energy Transport

12-1 OVERALL HEAT TRANSFER COEFFICIENTS

The concept of transfer coefficients was introduced in Chapter 10 where it was pointed out that they are employed as a semi-empirical approach to transport problems which are too complex to solve mathematically. Because heat exchange equipment is almost always complex and typically involves indirect contact between two fluids, one hot and one cold, heat exchangers are designed and analyzed in terms of another transfer coefficient – the overall heat transfer coefficient. While this leads to more simplified algebraic equations, the disadvantage is that these coefficients combine a number of energy transport steps and thus direct experimental correlations are not available or even appropriate.

The concept behind overall heat transfer coefficients is quite similar to the "two-film theory" in mass transport (see Section 10-3b) except that, in place of a fluid–fluid interface, there is usually a third phase separating the two fluids - the solid wall of the heat exchange equipment. Figure 12-1 shows the situation in a typical heat exchanger where heat is exchanged between a hot fluid and a cold fluid which are, in turn, separated from one another by the walls of a heat exchanger tube. Keep in mind that this sketch only shows one such contact point in the exchanger. In commercial heat exchangers, it is common to employ many tubes inside a "shell" (see Figure 8-6). For purposes of illustration, the sketch also shows a "fouling deposit" on the cold side of the tube wall. These deposits are typical of commercial heat exchangers and can arise as a result of corrosion, mineral deposits, or particulates carried by the fluid (the cold

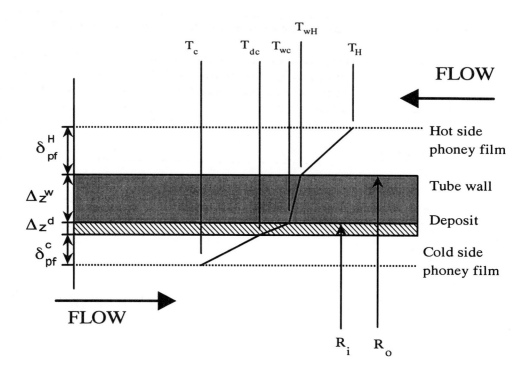

Figure 12-1 Convective Heat Transfer between a Hot and Cold Fluid

fluid, in this case). The temperature profiles imposed on this sketch indicate the magnitude of the resistance posed by each step in the transport of energy from the hot fluid to the cold fluid. These steps are: the convective transport of energy from the hot fluid to the tube wall, the molecular transport (conduction) through the tube wall and then through the fouling deposit and, finally, the convective transport from the edge of the deposit to the cold fluid. The steeper the temperature gradient in one of these steps, the larger the resistance offered by that step. Note that the convective transport steps are represented by "phoney" films; a concept introduced in Chapter 10 (see Section 10-4). As might be expected, the resistance of the tube wall is the smallest — usually heat-exchanger tubes are purposely thin and constructed of materials with high thermal conductivity. The magnitude of the resistance offered by the convective steps will depend on the state of mixing within these fluids and, in terms of the phoney film, this is represented by the thickness of the film; i.e., increased mixing results in thinner films.

　　　　If a differential balance is taken within any of the four regions shown in Figure 12-1, each would take the form

$$-\frac{1}{r}\frac{d}{dr}(rq_r) = 0$$

Then, upon integration

$$rq_r^H = rq_r^W = rq_r^f = rq_r^C = \text{constant} = C_1 \tag{12.1}$$

The physical meaning of the mathematics in Equation 12.1 is that the heat <u>rate</u> (i.e., $\dot{Q} = 2\pi L(rq_r)$) passing through each region is the same, as it should be in a steady-state system. Therefore, comparing Equation 12.1 to the expression for $\dot{Q}$, C_1 is

$$C_1 = \frac{\dot{Q}}{2\pi L} \tag{12.2}$$

where L is the length of the heat exchanger section being considered. Of the four regions, only the tube-wall thickness is very large (see Problem 10-15 for a typical value of the phoney film thickness). Thus cylindrical coordinates are only necessary within the tube wall and the variable, r, can be considered a constant in the other three regions.

Considering the phoney film on the cold side of the heat exchanger, the portion of Equation (12.1) which must be integrated is

$$rq_r^C = C_1$$

Applying Fourier's Law, $q_r^C \sim -k^C\dfrac{dT}{dz}$, and since r is essentially a constant in this region,

$k^C\dfrac{dT}{dr} = k^C\dfrac{\Delta T}{\Delta z}$, and with the boundary conditions

$$\text{at } z = z_1, \ T = T_c$$

$$\text{at } z = z_2, \ T = T_{fc}$$

direct integration yields an expression for C_1

$$C_1 = -r\left(\frac{k}{\Delta z}\right)^C (T_c - T_{fc}) \tag{12.3}$$

Substituting Equation (12.3) into (12.2), the heat rate can be calculated from

$$\dot{Q} = 2\pi r L \left(\frac{k}{\Delta z}\right)^c (T_{fc} - T_c)$$

(12.4)

Similar results are obtained in the fouling deposit and in the hot side phoney film,

$$\dot{Q} = 2\pi r L \left(\frac{k}{\Delta z}\right)^f (T_{wc} - T_{fc})$$

(12.5)

$$\dot{Q} = 2\pi r L \left(\frac{k}{\Delta z}\right)^H (T_H - T_{wH})$$

(12.6)

In general, the variability of r in the tube wall should be taken into account and the substitution of Fourier's Law into Equation (12.1) for this region takes the form

$$r\,q_r = r\left(-k^w \frac{dT}{dr}\right) = C_1$$

This equation can be integrated and solved for C_1 by using the boundary conditions

$$\text{at } r = R_i, \quad T = T_{wc}$$

$$\text{at } r = R_o, \quad T = T_{wH}$$

so that

$$\dot{Q} = 2\pi L \frac{k^w}{\ln(R_o/R_i)} (T_{wH} - T_{wc})$$

(12.7)

While any of the four expressions, Equations (12.4–7) could be used to calculate the heat rate, only two of the temperatures, T_c and T_H, are typically known. So, for practical reasons the <u>overall heat transfer coefficient</u>, U, is used and it is based on the <u>overall driving force</u>, $(T_H - T_c)$. Depending on whether it is based on the inside or outside tube surface area, the defining equation for U is the following:

$$\dot{Q} = U_o A_o (T_H - T_c) = U_i A_i (T_H - T_c)$$

(12.8)

where, for a single tube

$$A_o = 2\pi R_o L$$

$$A_i = 2\pi R_i L$$

(12.9)

It is important to keep in mind that U is a <u>defined</u> parameter, consequently it only indirectly relates to the state of mixing of the fluids and their thermal properties. To be totally useful, a relationship between U and these properties is needed and this can be done in one of two ways: by solving Equations (12.4-7) simultaneously to eliminate the unknown temperatures, T_{fc}, T_{wc}, T_{wH}, and T_{wc}, or by using the electrical analogy of resistances in series.

The simultaneous solution is left for a homework exercise (Problem 12-1) and the analogy approach is presented here. The electrical analogy concept was introduced in Chapter 2-2b for the phenomenological laws and the same can be done for the overall heat transfer coefficient. In this case the "current" is the heat rate, the "potentials" are the temperature driving forces, and the "resistances" are the reciprocals of the terms in front of the temperature driving forces in Equations (12.4-7). Thus in view of Equation 12.8, the total resistance, based on the outside tube area, is

$$R_{TOT} = \frac{1}{U_o A_o}$$

(12.10)

and for the phoney films they are

$$(R_{pf})^C = \frac{1}{h_i A_i}$$

(12.11)

$$(R_{pf})^H = \frac{1}{h_o A_o}$$

(12.12)

where individual heat transfer coefficients are used in place of the phoney film thicknesses (see Chapter 10-3). The comparable expressions for the tube wall conductance and the fouling deposit are

$$R_w = \frac{\ln(\frac{R_o}{R_i})}{2\pi L k^w}$$

$$R_f = \frac{1}{h_f A_i}$$

Note that since the fouling thickness is rarely known, <u>fouling factors</u> in terms of "pseudo" heat transfer coefficients, h_f, are typically employed.

The overall resistance is equal to the sum all of these resistances

$$\frac{1}{U_o A_o} = \frac{1}{h_i A_i} + \frac{1}{h_f A_i} + \frac{\ln(\frac{R_o}{R_i})}{2\pi L k^w} + \frac{1}{h_o A_o} \tag{12.13}$$

and the expressions for U_o and U_i are then:

$$U_o = \frac{1}{\frac{A_o}{A_i}\left[\frac{1}{h_i} + \frac{1}{h_f}\right] + \frac{A_o \ln(\frac{R_o}{R_i})}{2\pi L k^w} + \frac{1}{h_o}} \tag{12.14}$$

$$U_i = \frac{1}{\left[\frac{1}{h_i} + \frac{1}{h_f}\right] + \frac{A_i \ln(\frac{R_o}{R_i})}{2\pi L k^w} + \frac{A_i}{A_o}\frac{1}{h_o}} \tag{12.15}$$

Again, since the overall heat transfer coefficient is a parameter which encompasses a number of individual transport steps, there are no experimental correlations available which allow us to calculate U directly. On the other hand, there are "rules of thumb" which are based on typical experience and which give estimates of U for a variety of heat exchanger configurations. One such table of values is given in Table 12-1 for shell & tube exchangers.

Quite often only one of the individual transport steps governs the value of U. That is, since the resistances are in "series," if one of the resistances is much larger than the others, it is that resistance which determines the overall resistance. For example, if the heat transfer coefficient corresponding to the "hot" fluid in Figure 12-1 is very large (condensing steam, for example, see Section 12-2) and the tubes are clean, then $U \sim h_i$ and $\dot{Q} \sim h_i \, A_i \, (T_H - T_c)$.

Table 12-1 "Typical" Values of U: Shell and Tube Exchangers

Shell Side	Tube Side	U (BTU/hr-ft²-F)
Liquid – Liquid Systems		
Fuel Oil	Water	20
Gasoline	Water	60 – 100
Heavy Oils	Heavy Oils	10 – 40
Heavy Oils	Water	15 – 50
Gas Oil	Water	25 – 50
Water	Water	225
Condensing Vapor – Liquid Systems		
Alcohol Vapor	Water	100 – 200
Hydrocarbons (High Boilers)	Water	20 – 50
Hydrocarbons (Low Boilers)	Water	80 – 200
Steam	Water	400 – 1000
Steam	No. 2 Fuel Oil	60 – 90
Steam	No. 6 Fuel Oil	15 – 25
Steam	Gases	5 – 50
Gas - Liquid Systems		
Air	Water	10 – 50
Water	Air	20 – 40
Water	Hydrogen-laden Gases	80 – 125
Vaporizer Systems		
Anhydrous Ammonia	Condensing Steam	150 – 300
Light Ends (C$_3$- C$_5$)	Condensing Steam	200 – 300
Water	Condensing Steam	250 – 400

EXAMPLE 12-1: Effect of Tube Properties on Overall Heat Transfer Coefficient

It is proposed to replace the copper tube bundle of a shell and tube heat exchanger with a #304-stainless steel tube bundle. The tubes are 1" BWG # 14 and the heat transfer coefficients on the tube and shell side of the exchanger are 130 and 45 BTU/hr-ft^2-F, respectively. Assuming clean surfaces, calculate the percent change in the overall heat transfer coefficient if this proposal were adopted.

<u>Solution</u>

Basing the overall coefficient on the outside tube area, Equation (12.14) applies. From *Chemical Engineering Handbook* (R.H. Perry and C.H. Chilton, 5th ed., McGraw-Hill Co., N.Y., 1973, p.11-12), the dimensions of 1", BWG #14 tubing are: $\frac{A_o}{L} = 0.2618$, $\frac{A_i}{L} = 0.2047$ ft^2/ft and the thermal conductivities of copper and 304 stainless are 220 and 9.4 BTU/ft-hr-F, respectively. Therefore

$$U_o = \frac{1}{\frac{0.2618}{0.2047}\frac{1}{130} + \frac{0.2618\ln[.2618/.2047]}{2(3.14)k^w} + \frac{1}{45}} = \frac{1}{0.03206 + \frac{0.0102}{k^w}}$$

Substituting the thermal conductivities for copper and stainless, the overall heat transfer coefficients are calculated to be:

$$U_o(SS) = 30.16 \text{ BTU/ft}^2\text{-hr-F}, \quad U_o(Cu) = 31.1 \text{ BTU/ft}^2\text{-hr-F}$$

or, a drop of only 3%.

12-2 INDIVIDUAL HEAT TRANSFER COEFFICIENTS

In order to be able to calculate values for the overall heat transfer coefficient, it is necessary to be able to estimate values for the individual heat transfer coefficients. The word "estimate" is a purposeful choice of words since, as pointed out earlier in text, quantitative values for h are based on empirical correlations and thus are only as accurate as the data used to obtain the correlations. Over the years, many correlations have been published for a wide variety of heat exchange systems and only a representative sample will be offered here. This representation is restricted to only general applications, such as flow in ducts or in packed beds. There are many specific heat transfer coefficient correlations in the literature and if a problem requires that kind of specificity, a literature search may be a better alternative than attempting to apply the correlations listed here.

In any case, correlations are more generally applicable if they are presented in the form of similarity analysis. Thus the data are correlated in terms of dimensionless groups that were derived in Chapter 7, and vividly demonstrate how transport phenomena analyses can lead to results that eventually become useful in macroscopic calculations. The primary dimensionless groups used here are the Nusselt, Reynolds, and Prandtl numbers and all have been previously defined earlier in the text. However, experimental heat transfer coefficient correlations are often in terms of other dimensionless parameters, and so it is worthwhile to show how those other groups relate to the Nusselt number. One such group is the Stanton number. It is defined and related to the Nusselt number in the following manner

$$St = \frac{h}{\hat{c}_P G} = \frac{Nu}{Re\ Pr} \tag{12.16}$$

Another often used dependent dimensionless parameter is the *j-factor*. It has the following definition and relationships

$$j_H = \frac{h}{\hat{c}_P G} Pr^{2/3} = \frac{Nu\ Pr^{1/3}}{Re} \tag{12.17}$$

12-2.a Correlations for Tubes and Ducts: "Free" (Natural) Convection

Because fluid density is a function of temperature, differences in temperature can cause circulation currents. For example, if a horizontal flat plate is heated in air, the air density near the surface of the plate will be lower than the density of the air high above the plate. As a result of gravitational forces, the higher density air moves downward and the lower density air must move upwards to replace it. This causes circulation currents ("free" convection) normal to the plate and this mixing is more efficient than molecular conduction in removing energy from the plate. Even though these flow rates are low and can be described by differential momentum balances, the geometrical complexities are such that mathematical/computer solutions are rarely adequate to sufficiently predict energy transport rates. Consequently, empirical heat transfer coefficient correlations are used to predict free convection heat transfer rates. A typical application of these types of calculations is the prediction of heat losses from the surfaces of pipes and vessels (see Problems 12-4, 12-5).

The important dimensionless parameter in free convection situations is the Grashoff number which is defined as

$$Gr \equiv \frac{L_c^3 \rho^2 g \beta \Delta T}{\mu^2}$$

where L_c is the characteristic length of the surface, and β is the thermal expansion coefficient of the fluid, which is defined as

$$\beta \equiv \frac{\left(\dfrac{\partial \hat{V}}{\partial T}\right)_P}{\hat{V}} \tag{12.18}$$

where ΔT is the temperature difference between the surface and the bulk fluid and all properties are evaluated at the mean average temperature between the two. The thermal expansion coefficient for gases is easily determined by using an appropriate equation of state (e.g., the Ideal Gas Law). For a liquid, all that is needed is the change in density with respect to temperature; i.e., from Equation (12.18)

$$\beta = \rho \frac{\partial}{\partial T}(\frac{1}{\rho}) = -\frac{1}{\rho}\frac{\partial \rho}{\partial T}$$

Heat transfer coefficients in free convection can be correlated by the so-called <u>Nusselt equation</u>

$$Nu = a \ (Gr \ Pr)^m \tag{12.19}$$

The values of a and m are listed in Table 12-2 for various surfaces and their orientation and the Nusselt number is based on the horizontal or vertical characteristic length.

12-2.b Correlations for Tubes and Ducts: "Forced" Convection

While free convection heat transfer coefficients are useful for calculating ambient heat losses from heated vessels, forced convection heat transfer is the primary mode of energy transport when fluids must be heated or cooled. Although commercial heat exchange equipment can be quite complex, in a basic sense it involves flow within and around tubes and ducts. Dealing first with heat transfer to and from fluids within tubes and ducts, Table 12-3 lists correlation parameters for a generalized correlation of the form

$$(Nu)_{\Delta T} = C_1 + C_2 (Gz)^a (Re)^b (Pr)^c \Phi^d \Lambda_i \tag{12.20}$$

Table 12-2 Parameters for Equation (12.19)

Surface	Gr Pr	A	m
Vertical Surfaces	$< 10^4$	1.36	.20
	$10^4 - 10^9$	.59	.25
	$> 10^9$	.13	.33
Horizontal Cylinder ($L_c = D$)	$< 10^{-5}$	.49	0
	$10^{-5} - 10^{-3}$	.71	.04
	$10^{-3} - 1.0$	1.09	.10
	$1.0 - 10^4$	1.09	.20
	$10^4 - 10^9$	.53	.25
	$> 10^9$	.13	.33
Horizontal Plate (up)	$10^5 - 2 \times 10^7$	.54	.25
	$2 \times 10^7 - 3 \times 10^{10}$	.14	.33
Horizontal Plate (down)	$3 \times 10^5 - 3 \times 10^{10}$	.27	.25

where the Nusselt number, or the "dimensionless heat transfer coefficient" is defined as $Nu = \dfrac{h L_c}{k}$, and L_c is the characteristic dimension of the geometry (e.g., the tube diameter in tube flow). Keeping in mind that these correlations are derived from experimental measurements of the heat transfer coefficients, the ΔT subscript on the Nusselt number refers to the manner in which they were measured. Strictly speaking, they should also be applied in the same way. Thus, if the coefficients over an experimental heat exchanger were calculated from arithmetic mean temperature driving forces ("am"), then the heat transfer coefficients calculated from Equation (12.20) should be used in conjunction with an arithmetic mean temperature driving force when being used for design purposes. The function, Φ , corrects for the effect of temperature on viscosity at the wall of the duct and is defined as $\Phi = \left(\dfrac{\mu}{\mu_w} \right)^{.14}$. The function, Λ_i , is a miscellaneous correction for "entrance" and free convection effects and is defined in Table 12-3.

The state of mixing of the fluid determines the magnitude of the heat transfer coefficients and thus there are separate correlations for laminar and turbulent flow. Transition flow is treated in terms of a "merged" correlation between the two. However, all of the correlations can

Table 12-3 Heat Transfer Coefficient Correlations (Internal Flow in Ducts)

Flow	Geometry	Limitations	C_1	C_2	a	b	c	d	i	$\Delta T^{\#}$	REF
Laminar	Horiz Tubes	Re < 2100; Gz < 100	3.66	0.0	0	0	0	0	2	lm	1
	Horiz Tubes	Re < 2100; Gz > 100	0.0	1.86	.33	0	0	1	4	am	2
	Vertical Tubes	Re < 2000									
	Annuli	Re < 2000; D = D_{eq}	0.0	1.02	0	.45	.5	1	3	am	
	Rect.Ducts	Re < 2000; D = D_{eq}	0.0	1.85	.33	0	0	0		lm	
Transition	Tubes	$2000 < Re < 10^4$	0.0	.116	0	0	.33	1	5	am	
Turbulent	Smooth Tubes$^{\#\#}$	$Re > 10^4$; .7<Pr<700; L/D>60	0.0	.023	0	.8	.33	1	1	-	
	Smooth Tubes	$Re > 10^4$; L/D < 60	0.0	.023	0	.8	.33	1	$6^{\#\#\#}$	-	
	Rect. Ducts	D = D_{eq}; $L/D_{eq} > 60$	0.0	.023	0	.8	.33	1	1	-	

Equation Form: $(Nu)_{\Delta T} = C_1 + C_2 (Gz)^a (Re)^b (Pr)^c \Phi^d \Lambda_i$

$$\Lambda_1 = 1 \qquad \Lambda_2 = \frac{.085\,(Gz)}{1 + .047(Gz)^{.67}} \qquad \Lambda_3 = \left(\frac{D_{eq}}{L}\right)^{.4}\left(\frac{D_2}{D_1}\right)^{.8}(Gr)^{.05}$$

$$\Lambda_4 = .87[1.0 + .015(Gr)^{.33}] \qquad \Lambda_5 = \left[(Re)^{.67} - 125\right]\left[1 + \left(\frac{D}{L}\right)^{.67}\right] \qquad \Lambda_6 = \left(1 + F\frac{D}{L}\right)$$

be placed in the generalized form of Equation (12.20). With respect to *laminar* flow, the pertinent dimensionless parameter is the Graetz number, Gz, which is defined as

$$Gz = \frac{\dot{m}\,\hat{c}_p}{k\,L} = \frac{\pi}{4}\,Re\,Pr\,\frac{D}{L}$$

In principle, heat transfer coefficients in laminar flow can be theoretically calculated and "limiting" Nusselt numbers for constant heat flux at the walls of tubes and ducts (i.e., for fully developed temperature profiles) are relatively easy to calculate and examples are given in Chapter 8$^{\#\#\#\#}$. When temperature profiles are not fully developed, the heat transfer coefficient will be a function of distance down stream, L, and this is reflected in the definition of the Graetz

---

$^{\#}$ *am, lm*: arithmetic, log-mean driving forces.

$^{\#\#}$ Correction for rough tubes: $(Nu)_r = (Nu)_{sm}\dfrac{f_r}{f_{sm}}$.

$^{\#\#\#}$ See Table 12-4 for values of F.

$^{\#\#\#\#}$ Note that, when Gz < 100 (i.e., "long" tubes), the limiting Nusselt number, 3.66, is part of the correlation.

number. However, pure laminar flow is nearly impossible to achieve and free convection currents are often present. When these currents are normal to the laminar flow stream lines, mixing takes place and the heat transfer rates can be many times that calculated due to molecular transport. The general correlation of Equation (12.20) accounts for this through the use of an appropriate Λ function. For example, suppose we have flow in a horizontal tube with a Reynolds number of 1200 and a Graetz number of 300. Referring to Table 12-3, the appropriate form of the correlation is then

$$(Nu)_{am} = 1.86(Gz)^{.33} \, \Phi \, [\, 1.0 + .015(Gr)^{.33} \,] \tag{12.21}$$

where the Grashoff number is introduced in order to account for the increased heat transfer rates arising from free convection currents. Note that the same form of the correlation can also be used in annuluses and in rectangular ducts as long as the diameter is replaced by the "equivalent diameter," D_{eq}.

The heat transfer coefficient correlations for *turbulent* flow inside tubes and ducts are based upon the Seider-Tate correction to the Dittus-Boelter Equation for fully developed flow in smooth tubes, which from Table 12-3 is

$$Nu = .023 \, (Re)^{.8} \, (Pr)^{.33} \, \Phi \tag{12.22}$$

Note that this is valid only for $L/D > 60$ (i.e., fully developed flow where Nu is independent of distance downstream). For short tubes, a correction factor must be added

$$\Lambda_6 = 1 + F \, \frac{D}{L}$$

where F accounts for irregularities in the tube entrance geometries and can be obtained from Table 12-4. Since a rough tube surface increases turbulence, the Nusselt numbers in rough tubes are larger and can be estimated by multiplying the smooth tube value by the ratio of the friction factor for the rough tube to the friction factor for smooth tubes; i.e,

$$(Nu)_r = (Nu)_{sm} \, \frac{f_r}{f_{sm}} \tag{12.23}$$

Again, these same correlations can be used in other closed duct geometries by using D_{eq} in place of the diameter, D.

Finally, heat transfer coefficients in transition flow can be calculated by merging the correlations in laminar and turbulent flow. An empirical equation for doing this is presented in Table 12-3 where the correction function, Λ_5, links the two correlations.

Table 12-4 Values For F In Equation (12.22)

Entrance Conditions	F
Fully Developed Velocity Profile	1.4
Abrupt Contraction	6
90 degree Right Angle Bend	7
180 degree Round Bend	6

EXAMPLE 12-2: Calculation of Overall Heat Transfer Coefficient

280,000 lbs/hr of a hot, heavy oil is fed to the tube side of a shell and tube heat exchanger where it is cooled with cooling water which has a geometric mean mass velocity of 300,000 lbs/hr-ft^2. The exchanger consists of 150, 3/4" BWG 14, 12 ft long copper tubes (k = 220 BTU/ft-hr-F). The properties of the cooling water correspond to a temperature of 70 F and the oil properties are as follows: $\hat{c}_P = 0.49, k = 0.077, \rho = 53, \mu = 5.2$ where all the properties have AES units (BTU, lbm, F, ft, hr) except for the viscosity which is in centipoise. Calculate the overall heat transfer coefficient, based on the inside area, for this exchanger.

Solution

The tube dimensions are taken from *Chemical Engineering Handbook* (R.H. Perry and C.H. Chilton, 5th ed., McGraw-Hill Co., N.Y., 1973, p.11-12), which for, 1", BWG #14 tubing are: $\dfrac{A_o}{L} = 0.1963, \dfrac{A_i}{L} = 0.1529$ ft^2/ft, $D_i = 0.584", A_{x_i} = 0.00186$ ft^2

The mass velocity on the tube side is calculated from

$$G = \frac{(280,000)/150}{0.00186} = 1.0 \times 10^6 \ \text{lbm/ft}^2\text{-hr}$$

so that the tube side Reynolds number is

$$Re = \frac{(0.584/12)(1.0 \times 10^6)}{(5.2)(2.42)} = 3,900$$

which is in transition flow. The Prandtl number for the oil is,

$$Pr = \frac{(0.49)(5.2)(2.42)}{0.077} = 80$$

The tube side heat transfer coefficient can be calculated from the correlation for transition flow in a tube which, from Table 12-3, is

$$Nu = 0.116(80)^{1/3}[(3,900)^{0.67} - 125]\left[1 + \left(\frac{0.584/12}{12}\right)^{.67}\right] = 65$$

and $\quad h_i = \dfrac{(65)(0.077)}{0.584/12} = 102 \ \text{BTU/ft}^2\text{-hr-F}$

The heat transfer coefficient on the shell side is calculated from the Donahue equation, Equation (12-27)

$$Nu = \frac{h_o D_o}{k} = 0.20\left[\frac{(0.75/12)(300,000)}{(1)(2.42)}\right]^{0.6}(7.7)^{1/3} = 85.1$$

$$h_o = \frac{(85.1)(0.077)}{0.75/12} = 105 \ \text{BTU/ft}^2\text{-hr-F}$$

The overall heat transfer coefficient is calculated from Equation (12.15)

$$U_i = \frac{1}{\dfrac{1}{102} + \dfrac{(0.1529)\ln[.1963/.1529]}{2(3.14)12(220)} + \dfrac{.1529}{.1963}\dfrac{1}{105}} = 58 \ \text{BTU/ft}^2\text{-hr-F}$$

This value is very nearly equal to the high end of the range given in Table 12-1 for a shell & tube exchanger in this type of service.

The fluid flow external to the tubes in a shell and tube heat exchanger can be in parallel flow, cross flow, or a mixture of the two. The case of parallel flow is easily handled by using the correlations for internal flows with the diameter replaced by the equivalent diameter. There have been many studies of heat transfer normal to single tubes and one correlation that can be used for liquids and gases (up to $Re = 10^4$) is the following

Heat Transfer Normal to Single Tubes

$$Nu = Pr^{.30}[0.35 + 0.56 \ Re^{0.52}] \tag{12.24}$$

where the characteristic dimension is the outside tube diameter.

Correlations for flow normal to banks of tubes tend to be very specific because of the flow complexities introduced by tube layout, baffles and the like. However, Equation (12.25), while recommended for gases, should also be applicable to liquids since it is in dimensionless form.

Heat Transfer Normal to Tube Banks

$$Nu = 0.33 \ Re^{0.60} \ Pr^{1/3} \tag{12.25}$$

where the outside tube diameter is the characteristic dimension. More often, the flow external to tube banks in a shell and tube heat exchanger is a mixture of parallel and cross-flow. In this case, the Donahue equation is applicable. Basically, this correlation uses a geometric mean of the cross flow and parallel flow mass velocities

$$\overline{G} = \sqrt{G_C \ G_P} \tag{12.26}$$

The cross-flow and parallel mass velocities, G_C and G_P, are calculated from a knowledge of the tube layout and the baffle spacing. The specifics of these calculations will be discussed in connection with shell and tube heat exchanger design, later in this chapter. The Donahue correlation is given by Equation (12.27)

Heat Transfer to Baffled Tube Banks

$$Nu = 0.2 \ Re^{0.6} \ Pr^{1/3} \tag{12.27}$$

where, again, the characteristic dimension is the outside tube diameter and the Reynolds number is calculated from the geometric mean mass velocity given in Equation (12.26).

12-2.c Correlations in Packed and Fluidized Beds

Many industrial processes involve packed or fluidized beds; very often for the purpose of carrying out reactions which are accompanied by large enthalpy changes. Consequently, there is a need to compute heat transfer rates between the packing and the fluid since high temperatures on the surface of the packing (catalyst) can lead to catalyst deactivation. The same is true for heat transfer between the packing and the walls of the vessel, since it may be necessary to cool (or heat) the reactor. Since, as pointed out in Chapter 11, a fluidized bed is a packed bed operated at velocities which cause the packing to become suspended, the same needs apply equally well to fluidized beds. Only the so-called, dense phase, "bubbling" fluidized beds are reviewed here. More recent developments of "fast" fluidized beds are discussed in the literature [3] and should be consulted for prediction of heat transfer coefficients in those suspended solid systems.

In general, the heat transfer rates between the fluid and the particles in a fluidized bed are very high (one of the reasons for its use) and so only fluid-solid heat transfer in packed beds will be discussed here. Many of the early correlations were based on the assumption that the limiting Nusselt number (as the Reynolds number approaches zero) for heat transfer between a fluid and spherical packings would be 2.0, as predicted for molecular transport between a single sphere and a fluid (see Problem 3-27). Thus the correlations were usually fit to equations of the form

$$Nu_P = 2.0 + a\ Re_P^m\ Pr^n$$

where, for a single sphere, $a = 0.6$, $m = 1/2$, $n = 1/3$

However, once experimental data were available at very low Reynolds numbers, it became apparent that Nusselt numbers could be much lower than 2.0. These data are not easy to come by and thus correlations at Reynolds numbers much below 5 should be used with caution. One correlation that appears to work reasonably well is that proposed by Whitaker [4]

Fluid-Solid Heat Transfer in Packed Beds

$$Nu' = [\ 0.5\ (R'e)^{1/2} + 0.2\ (R'e)^{2/3}\ ]\ Pr^{1/3} \tag{12.28}$$

where the Nusselt and Reynolds numbers are defined in terms of the specific area of the particle and the void fraction in the bed as

$$Re' = \frac{\dfrac{6}{a_P}\dfrac{\varepsilon}{1-\varepsilon}\dfrac{G_o}{\varepsilon}}{\mu}$$

$$Nu' = \frac{h\,\dfrac{6}{a_P}\dfrac{\varepsilon}{1-\varepsilon}}{k}$$

This correlation has the advantage of being applicable to shaped particles and, for spheres,
$\dfrac{6}{a_P} = d_P$.

   One of the most important aspects of heat transfer between packed beds and the walls of a vessel is the ratio of the particle diameter to the bed diameter, d_P / D. While the presence of a packing tends to promote wall turbulence and increase heat transfer coefficients, a significant fraction of the fluid begins to channel near the wall as d_P / D approaches about 0.25. Consequently, Leva [5] proposed two correlations

Heat Transfer between a Packed Bed and the Wall

$$\frac{d_P}{D} \le 0.35 \; : \; Nu = 0.813 \; Re_P^{0.90} \exp\left(-\frac{6\,d_P}{D}\right)$$

(12.29)

$$\frac{d_P}{D} > 0.35 \; : \; Nu = 0.125 \; Re_P^{0.75}$$

where the Nusselt number is based on the wall diameter and the Reynolds number is based on the particle diameter.

   In a number of fluidized bed applications, heat exchanger bundles are placed in the beds to take advantage of the higher heat transfer coefficients. Thus heat transfer coefficient correlations exist for both immersed objects as well as between the bed and the wall. One correlation for the wall coefficient is given in Equation (12.30)

Heat Transfer between a Fluidized bed and the Wall

$$Nu = 2.53 \left[(Re_P)_{\text{mf}} \; \ln\left(\frac{v_0}{v_{\text{mf}}}\right) \right]^{0.47}$$

(12.30)

where the particle Reynolds number is calculated at minimum fluidizing conditions (see Chapter 11) and v_0 is the superficial velocity.

   For fluidized heat transfer to immersed tube bundles, Gelperin and Einstein [6] suggest the following correlations (staggered tubes):

Heat Transfer from Fluidized Beds to Vertical Tube Bundles

$$Nu = 0.75 \ Ar^{0.22} \left[1 - \frac{D}{S_H} \right]$$ (12.31)

where D is the tube diameter, S_H is the horizontal tube spacing and Ar is the <u>Archimedes number</u>, defined by

$$Ar = \frac{d_p^3 \rho_f \ (\rho_s - \rho_f) g}{\mu^2}$$

where ρ_f is the density of the fluid and ρ_s is the density of the solid.

Heat Transfer from Fluidized Beds to Horizontal Tube Bundles

$$Nu = 0.75 \ Ar^{0.22} \left[1 - \frac{D}{S_H} (1 + \frac{D}{D + S_V}) \right]^{0.25}$$ (12.32)

where S_V is the vertical tube spacing.

EXAMPLE 12-3: Temperature on A Catalyst Surface

A 15-mm spherical Ni/Al_2O_3 catalyst is used to catalyze the endothermic, steam reforming of methane

$$CH_4 + 2H_2O = 4H_2 + CO_2 \qquad \Delta \tilde{H}_r = 40 \ Kcal/mol$$

If the gas at a point in the packed bed reactor has a superficial mass velocity of 3500 kg/m²-s and is at 800 °C, what will be the steady-state temperature of the catalyst surface if the methane reaction rate is 0.05 mol/m²-s? The properties for the reaction mixture can be taken as follows: $\hat{c}_P = 3.2 \ cal/g - K$, $\mu = 0.03 \ cp$, $k = 0.33 \ J/m-s-K$.

Solution

A steady-state energy balance between the reaction mixture and the catalyst particle is

$$R_{AS} \ \Delta \tilde{H}_r = h (T_f - T_{cat})$$

where T_{cat} is the catalyst surface temperature and T_f is the fluid temperature.

The next step is to calculate the heat transfer coefficient between the fluid and the catalyst particle. Equation (12.28) is applicable and so,

$$Re' = \frac{d_p \dfrac{G_o}{\varepsilon} \dfrac{\varepsilon}{1-\varepsilon}}{\mu} = \frac{(.015)\dfrac{(3500)}{0.46}\dfrac{(0.46)}{(1-0.46)}}{(0.03)(1 \times 10^{-3})} = 902$$

and

$$Nu' = \left[0.5 \ (Re)^{1/2} + 0.2 \ (Re)^{0.67}\right](Pr)^{1/3}$$

$$Nu' = \left[0.5 \ (902)^{0.5} + 0.2 \ (902)^{0.67}\right]\left(\frac{(3.2)(4.183)(10^3) \ (0.03)(1 \times 10^{-3})}{0.33}\right)^{0.33} = 6.57$$

$$h = 6.57 \ \frac{\dfrac{0.33}{(0.015)(0.46)}}{0.54} = 169 \quad J/m^2\text{-}s\text{-}K$$

Using this value of h in the energy balance, the catalyst temperature is calculated to be

$$T_f - T_{cat} = \frac{(0.05)\,4\,(10^4)\,(4.183)}{169} = 49.5 \quad K$$

$$T_{cat} = 800 - 49.5 = 750.5 \quad {}^0C$$

In effect, this means that the catalytic reaction is actually taking place at a temperature which is 50 ^{0}C lower than that measured in the reacting gas.

12-2.d Correlations for Boiling and Condensation

Boiling heat transfer can be an unpredictable and sporadic phenomenon. If nucleate boiling takes place, the heat transfer rates are usually high and the limiting heat transfer resistance lies elsewhere. However, if "film" boiling takes place, heat transfer rates are lower and the ΔT driving force for heat transfer can become unacceptably large. For these reasons, it is important to have a knowledge of these two boiling mechanisms and be able to predict the heat transfer coefficients associated with each.

When the liquid adjacent to a hot surface is at its saturation temperature and the temperature of the surface is raised above that value, the resulting heat flux varies with the ΔT driving force in the manner sketched in Figure 12-2. At ΔT's less than 10, heat is transferred due to natural convection currents in the liquid and the heat flux dependency on ΔT is, $q_w \sim (\Delta T)^{1.25}$. As ΔT increases, discrete bubble formation occurs and the heat transfer rate increases dramatically in a process known as "nucleate" boiling. The bubble formation is difficult to predict since it results from microscopic nucleation sites at the surface, but the dependency of flux on ΔT is generally of the form, $q_w \sim (\Delta T)^{3-4}$. The nonlinearity of nucleate boiling is evident in the following correlation

Heat Transfer Coefficient — Nucleate Boiling

$$h = 0.225 \left(\frac{q_w \hat{c}_P^{\,L}}{\Delta \hat{H}_V} \right)^{0.69} \left(\frac{144 P\, k_L}{\sigma} \right)^{0.31} \left(\frac{\rho_L}{\rho_G} - 1 \right)^{0.33} \tag{12.33}$$

where σ is the surface tension (lb$_f$/ft), the total pressure, P, is in psia and all other variables are in the AES of units (ie., BTU, lb$_m$, ft, hr, °F). Since the wall heat flux appears in the correlation, prediction of the heat transfer rates is an iterative process.

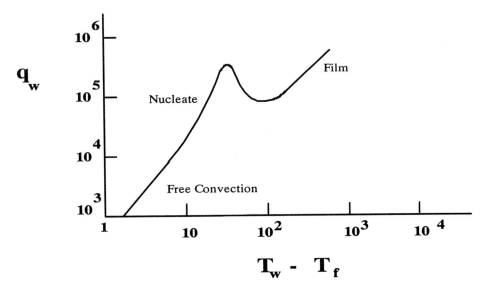

Figure 12-2 Boiling Heat Flux Curve

At still higher ΔT, some of the bubbles coalesce to form gas pockets at the surface, resulting in higher thermal resistance. This continues until there is a continuous film, separating the surface from the liquid. In this situation the boiling is more uniform and predictable although unacceptably high ΔT's can often result. One correlation that can be used in this region for tubes was suggested by Bromley [7]

<u>Film Boiling from Tubes</u>

$$h = 0.62 \left[\frac{k_G^3 (\rho_L - \rho_G) \rho_G g}{\mu_G D \Delta T} \right]^{1/4} \qquad (12.34)$$

where the subscripts G and L refer to the gas and liquid and D is the tube diameter.

Condensers are used to either remove sufficient heat to promote condensation or to use the enthalpy of a condensing vapor to supply heat to a colder fluid. In either case, the liquid condensate serves as a barrier to heat transfer and thus it is necessary to account for its accumulation as condensation takes place. The simplest view of this process is to consider condensation on a vertical surface, as in Figure 12-3.

Here the condensate film grows in the direction normal to the surface and has a thickness, δ_f, at any point along the surface. This is identical to the situation already described in

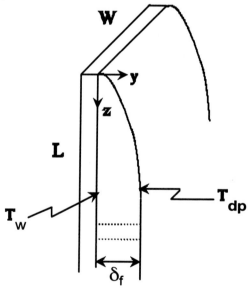

Figure 12-3 Heat Transfer to a Condensate Film

Section 5-3 and in Problem 5-8. There it was shown that the volumetric flow rate of the liquid could be related to the film thickness and, for a flat surface,

$$\dot{V} = \frac{\delta_f^3 \rho\, g\, W}{3\mu}$$

A useful parameter in these flow systems is the "hydraulic loading," Γ, the mass flow rate per unit width of the surface. It can be derived from the above equation as

$$\Gamma = \frac{\delta_f^3 \rho^2 g}{3\mu} \tag{12.35}$$

In a condenser, the driving force for heat transfer is the temperature difference between the edge of the condensate film (the dew point of the vapor) and the wall temperature. The heat transfer resistance is entirely across the liquid film and since the film flow is taken to be laminar, this is strictly due to molecular transport, or

$$q_W = \frac{k}{\delta_f}(T_{dp} - T_W) \tag{12.36}$$

Thus, as the film thickness grows, the heat transfer rate diminishes. Since the film thickness depends on the condensation rate and it, in turn, depends on the heat transfer rate which is a function of the film thickness, the two phenomena must be solved simultaneously. This can be done by taking a differential mass balance across a DVE of thickness, δz, and width, W, while recognizing that there is a "source" of mass due to the condensation rate. Mathematically, this is can be expressed as

$$\dot{m}_{|z} - \dot{m}_{|z+\delta z} - \frac{q_W\, W}{\Delta \hat{H}_v}\, \delta z = 0$$

$$-\frac{d\dot{m}}{dz} = \frac{q_W\, W}{\Delta \hat{H}_v}$$

The differential equation can be written in terms of Γ by dividing through by W and then in terms of the film thickness by substituting for q_W and Γ from Equations (12.35, 12.36) to obtain

$$\delta_f \frac{d\delta_f^3}{dz} = \frac{3\mu k (T_{dp} - T_W)}{\rho^2 g\, \Delta \hat{H}_v}$$

which can be integrated to give

$$\delta_f = \left[\frac{4\mu k \, (T_{dp} - T_W)}{\rho^2 g \, \Delta\hat{H}_V} \right]^{1/4} z^{1/4}$$

The local heat transfer coefficient is just $\dfrac{k}{\delta_f}$, and the average heat transfer coefficient over a vertical distance, L, is

Condensation on Vertical Tubes – Laminar Film

$$\bar{h} = \frac{1}{L} \int_0^L h \, dz$$

(12.37)

$$\bar{h} = 0.943 \left[\frac{k^3 \rho^2 g \, \Delta\hat{H}_V}{\mu \, (T_{dp} - T_W)L} \right]^{1/4}$$

Equation (12.37) is only valid in laminar film flow; that is, when the film is smooth and without ripples. The usual criterion for this restriction is that, at the bottom of the tube, $\dfrac{4\Gamma}{\mu} \leq 2100$, where for cylindrical tubes, Γ is equal to the mass flow rate divided by the perimeter of the tube ($\dot{m} \, / \pi D$), which is often called the *hydraulic loading*.

The heat transfer coefficients in <u>turbulent</u> condensate films vary with mass flow rate and the experimental data can be empirically correlated by Equation (12.38)

Condensation in Vertical Tubes – Turbulent Films

$$\bar{h} \left[\frac{\mu^2}{k^3 \rho^2 g} \right]^{1/3} = 0.0076 \left(\frac{4\Gamma}{\mu} \right)^{0.4}$$

(12.38)

where again, Γ is calculated at the bottom of the tube.

An equation similar to (12.37) applies to single horizontal tubes

Condensation on a Single Horizontal Tube

$$\bar{h} = 0.725 \left[\frac{k^3 \rho^2 g \Delta\hat{H}_V}{\mu \, (T_{dp} - T_W) D} \right]^{1/4} \tag{12.39}$$

where D is the tube diameter and Γ is the mass flow rate divided by the length of the tube. Horizontal tubes are almost always in laminar flow since the hydraulic loadings are typically one-tenth of the values for vertical tubes.

Equation (12.39) can also be used for vertical banks of horizontal tubes since condensate falls cumulatively from one bank to another. However, the average heat transfer coefficient for each bank is successively lower than the previous bank. If there are N banks of tubes, the equation becomes

Condensation on Banks of Horizontal Tubes

$$\bar{h} = 0.725 \left[\frac{k^3 \rho^2 g \Delta\hat{H}_V}{N \mu \, (T_{dp} - T_W) D} \right]^{1/4} \tag{12.40}$$

12-3 HEAT EXCHANGER DESIGN

12-3.a Double-Pipe Heat Exchangers

The basis for the design of shell and tube heat exchangers is the so-called, "double-pipe" heat exchanger which consists of one tube surrounded by another. When the fluids do not undergo phase changes, the flow pattern is countercurrent and the temperature profiles are similar to those shown in Figure 12-4. As can be seen, the temperature profiles for the cold and hot fluids are essentially parallel and almost linear. The purpose for countercurrent flow is to keep a high temperature driving force over the length of the exchanger. These profiles can be quantitatively predicted by taking differential energy balances on the two streams and solving them with the boundary conditions given in the figure. A differential balance over a cylindrical DVE for the cold fluid is

$$\dot{m}_c \hat{c}_{P_c} \left(T_{c_z} - T_{c_{z+\delta z}} \right) \delta t + U \left(T_h - T_c \right) \pi D \delta z \, \delta t = 0$$

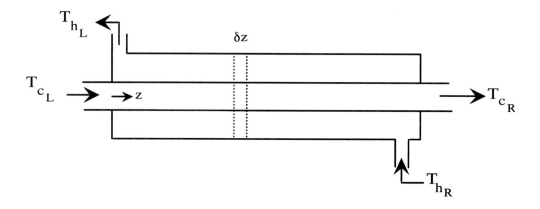

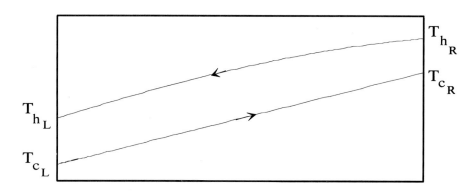

Figure 12-4 Countercurrent Double Pipe Heat Exchanger

Note that the cold fluid receives heat as a source and its rate is expressed in terms of the overall heat transfer coefficient, U. The corresponding differential equation is

$$\frac{dT_c}{dz} = \frac{U\pi D}{\dot{m}_c \hat{c}_{P_c}} (T_h - T_c) \tag{12.41}$$

An identical equation could be written for the hot fluid (with an energy "sink") and the two equations could be solved simultaneously, using MATLAB. However, an analytical solution can be found in this case by relating the temperature of the hot fluid to the temperature of the cold fluid by means of an overall energy balance between the right-hand side of the exchanger and any point in the exchanger (assuming an adiabatic exchanger); viz.

$$T_h = \left[\frac{\dot{m}_c \hat{c}_{P_c}}{\dot{m}_h \hat{c}_{P_h}} \right] (T_c - T_{cR}) + T_{hR} \tag{12.42}$$

If Equation (12.42) is substituted for T_h in Equation (12.41), it is separable and of the form

$$\frac{U \pi D}{\dot{m}_c \hat{c}_{P_c}} \int_0^L dz = \int_{T_{cL}}^{T_{cR}} \frac{dT_c}{aT_c + b}$$

This can be integrated and solved for the area of the heat exchanger

$$\pi DL = \frac{\dot{m}_c \hat{c}_{P_c}}{U} \left[\frac{T_{cR} - T_{cL}}{(T_{hR} - T_{cR}) - (T_{hL} - T_{cL})} \ln \left(\frac{T_{hR} - T_{cR}}{T_{hL} - T_{cL}} \right) \right] \tag{12.43}$$

The heat exchanger duty can also be calculated by recognizing that the numerator in Equation (12.43) is the total heat rate, $\dot{Q}$, i.e.,

$$\dot{Q} = \dot{m}_c \hat{c}_{P_c} (T_{cR} - T_{cL}) \tag{12.44}$$

If Equation (12.43) is divided by (12.44) and solved for $\dot{Q}$

$$\dot{Q} = U \pi DL \left[\frac{(T_{hR} - T_{cR}) - (T_{hL} - T_{cL})}{\ln \left(\dfrac{T_{hR} - T_{cR}}{T_{hL} - T_{cL}} \right)} \right]$$

The bracketed term in the above equation is defined as the <u>log mean driving force</u>, $(\Delta T)_{LM}$, and is the basis for the driving force for all exchangers, so that the general expression for the heat duty is

$$\dot{Q} = U A (\Delta T)_{LM} \tag{12.45}$$

where

$$(\Delta T)_{LM} = \left[\frac{(T_{hR} - T_{cR}) - (T_{hL} - T_{cL})}{\ln\left(\dfrac{T_{hR} - T_{cR}}{T_{hL} - T_{cL}}\right)} \right]$$

(12.46)

12-3.b Shell and Tube Heat Exchangers

While double-pipe heat exchangers are sometimes used in an industrial setting, usually as a series of "hairpin" U-tubes, the most common heat exchangers are of the shell and tube type since much larger surface areas can be placed in a single vessel. Figure 12-5 shows three typical shell and tube configurations. Whereas these exchangers may have hundreds of tubes, only four are shown here in the interest of clarity. The classifications of these exchangers are given in terms of the number of shell "passes" and tube "passes". Thus the exchanger at the top of Figure 12-5 is a 1-1 exchanger since both the fluid in the shell and in the tubes makes a single pass through the exchanger. In the 1-2 exchanger, the shell-side fluid still makes one pass through the exchanger but the tube-side fluid makes two passes through the shell. The bottom sketch shows a configuration where the shell-side fluid makes two passes from one end of the shell to the other, and the tube-side fluid passes through the shell four times (a 2-4 exchanger).

There are advantages and disadvantages to all of these configurations. Using many tubes in parallel provides high surface area but the velocities in each tube are lower. The latter can be a problem because heat transfer coefficients depend on the fluid velocity and multiple tubes will therefore produce lower heat transfer coefficients. This can be partially offset by using "multi-passing", so that the fluid does not go through all of the tubes simultaneously. In the 1-2 exchanger, the tube-side velocity would be twice the value in the 1-1 exchanger and it would be four times higher in the 2-4 exchanger. Using segmented baffles in the shell also increases the velocity of the shell-side fluid and promotes more uniform contacting of the shell-side fluid with the tubes. For example, if there were no baffles in the 1-1 exchanger, there is a possibility that the shell side-fluid would have a diagonal path between the inlet and outlet ports. However, multipassing produces higher pressure drops and thus heat exchanger design involves trade-offs between surface area, heat transfer coefficients, and pressure drop. These will be discussed later in this chapter in connection with the optimization of heat exchanger design.

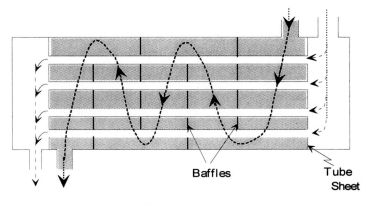

Baffles   Tube
                 Sheet

1-1 Exchanger

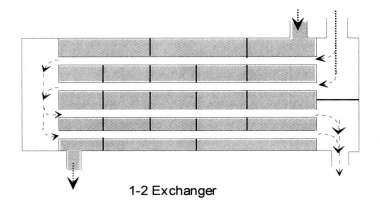

1-2 Exchanger

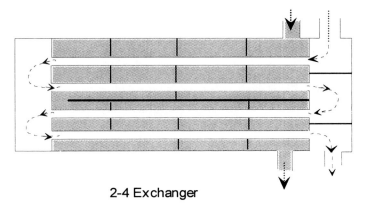

2-4 Exchanger

Figure 12-5 Shell and Tube Heat Exchangers

The next issue to address is the design calculations for shell and tube exchangers. Obviously the flow patterns are quite complex in these exchangers, and this will affect the heat transfer coefficients as well as the thermal driving force. The choice of the tube diameter (and number of tube passes, for a given surface area) and the baffle spacing will affect the heat transfer coefficients and the number of shell and tube passes will influence the driving force. For example, in the 2-4 exchanger, the shell-side flow is, at times, both parallel and normal to the tube flow. In addition, the flow is co-current and countercurrent to the tube-side flow at different locations in the exchanger.

In any case the basic design equation is still Equation (12.45). This equation can be adapted to shell and tube exchangers by calculating more appropriate heat transfer coefficients and correcting the log-mean driving force for the flow patterns. The tube-side heat transfer coefficients are still calculated from equations such as those listed in Table 12-3; the only additional consideration is the influence of the number of tube passes on the fluid velocity. The velocity of the shell-side fluid is also the primary factor for the shell-side heat transfer coefficient and is determined by the "free" cross-sectional area in the shell (dependent on the tube lay-out) and the baffle spacing. However, it is also influenced by the alternating flow pattern (parallel or normal to the tubes). The Donahue equation, Equation (12.27), which was presented earlier, attempts to account for all of these factors.

$$Nu = 0.2 \ Re^{0.6} \ Pr^{1/3} \tag{12.27}$$

Parallel versus cross-flow is taken into account by basing the Reynolds number on the outside tube diameter and the geometric mean of the cross flow and parallel flow mass velocities, (see Equation (12.26)). Figure 12-6 shows a sketch of the tube-baffle arrangement that can be used to calculate the flow cross sectional areas and then, the two mass velocities. For cross flow (Figure 12-6a), if N_T is the number of tubes at the <u>center</u> of the shell, P_T is the tube "pitch" (center-to-center), c' is the tube spacing and B is the baffle spacing, then the mass velocity is the shell-side mass flow rate divided by the free cross-sectional area available for cross flow, or

$$G_c = \frac{\dot{m}_s}{(Ac)_f} = \frac{\dot{m}_s}{B \ N_T \ c'} = \frac{\dot{m}_s}{B \ (\frac{Ds}{P_T}) \ c'} \tag{12.47}$$

where D_s is the inside diameter of the shell.

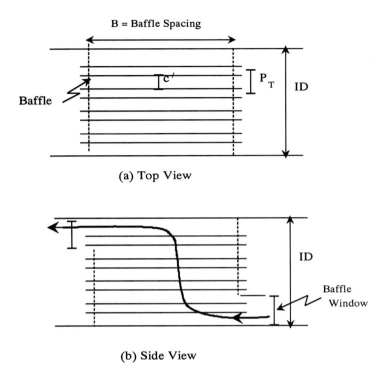

(a) Top View

(b) Side View

Figure 12-6 Tube / Baffle Layout

For parallel flow, the fluid enters a baffle segment by flowing horizontally through the "baffle window" (Figure 12-6b) and the "free" cross-sectional area available for parallel flow can be calculated from

$$(A_P)_f = f_B \frac{\pi}{4} \left[D_S^2 - N_T \ D_T^2 \right]$$

(12.48)

where f_B is the fraction of shell cross section occupied by the baffle window (0.1955 is common) and D_T is the outside diameter of the tubes. The next step is the correction for the log-mean temperature driving force. The procedure is to calculate $(\Delta T)_{LM}$ as if the shell and tube exchanger were a double-pipe exchanger in countercurrent flow

$$(\Delta T)_{LM} = \left[\frac{(T_{h_i} - T_{c_o}) - (T_{h_o} - T_{c_i})}{\ln\left(\dfrac{T_{h_i} - T_{c_o}}{T_{h_o} - T_{c_i}} \right)} \right]$$

(12.49)

where the i and o subscripts refer to the inlet and outlet streams. Equation (12.45) is then modified by multiplying by a correction factor, F_T, so that it is re-written as

$$\dot{Q} = U\, A (\Delta T)_{LM}\; F_T \qquad (12.50)$$

The F_T correction factors have been empirically determined for various shell and tube configurations and a complete set can be found in the Chemical Engineering Handbook [8]. For illustration purposes, Figure 12-7 shows one example from this set of correction factors. As can be seen, the F_T values are correlated with two dimensionless temperature parameters, S and R, which are defined as:

$$R = \frac{T_{hi} - T_{ho}}{T_{co} - T_{ci}} \;,\; S = \frac{T_{co} - T_{ci}}{T_{hi} - T_{ci}} \qquad (12.51)$$

Notice that R is simply the ratio of the heat capacities of the hot and cold streams and S, sometimes referred to as an "effectiveness", is the ratio of the heat actually absorbed by the cold fluid to the heat that would be absorbed if the cold stream were to be heated to the temperature of the inlet hot stream.

As mentioned earlier, heat-exchanger design involves a series of trade-offs between heat transfer coefficients, pressure drop and the required surface area for a given heat duty. The dependence of these variables on tube diameter, the number of tubes, and the number of tube passes can be illustrated by assuming turbulent tube flow and that the heat transfer rate is limited by the resistance on the tube-side of a shell and tube exchanger; i.e., $U \sim h_T$. From Table 12-3, the Nusselt number is dependent on the Reynolds number raised to the 0.8 power. Therefore:

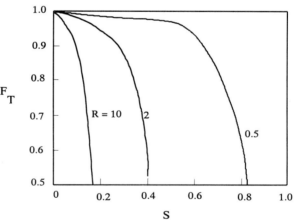

Figure 12-7 F_T Correction Factors [One Shell Pass, Two or More Tube Passes]

$$Nu = \frac{hD}{k} \propto (D_t \bar{v})^{0.8}$$

(12.52)

$$h \propto \frac{\bar{v}^{0.8}}{D_t^{0.2}}$$

where D_t is the tube diameter. For a fixed volumetric flow rate of fluid, the tube velocity can be calculated from a knowledge of the tube diameter, the number of tubes, N_t, and the number of tube passes, P_t,

$$\bar{v} = \frac{\dot{V} P_t}{\frac{\pi D_t^2}{4} N_t} \propto \frac{P_t}{D_t^2 N_t}$$

(12.53)

Using the definition of the Reynolds number and substituting $\bar{v}$ from Equation (12.53) into (12.52), h is found to be proportional to

$$h \propto \left[\frac{P_t}{N_t} \right]^{0.8} \frac{1}{D_t^{1.8}}$$

(12.54)

A similar dependency can also be derived for the pressure drop in the tubes. From Chapter 11, the friction factor for smooth tubes in turbulent flow was shown to be dependent on the Reynolds number raised to the negative 0.2 power

$$f = \frac{L}{D} \frac{\Delta P}{\rho \bar{v}^2} \propto (\text{Re})^{-0.2}$$

$$\Delta P \propto \frac{\bar{v}^2 (\text{Re})^{-0.2}}{D}$$

Substituting for $\bar{v}$ from Equation (12.53),

$$\Delta P \propto \frac{P_t^{1.8}}{N_t^{1.8}} \frac{1}{D_t^{4.8}}$$

(12.55)

Finally, the surface area required for a given heat duty, $\dot{Q}$, can be calculated from Equation (12.50) with U replaced by h

$$As = \frac{\dot{Q}}{h \, (\Delta T)_{LM} F_T}$$

The iterative nature of the design calculations can be seen from Equation (12.56).

$$N_t \, D_t = \left[\frac{1}{h(\Delta T)_{LM} F_T} \right] \frac{\dot{Q}}{\pi \, L} \qquad (12.56)$$

That is, the bracketed term depends on the number of tubes, the number of tube passes, and the tube diameter. Because of the large dependency of pressure drop on the tube diameter [Equation (12.55)], the tube diameter is usually chosen on the basis of pressure drop considerations. Then the number of tube passes is chosen in an effort to maximize h [Equation (12.54)] and to calculate F_T (Figure 12-7, or see [8] for other configurations). The total number of tubes, N_t, is then calculated from Equation (12.56).[#] However, since this choice will also influence the value of h, iterations are necessary to arrive at the optimum choice of the three parameters. It should be kept in mind that while increasing the number of passes will increase the heat transfer coefficient (at a cost of pressure drop), it will also *decrease* the log mean driving force. For example, if a 1-2 exchanger is used instead of a 1-1 exchanger and R = 10 and S = 0.15, the correction factor drops $(\Delta T)_{LM}$ by about 20% (from Figure 12-7).

EXAMPLE 12-4: Heat-Exchanger Design

7,000 lbs/hr of a lube oil is to be cooled from 200 to 100 F using 62,000 lbs/hr of cooling water available at 68 F. Specify a heat exchanger to meet these specifications. The following properties apply to the lube oil: $\hat{c}_P = 0.62$ BTU/lbm-F, $\mu = 5$ cp, $k = 0.067$ BTU/ft-hr-F. Typical practice in this case is to employ 20' "hairpin" U-tubes for double-pipe exchangers and 16', 1"-BWG 12 tubes laid out in a 1.25"-square pitch with 5" baffle-spacing for shell and tube exchangers.

Solution

The heat duty of this exchanger is

$$\dot{Q} = \dot{m} \, \hat{c}_P (T_{hR} - T_{hL}) = (7000)(0.62)(200 - 100) = 4.34 \times 10^5 \text{ BTU/hr}$$

which, from an energy balance on the cooling water, gives the exit water temperature as

---

[#] Heat-exchanger tubes are usually supplied in "off-the-shelf" lengths (12' ft. is typical).

$$T_{CR} = \frac{(4.34)10^5}{(1.0)(62,000)} + 68 = 75 \text{ F}$$

To start, let's try a shell and tube heat exchanger. From Table 12-1, a midrange overall heat transfer coefficient for this type of service is about 35 BTU/ft^2-hr-F. The log mean driving force can be calculated from Equation (12.46) as

$$(\Delta T)_{LM} = \frac{(200 - 75) - (100 - 68)}{\ln\left[\dfrac{125}{32}\right]} = 68 \text{ F}$$

Therefore, the required heat exchanger area from Equation (12.45) is

$$A = \frac{\dot{Q}}{U(\Delta T)_{LM}} = \frac{(4.34)10^5}{(35)(68)} = 182 \text{ ft}^2$$

This is not a large heat transfer area for a shell and tube exchanger, so let's explore the possibility of a double pipe exchanger. However, we must first pay attention to the pressure drop that might be experienced with this type of exchanger, since the water flow is rather large. As a first guess, choose a configuration consisting of 2" x 1 1/4" schedule-40 pipes. With water in the tube, the Reynolds number is

$$Re = \frac{4\dot{m}}{\pi D_i \mu} = \frac{4(62000)}{(3.14)(\dfrac{1.38}{12})(1.0)(2.42)} = 2.84 \times 10^5$$

The friction factor can be calculated from Equation (11.12)

$$f = \frac{0.046}{(Re)^{.20}} = \frac{0.046}{[(2.84)10^5]^{.20}} = 0.00373$$

The pressure drop per foot of pipe can then be calculated from Equation (11.10)

$$\frac{\Delta P}{L} = \frac{2 f \rho \bar{v}^2}{D} = \frac{2(0.00373)(62.4)\left[\dfrac{62000}{7.48\dfrac{\pi}{4}\left(\dfrac{1.38}{12}\right)^2}\right]^2}{\left(\dfrac{1.38}{12}\right)} = 2.5 \times 10^{12} \ (\text{lbm/ft-hr}^2)/\text{ft}$$

which converts to about 45 psi/ft of pipe length. Obviously this is unacceptable and so a shell and tube heat exchanger must be used.

For a single pass exchanger, the number of 1" BWG 12 tubes required is

$$N_T = \frac{182}{\pi D_o L} = \frac{182}{(3.14)(\frac{1}{12})(16)} = 44$$

The inside diameter of 1" BWG 12 tubes is 0.75" and with water flowing in the tubes, the Reynolds number is

$$Re = \frac{4\dfrac{62000}{44}}{(3.14)(\frac{0.75}{12})(1)(2.42)} = 11,870$$

The Reynolds number based on the geometric mean mass velocity on the shell side is calculated from Equations (12.47) and (12.48), using the dimensions described in Figure 12-6. With $P_T = 1.25$", $c' = 0.25$" and $B = 5$" and $f_B = 0.1955$, the diameter of the shell is about 9"

$$G_C = \frac{7000}{(\frac{5}{12})(\frac{9}{1.25})(\frac{0.25}{12})} = 90,720$$

$$G_P = \frac{\dot{m}_S}{(A_P)_f} = \frac{7000}{(.1955)(\frac{3.14}{4})\left(1^2 - 44(\frac{1}{12})^2\right)} = 66,000$$

$$\overline{G} = \sqrt{(90720)(66000)} = 77,380 \text{ lbm/ft}^2\text{-hr}$$

The Prandtl numbers for the oil and water are

$$Pr(oil) = \frac{(0.62)(5)(2.42)}{0.067} = 112$$

$$Pr(water) = \frac{(1.0)(1)(2.42)}{0.36} = 6.7$$

The tube side heat transfer coefficient is calculated from the correlation for turbulent flow in a smooth tube and the shell-side coefficient is calculated from Equation (12.27)

$$h_o = \frac{0.067}{1/12} (0.2) \left[\frac{(1/12)(77380)}{(5)(2.42)} \right]^{0.6} (112)^{.33} = 33 \ \text{BTU/ ft}^2\text{-hr-F}$$

$$h_i = \frac{0.36}{0.75/12} (0.023)(11{,}870)^{.80}(6.67)^{.33} = 451 \ \text{BTU/ ft}^2\text{-hr-F}$$

Neglecting conduction through the tube walls, the overall heat transfer coefficient is

$$U_o = \frac{1}{\dfrac{1}{.75}\dfrac{1}{451} + \dfrac{1}{33}} = 30 \ \text{BTU/ ft}^2\text{-hr-F}$$

which is very close to the originally assumed value of 35. Finally, the pressure drop for the water stream should also be calculated. Following the same set of calculations as listed above, except that, since the flow is in the transition region, Equation (11.13) is used in place of (11.12), the pressure drop is about 0.95 psi/ft pipe, a much more tolerable value. Note that, since the primary heat transfer resistance is due to the shell side convection, if a lower pressure drop is desired, the number of tubes could be increased without seriously affecting the heat transfer rate.

12-3.c Extended Area Heat Exchangers

There is one more heat exchanger configuration which needs to be discussed in the context of macroscopic heat transfer calculations and that is the "extended area" heat exchanger. These exchangers are used when heat must be exchanged between gases and liquids; for example, an air cooled heat exchanger. In these cases the overall heat transfer coefficients are totally dominated by the gas-side ("outside") heat transfer coefficients; i.e., $U \sim h_o$. When this occurs, one strategy is to "extend" the surface area on the gas-side of the exchanger and consequently these are sometimes called extended area heat exchangers. The surface area can be extended by attaching "fins" to the outside tubes of the heat exchanger and the magnitude of the increase will be a function of the length of the fins as well as the number of fins that are placed on the tube. There are two limitations here. If the fins are placed too close to one another, the convective heat transfer coefficient between the fins and the surrounding gas will decrease as a result of stagnant zones of fluid in between the fins. The other limitation has been discussed at length in Section 3-3. That is, the local temperature along the fin decreases due to molecular transport limitations. If the fin is too long, the temperature driving force between the fin and the surrounding gas will drop significantly. The magnitude of this effect can be accounted for in terms of the "effectiveness factor", η, (Equation 3-25). The variation of η with the parameter, $\Gamma_h L$ (a dimensionless parameter relating the thermal properties and the geometry of the fin) was illustrated in Figure 3-10.[#] Once a value of η is obtained, the outside resistance of the tube is corrected according to Equation (12.57)

$$R_o = \frac{1}{h_o \left(\eta \, A_F + A_o \right)}$$

(12.57)

where A_F is the fin area and A_o is the outside area of the *bare* tube.

REFERENCES

[1] Kern, D.Q., *"Process Heat Transfer"*, McGraw-Hill, N.Y., 1965

[2] Kern, D.Q., *"Process Heat Transfer"*, McGraw-Hill, N.Y., 1965, Figures 4.7-4.9

[3] Dry, R.J. and R.D. La Nauze, Chem. Engr. Progr., July, 1990, p. 31

[4] Whitaker, S., AIChE J., 18, p. 361 (1972)

[5] Leva, M, Ind. & Eng. Chem., 42, p.2498 (1950)

[6] Gelperin, N.I. and V.G. Einstein, in *"Fluidization"*, (ed., J.F. Davidson and D. Harrison), Academic Press, N.Y., (1971), p. 517

[7] Bromley, F., Chem. Engr. Progr., 46, p. 221 (1950)

[8] Perry, R.H and C.H. Chilton, "Chemical Engineer's Handbook", 5th ed., McGraw-Hill, N.Y., 1973

PROBLEMS

12-1   Solve Equations (12.4-12.7) simultaneously to eliminate the temperatures, T_{fc}, T_{wc}, and T_{wH} and obtain an expression for $\dot{Q}$ in terms of the driving force, $(T_H - T_c)$. Then compare this result to Equation (12.8) to obtain an expression for $U_{,}$. Finally, express this result in terms of the heat transfer coefficients , using the definition that $h = \dfrac{k}{\delta_{pf}}$

12-2   Kato et al. (*Int. J. Ht. & Mass Transfr*, v 11, p.1117, 1968) recommend the following heat transfer coefficient correlations for free convection from vertical plates and cylinders:

[#]   This was for a cylindrical fin. Other solutions are available for radial and rectangular fins.

For $Gr > 10^9$: $Nu = 0.138 \ Gr^{.36} \ (Pr^{0.175} - 0.55)$

For $Gr < 10^9$: $Nu = 0.683 \ (Gr \ Pr)^{.25} \left[\dfrac{Pr}{0.861 + Pr} \right]^{.25}$

Calculate the heat transfer coefficient using the appropriate correlation for air surrounding a vertical, 2-in schedule-40 pipe that is 12-ft long, if the pipe wall is at 115 °C and the surrounding air is at 3 °C. Compare these results using the Nusselt equation [Equation (12.19)] together with the appropriate parameters from Table 12-2.

12-3 The Handbook of Chemical Engineering (R.H. Perry and C.H. Chilton, 5ᵗʰ ed., McGraw-Hill Co., N.Y., 1973, p.10-11) claims that, in free convection, the Nusselt number (Equation (12.19)) reduces to

$$h = b \, (\Delta T \,)^m \, L^{3m-1}$$

Specifically, it is stated that for air at 70 °F and a vertical surface where $10^4 < Gr \ Pr < 10^9$, the value of b = 0.28. Show that the form of the equation is valid and then calculate the value of b (give the units). Are there any restrictions on ΔT and L?

12-4 A distillation column, 8-m high and 1 m in diameter has a surface temperature of 140 °C.

(a) Calculate the heat loss rate from the column under ambient air conditions at 22 °C.

(b) It is proposed to add 2" of insulation to the column, which will reduce the surface temperature to 40 °C. What will be the payout time if the installed insulation cost is 12 $/ft² and the cost of energy is $2.50/ "therm" ($10^6$ BTU)?

12-5 The Alaska pipeline is 800 miles long and 420 miles of it are aboveground so as not to damage the permafrost. Crude oil from the north slope flows at about 88,000 barrels per day, entering the pipeline at 116 °F. The pipe is stainless steel, has a 48" inside diameter, a 0.462" wall thickness and is covered with 3.75" of insulation (thermal conductivity = 0.01 BTU/hr-ft-F). Assume that the crude oil has a viscosity of 6 cp, a thermal conductivity of 0.07 BTU/hr-ft-F, a specific heat of 0.76 BTU/lb-F and a specific gravity of 0.89. Assume also that the heat losses from the pipe during its below ground transit can be ignored.

(a) Derive and solve the differential equation which describes the temperature of the oil as a function of distance along the pipe line.

(b) Will the pipe need to be heated at periodic intervals in order to insure that the temperature does not fall below 82 °F? If so, at what intervals? (Assume the 420 miles are continuous).

(c) Assuming that each station will heat the oil to 116 °F, calculate the cost per year of heating the pipeline if the energy cost is $2.00/ 10^6 BTU.

12-6   Table 12-3 lists heat transfer coefficient correlations for both laminar and turbulent flows in tubes. Using this table, calculate the value for the Nusselt number in a horizontal pipe at a Reynolds number of 10^4 and also at a Reynolds number of 1000. In each case use <u>BOTH</u> the turbulent and laminar correlations, assuming that the Prandtl number is 8 and that the L/D of the pipe is 200.

12-7   The parameter, $\Phi = \left(\dfrac{\mu}{\mu_w}\right)^{.14}$ appears in many of the correlations listed in Table 12-3 and attempts to correct for the dependency of viscosity on temperature when the wall temperatures are different than the bulk fluid temperatures; i.e., when $\Delta T = T_w - \overline{T}$ is large. Given the fact that heat transfer coefficient correlations are rarely better than 10-20%, estimate the value of T_w which would make this correction significant (i.e., = 20%) for both air and water at 60 °F being heated by a pipe wall.

12-8   A new house is located on hill where the average wind velocity is 20 mph. The owners specifically designed the house to take advantage of the view and therefore 40% of the external surface area of the house consists of thermal pane windows without window coverings. The house is heated by a natural gas-fired (80% efficient) furnace and during one winter month when the average outside temperature was 28 °F and the average inside temperature was 65 °F, the gas bill was $105. The cost of the gas was $3.00/$10^6$ BTU. The owners are now considering whether to cover the windows in order to minimize radiation losses from the windows. The following properties of the house apply: Total external surface area = 1700 ft^2, effective "R" Value of walls (including wallboard and insulation) = 41 ft^2-hr-°F /BTU. The thermopane windows consist of 1/4"-plate glass (k = 0.5 BTU/hr-ft-F) sandwiched around a 1/4"-air gap.

12-9   Assume that an oil, at 210 lb/min, is being cooled from 250 °F to 125 °F in a counter-current, double pipe heat exchanger (carbon steel, inside pipe = 2", schedule-40, outside pipe = 3", schedule- 40). Cooling water enters the annulus of the double pipe exchanger at 45 gpm and 60 °F.

The oil has a viscosity of 2.7 cp, a specific heat of 0.42 BTU/lbm-F, a specific gravity of 0.89, a Prandtl number of 19, and a coefficient of thermal expansion of 9 x 10^{-4} °F^{-1}

(a) Calculate the exit water temperature.

(b) Calculate the length of pipe required.

12-10 For the oil-water system described in Problem 12-9, to what temperature will the oil be cooled if the oil and water flow rates are reduced to 20 lbm/min and 8 gpm, respectively and the length of the exchanger is 20 ft?

12-11 An experiment to determine the inside heat transfer coefficient in a 4' long double pipe heat exchanger yields the following data:

Pipes: Inside = 5/8" 12 BWG copper tube
            Outside = 2" schedule 40 pipe

Fluids: Inside = orthophenyl phenophenate (OOP) at 100 lb/hr , $\hat{c}_P$ = 0.68 BTU/lb-F
            Outside = saturated steam at 75 psig

Temperatures: Inlet OOP = 60 F
                   Outlet OOP = 130 F.

Calculate the inside heat transfer coefficient.

12-12 A 1-1 shell and tube heat exchanger with a total inside surface area of 51 ft^2 (1" BWG 12 tubes, 12' long) operates with a petroleum fraction on the tube side and saturated steam on the shell side (temperature = 300 °F). The petroleum fraction flows at 10,000 lb/hr, has a $\hat{c}_P$ = 0.57 BTU/lb-F, a viscosity of 1 cp, enters at 60 °F and leaves at 250 °F. It is proposed to increase the petroleum flow to this exchanger to 17,500 lb/hr and maintain the same outlet temperature. What must the steam temperature be raised to in order to operate under these proposed conditions? It can be assumed that the limiting heat transfer coefficient is essentially the tube side value.

12-13 Kerosene, at a flow rate of 43,000 lb/hr, leaves the bottom of a distillation column at 390 °F and is to be cooled to 200 °F by interchanging heat with 149,000 lb/hr of a crude oil stream delivered from storage at 100 °F. Specify a shell and tube heat exchanger to meet this requirement. The shell is to be equipped with baffles spaced at 5" intervals and the tubes are to be 1"-copper, BWG 13,

16' long and laid out in a 1.25" square pitch. The pressure drop for the crude stream is restricted to be below 10 psi. The following constant fluid properties apply:

	Kerosene	Crude
$\hat{c}_P$	0.605	0.49 BTU/lb-F
k	0.0765	0.077 BTU/ft-hr-F
ρ	45.6	51.8 lb/ft^3
μ	0.4	3.6 cp

CHAPTER 13

Macroscopic Calculations: Mass Transfer

In this chapter, the basics of transport phenomena are combined with the macroscopic approach of mass transfer coefficients to calculate the rates of interphase transport for selected multi-phase systems. The goal here is to illustrate how a knowledge of transport phenomena can be used to <u>size</u> mass transfer equipment; i.e., how to determine the residence time necessary to achieve separation of one or more species from one phase to another. There are many applications of this approach [1] and the illustrations below are not meant to be an exhaustive reference. Rather, only the more common applications are analyzed and only one application is chosen for each of the specific multiphase systems. Thus, gas absorption (or "stripping"), a gas-liquid mass transfer system is analyzed, but distillation, another common gas-liquid separation system, is not covered. Liquid-liquid extraction is also not covered because it (and distillation) are commonly analyzed in terms of the prevailing equilibria and then corrected for finite mass transfer rates by the use of *stage efficiencies*. The latter is largely empirical and thus does not (yet) lend itself to the transport phenomena approach. However, gas adsorption, an example of separation in a gas-solid system, <u>is</u> compatible with this approach and is used to illustrate how these principles can be applied to mass transfer between fluids and solids. More complex applications of mass transport phenomena — cooling towers, gas-stripping with chemical reactions, and the membrane separations of gases are covered in the final portion of this chapter.

13-1 INTERFACIAL MASS TRANSFER COEFFICIENTS

It has already been pointed out that mass transport rates epend on both the degree of mixing in a system as well the magnitude of the interfacial area. In many gas and liquid systems, the degree of mixing is directly related to the creation of interfacial surface area. That is, energy is expended in order to create an intimate gas-liquid mixing state, either through the use of spargers or solid particles ("packings"). Since mass transfer operations purposely operate with a high degree of mixing, there is little chance of being able to calculate the mass transport rates from the methods employed in Chapter 4. Instead, the approach is necessarily empirical and thus the methods of Chapter 10 must be used. That is, mass transfer coefficients are defined and then experimental data combined with similarity analysis, are used to correlate the data.

13-1.a Overall Mass Transfer Coefficients

Keep in mind that when we use mass transfer coefficients we have abandoned the goal of calculating concentration profiles and are focusing instead on the calculation of mass transfer rates between adjacent phases. Recall that the mass transfer coefficient can be defined in terms of a variety of driving forces, and in Chapter 10, it was shown that

$$k_{mx} = k_{mc} C = \frac{k_{mc} P}{R_g T} = k_{mp} P \tag{10.8}$$

It was also shown in Chapter 10 that an "overall" mass transfer coefficient can be defined in terms of the <u>difference</u> in the bulk average mole fractions in the two phases, so that the molar flux can be calculated from

$$N_A = K_{OGx} \Delta X_A \tag{13.1}$$

and where K_{OGx} is defined by Equation (10.28)

$$K_{OGx} = \left[\frac{1}{\dfrac{m_A}{k_{mx}^L} + \dfrac{1}{k_{mx}^G}} \right] \tag{10.28}$$

Equation (10.28) can also be derived by using "resistances in series," similar to what was done in Chapter 12 for the overall heat transfer coefficient (see Section 12-1). Referring to the phoney film concept discussed in Section 10-3 and shown in Figure 13-1, there are two resistances; the phoney films on either side of the gas-liquid interface. The molar flux can be

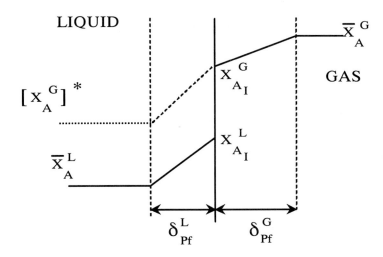

Figure 13-1 Two-Film Theory

calculated in terms of the mole fraction driving force across either of the films, or in terms of the overall driving force between the two bulk phases

$$N_{A_I} = k_{mx}^G (\overline{X}_A^G - X_{A_I}^G) = k_{mx}^L (X_{A_I}^L - \overline{X}_A^L) = K_{OG_x} \Delta X_{OG} \tag{13.2}$$

where ΔX_{OG} is an overall gas-phase driving force, yet to be determined

Before we can combine the resistances in Equation (13.2), we must first get all the driving forces in a consistent set of units. As it now stands, this equation has both gas-phase and liquid-phase mole fractions as driving forces and the two are not really compatible because of the differences in the density of the two phases.[#] In order to get both sets of driving forces in the same set of units, we can use the equilibrium relationship between the two phases; i.e.,

$$X_A^G = f \ (X_A^L) = m \ X_A^L \tag{13.3}$$

where, for simplicity, we assume that the equilibrium relationship is a linear one.[##]

The usual assumption is that equilibrium exists at the interface, so that

---

[#]   If chemical potential (or fugacity) were used as the driving force, this would not be a problem since they would be continuous across the interface.

[##]   This is usually valid at low liquid concentrations.

$$X_{A_I}{}^G = m_A \, X_{A_I}{}^L$$

$X_{A_I}{}^L$ can now be written in terms of $X_{A_I}{}^G$ and $\overline{X}_A^L$ can be multiplied and divided by m_A , so that Equation (13.2) becomes

$$N_{A_I} = k_{mx}^G (\overline{X}_A^G - X_{A_I}{}^G) = k_{mx}^L \left(\frac{X_{A_I}{}^L}{m_A} - \frac{m_A}{m_A} \, \overline{X}_A^L \right) = K_{OG_x} \, \Delta X_{OG}$$

(13.4)

$$N_{A_I} = k_{mx}^G (\overline{X}_A^G - X_{A_I}{}^G) = \frac{k_{mx}^L}{m_A} (X_{A_I}{}^G - [X_A^G]^*) = K_{OG_x} \, \Delta X_{OG}$$

where $[X_A^G]^* \equiv m_A \, \overline{X}_A^L$. We can use this same definition to express the overall driving force in terms of gas-phase concentrations. That is

$$\Delta X_{OG} = \overline{X}_A^G - [X_A^G]^*$$

Even though m_A is an equilibrium parameter, it is really used here as a conversion factor so that the bulk liquid-phase mole fraction can be expressed in terms of a gas-phase mole fraction. Stated in words, $[\overline{X}_A^G]^*$ is the gas-phase mole fraction which <u>would</u> be in equilibrium (of course it really isn't in equilibrium) with the actual liquid-phase mole fraction. So really, $[\overline{X}_A^G]^*$ is a "psuedo equilibrium concentration," It's effect on the gas-liquid concentration profile can be seen graphically in Figure 13-1 where the "adjusted" concentration profile through the liquid side phoney film (dashed line) is now continuous through the interface. Defining ΔX_{OG} in this manner now allows for the use of series resistances since the mole fractions at each point in the overall concentration profile are self-consistent [see Equation (2.14) in Section 2-2].

The overall mass transfer coefficient can now be expressed in terms of the individual mass transfer coefficients by identifying the resistances in Equation (13.4); namely

$$R_G = \frac{1}{k_{mx}^G} \; , \; \; R_L = \frac{m}{k_{mx}^L} \; , \; \; R_{Total} = \frac{1}{K_{OG_x}}$$

and since, for resistances in series, $R_{Total} = R_G + R_L$,

$$\frac{1}{K_{OG_x}} = \frac{1}{k_{mx}^G} + \frac{m}{k_{mx}^L}$$

or

$$K_{OG_x} = \frac{1}{\left[\dfrac{m}{k_{mx}^L} + \dfrac{1}{k_{mx}^G}\right]} \tag{13.5}$$

which is the same as Equation (10.28). With this definition of the overall mass transfer coefficient, the molar flow rate across the interface can be calculated from

$$N_I = K_{OG_x}\left(\overline{X}_A^G - [\overline{X}_A^G]^*\right)$$

Ideally, attempts are made to correlate data for the individual gas or liquid-side mass transfer coefficients. However, all that can be measured in complex gas-liquid systems is the overall mass transfer coefficient and therefore much of the collected experimental data utilized conditions where one "side" (gas or liquid) was presumably controlling. For example, in order to obtain correlations of gas-side mass transfer coefficients, Equation (10.28) shows that it is necessary that $K_{OG} \sim k_{mx}^G$, or, $\dfrac{m}{k_{mx}^L} \gg \dfrac{1}{k_{mx}^G}$. Since the degree of mixing generally affects the liquid and gas side mass transfer coefficients to the same degree, the value of the equilibrium parameter, m, determines which side will be controlling. That is, high solubility gases (low m) have been used to obtain correlations of gas-side mass transfer coefficients and low solubility gases (high m)[*] for liquid-side mass transfer coefficient correlations.

Correlations of experimental data are necessary in order to obtain numerical values of the mass transfer coefficient and the complexity of these correlations depend on whether or not the interfacial area can be independently determined. When the interfacial area is regular and known, the correlations are amenable to similarity principles and are remarkably similar to those used for heat transfer coefficients. However, when there is intimate mixing of two phases, the effect of interfacial area cannot be readily distinguished from the mass transfer coefficient. In these situations the mass transfer coefficients are "lumped" together with the interfacial area and the experimental data are often correlated in terms of "volumetric" mass transfer coefficients, $k_{mx}\,a_I$, where a_I is the interfacial surface area per unit volume of the experimental device (e.g., the volume a packed bed). Unfortunately these correlations tend to be very specific to the methods of producing the interfacial area (packings, spargers, etc.) and thus there are precious few generalized correlations available. For these reasons, the review of mass transfer

---

[*] Another method is to add a liquid component that reacts rapidly with the gas component.

correlations is presented here in two parts; the first for known interfacial areas and the second for systems where the area is not known *a priori.*

13-1.b Correlations for Known Interfacial Areas

Single-Phase Mass Transfer Coefficients – Internal Flows

It is sometimes necessary to calculate mass transfer rates in systems with well-established phase boundaries. Examples here would include membrane separators (see Section 6 of this chapter), fluid-solid reaction systems and catalytic wall reactors. In these applications, there is single-phase convective transport to well-defined "surfaces" and the flow can be either laminar or turbulent. Although "wetted wall columns" (see Problem 8-5) would appear to fit this category, disturbances at the gas-liquid-phase boundary produce sufficient disruption in the convective patterns that different correlations are needed. Consequently, these correlations are presented separately.

Due to the fact that the transport fluxes of both energy and mass are vectors, there are strong analogies between the two and thus the mass transfer coefficient correlations which apply to tubes/ducts are quite similar to those discussed in Chapter 12 for heat transfer in tubes and ducts. For example, in *laminar flow*, mass transfer coefficients can be correlated by a modification of the "Graetz" heat transfer correlation, Equation (12.21)

$$Sh = 1.86 \, (Gz_m)^{1/3} \tag{13.6}$$

Where the Sherwood number, Sh, is defined in the usual manner,[#] $Sh = \dfrac{k_{mc} D}{D_{G,L}}$ And the "mass transfer" Graetz number is

$$Gz_m = \frac{\pi D}{4L} \, Re \; Sc = \frac{\dot{m}}{D_A L \, \rho} \tag{13.7}$$

To apply this equation to annuli or rectangular ducts, the tube diameter, D, is replaced by the "equivalent diameter", D_{eq}, as defined in Equation (10.14).

In *turbulent flow,* the basis for mass transfer coefficient correlations is the "j-factor" correlation discussed in Chapter 12 which is used in terms of the Chilton-Colburn, analogy. That is,

---

[#]   The G, L subscripts on diffusivity indicate a gas or liquid value.

$$j_M = j_H = \frac{f}{2} = .023 \ Re^{-.2} \tag{13.8}$$

where

$$j_M = \frac{M \ k_{mx}}{G_{G,L}} \ Sc^{2/3} \tag{13.9}$$

and which only holds for long, smooth tubes with little "form drag" (i.e., minimal expansion, contraction, fittings etc.). This equation can be placed in terms of the Sherwood number by combining the definition of the Sherwood number with the relationship between k_{mx} and k_{mc} [Equation (10.8)] to give

$$Sh = .023 \ Re^{.8} \ Sc^{1/3} \tag{13.10}$$

which is identical to the Dittus-Boelter Equation discussed in Chapter 12 with the Prandtl Number replaced by the SChmidt number. A further restriction on Equation (13.10) is that the SChmidt number should be low; specifically, $.6 < Sc < 2.5$. For high SChmidt numbers ($430 < Sc < 10^5$) the following equation is recommended [2]

$$Sh = .0096 \ Re^{.91} \ Sc^{.35} \tag{13.11}$$

Single Phase Mass Transfer Coefficients – External Flows

The rates of many catalytic and noncatalytic reactions are limited by the rates of reactant transport to and from solid particles. In these situations it is necessary to be able to predict the influence of mass transfer on the reaction rates and therefore, to have a knowledge of the prevailing mass transfer coefficients. Although the j-factor analogy is not as strong when there is convective transport normal to single particles (objects), it is still applicable. For mass transfer to *single particles*, Ranz and Marshall [3] proposed the following correlation for *spheres*

$$Sh_p = 2.0 + .61 \ Re^{1/2} \ Sc^{1/3} \tag{13.12}$$

which is applicable for $Re_p < 10^3$, and the Sherwood number approaches the limiting value of 2 at very low Reynolds numbers.[#] Equation (13.12) can also be applied to situations with flow normal to *"long" cylinders (L/D > 1)*, in the form

---

[#]   See Problem 3-27 in energy transport and Section 4-5 in mass transport.

$$Sh_p = .61 \, Re_p^{1/2} \, Sc^{1/3} \tag{13.13}$$

for $10 < Re_p < 10^4$ and with the Sherwood number based on cylinder diameter.

The applications of *mass transfer coefficients in packed and fluidized beds* are similar to those listed above for single particles and they are used to correct catalytic and non-catalytic reaction rates for the influence of mass transport effects in packed or fluidized bed reactors.[#] There are many correlations available in the literature, depending on flow characteristics, particle size and shape, but one illustrative correlation for both packed and dense phase "bubbling" fluidized beds (see Chapter 11-3) has been proposed by Chu et al. [5] and is presented here. For low Reynolds numbers, $1 < \dfrac{Re_p}{1 - \varepsilon} < 30$

$$Sh_p = 5.7 \, Re_p^{.22} \, Sc^{1/3} \, (1 - \varepsilon)^{.78} \tag{13.14}$$

and for high Reynolds numbers, $30 < \dfrac{Re_p}{1 - \varepsilon} < 10^4$

$$Sh_p = 177 \, Re_p^{.56} \, Sc^{1/3} \, (1 - \varepsilon)^{.44} \tag{13.15}$$

These equations are valid for both gases and liquids over a Schmidt number range of .6 $< Sc < 1400$ and where ε is the void fraction in the bed or the void fraction of a fluidized bed at minimum fluidizing conditions.

Mass Transfer Coefficients – Simple Two-Phase Fluid Systems

Two examples of simple two-phase fluid systems are mass transfer between a liquid and well-defined gas bubbles and mass transfer between a gas and a thin film of liquid flowing down the walls of a wetted wall column. The former situation is important in many biological "aeration" processes and the latter has often been used in an attempt to separate the effects of interfacial area from convective mass transport in gas-liquid systems. For *bubble columns* where the liquid-side mass transfer is controlling, the Sherwood number based on the bubble diameter is correlated by

$$Sh_{d_B} = 1.13 \, Pe^{1/2} \left[\frac{d_B}{.45 + .2 \, d_B} \right] \tag{13.16}$$

[#] Another application is "fluidized bed driers" [4].

where d_B is the bubble diameter and the Peclet Number is related to the liquid-phase velocity and diffusivity by

$$Pe = \frac{d_B v^L}{D^L} = Re_{d_B} \; Sc$$

There are separate correlations for gas and liquid side mass transfer coefficients in *wetted wall columns* but there are very few wetted wall systems where the liquid-side is controlling. Consequently the correlation for the liquid-side coefficient should be used with caution. For the *gas-side mass transfer coefficient*, the correlation proposed by Gilliland and Sherwood [6] is most commonly used

$$Sh = .023 \, Re^{.83} \, Sc^{.44} \tag{13.17}$$

The similarity between this correlation and that for heat and mass transfer inside tubes is not unexpected. For the *liquid-side mass transfer coefficient*, Vivian and Peaceman [7] proposed

$$Sh_H = .433 \left(\frac{4\Gamma}{\mu} \right)^{.4} \left(\frac{\rho^2 g H^3}{\mu^2} \right)^{1/6} Sc^{2/3} \tag{13.18}$$

where the Sherwood number is based on the height of the column, H, and Γ is the "hydraulic loading", $\dfrac{\dot{m}}{\pi D}$, (see Section 12--2d).

13-1.c Correlations for Complex Interfacial Areas

Interfacial Surface Area Correlations

Ideally, the correlations of mass transfer coefficient data would be in dimensionless form and would be separate from correlations of interfacial area. As already mentioned, most mass transfer coefficient data in complex gas-liquid systems are in the form of "volumetric" mass transfer coefficients, $k_{mc} \, a_I$, which are the product of the mass transfer coefficient and the interfacial area/volume. However there have been a few attempts to obtain separate correlations of the interfacial surface area. The most recent correlation[#] is due to Onda et al. [8] who correlated the wetted surface area/volume of packing by

---

[#]   See Problems 13-3 and 13-4 for other measurements.

$$\frac{a_I}{a_p} = 1 - \text{Exp}\left[-1.45 \left(\frac{\sigma_c}{\sigma} \right)^{.75} Re_L^{.10} Fr^{-0.05} We^{0.20} \right] \qquad (13.19)$$

Where a_p is the surface area per unit volume of the dry packing (see Table 11-3), $\frac{\sigma_c}{\sigma}$ is the ratio of the critical surface tension of the packing to the surface tension of the liquid, and the Reynolds, Froude, and Weber numbers are defined as

$$Re_L = \frac{G_L}{a_p \mu_L} \quad , \quad Fr = \frac{G_L^2 a_p}{\rho_L^2 g} \quad , \quad We = \frac{G_L^2}{\rho_l \sigma a_p}$$

Values of σ_c for some representative packing materials are: ceramics = 61, polyethylene = 33, PVC = 40 dynes/cm. Note that the correlation is independent of the gas mass velocity since it is the liquid that typically covers portions of the packing,[#] thus reducing its effective surface area.

Liquid-Side Mass Transfer Coefficients

Liquid-side mass transfer coefficients are usually obtained from data on systems with slightly soluble gases [high values of m; see Equation (13.3)]. This is due to the fact that gas-phase characteristics do not influence liquid-side mass transfer until gas flows approach the flooding point. All of the following correlations are in terms of liquid properties unless otherwise stated.

Both Onda et al. [8] and Schullman et al. [9] obtained correlations which were independent of the interfacial area. Schullman's correlation is given in Equation (13.20)

$$Sh_p = 25.1 \ Re_p^{.45} \ Sc^{.50} \qquad (13.20)$$

and has the advantage that it is dimensionless. However the data were collected only for Raschig rings and Berl saddles. Onda et al. [8] correlated data for a wide range of packings and liquids and obtained Equation (13.21) which correlated the data to within 20%

$$\left(\frac{k_{mc}^3 \rho}{\mu g} \right)^{1/3} = 0.0051 \ (a_p d_p)^{.40} \left(\frac{G_L}{a_I \mu} \right)^{2/3} Sc^{-.50} \qquad (13.21)$$

[#] This is called *Hold-up*, which is defined as the fraction of the packing that is always occupied by liquid under steady-state conditions.

where the interfacial area, a_I, is obtained from Equation (13.19).

There are many other correlations in the literature which are in terms of the volumetric mass transfer coefficient. Typically these correlations are not dimensionless, but are given in terms of the <u>height of a transfer unit</u>, which is defined for the liquid-side as:

$$H^L = \frac{G_L}{k_{mc}\, a_I\, \rho} \tag{13.22}$$

The physical meaning of this term will become apparent in the next section of this chapter when we address the design of gas-liquid separators. For now it is only important to appreciate that it is *inversely proportional* to the volumetric mass transfer coefficient.

One of the earliest correlations of this type which is still in wide use is that due to Sherwood and Holloway [10]

$$H^L = \alpha \left(\frac{G_L}{\mu} \right)^n Sc^{.5} \tag{13.23}$$

where values of α and n are a function of the packing and size. For 1–2" Raschig rings, n = .22 and α = 0.011, and for 0.5 - 1.5 " Berl saddles, n = .28 and α = 0.0063. This equation is not dimensionless and thus all the parameters with the exception of the SChmidt number are accompanied by a specific set of units; viz., $H^L \sim$ [ft], $G^L \sim$ [lbm/ft²-hr], $\mu \sim$[lbm/ft-hr].

Gas-Side Mass Transfer Coefficients

Gas-side mass transfer coefficients are obtained from either the vaporization of pure liquids into gases or from highly soluble (or reacting) gases in liquids. Unfortunately all of these systems have heat effect problems (heats of vaporization, solution and reaction) which detract from obtaining good reliable data. Unlike liquid-side mass transfer coefficients where the coefficients are primarily dependent on the liquid mass velocity, the gas-side coefficients are dependent on both. The earliest attempt at the correlation of gas-side mass transfer coefficients was by Sherwood and Holloway [10] who correlated liquid vaporization data within 20 % in terms of the gas-side height of a transfer unit, H^G

$$H^G = \frac{G_G}{M\, k_{mx}^G\, a_I} = 1.01 \left(\frac{G_G}{G_L} \right)^{.32} \tag{13.24}$$

where the mass velocities are in lbm/ft²-hr units and H^G is in ft. Note that this correlation is essentially dependent on the ratio of the gas and liquid mass velocities. In another approach,

Schullman and his co-workers attempted to account for the liquid properties in terms of the liquid holdup in the bed by using

$$\frac{k_{mx}^G M}{G_G} = 1.195 \left[\frac{d_p G_G}{\mu (1 - \varepsilon_L)} \right]^{-0.36} Sc^{-2/3} \tag{13.25}$$

where ε_L is the void fraction corrected for liquid holdup which is, in turn, a function of G_L and the liquid properties [11]. Onda et al. [8] also developed a gas-side mass transfer coefficient correlation which was independent of the interfacial area

$$\frac{k_{mx}^G}{a_p D^G} = C_1 \left(a_p d_p \right)^{-2.0} \left(\frac{G_G}{a_p \mu} \right)^{0.7} Sc^{1/3} \tag{13.26}$$

where $C_1 = 5.23$ for Raschig rings and saddle packings greater than ½" and is equal to 2.0 for smaller packings. Note that this correlation is independent of the liquid flow or properties and consequently is associated with a rather large uncertainty ($\pm$ 20%).

Unless otherwise designated, the properties in these three correlations are gas properties. It should also be pointed out that k_{mx}^G should be corrected for the log-mean mole fraction difference of the non transferring components if stagnant film diffusion takes place (see Equation (10.26) in Chapter 10-3b).

13-2 MASS TRANSFER IN GAS ABSORBERS AND STRIPPERS

As pointed out early in the text, convective mixing is advantageous to transport rates because it serves to bring fluid elements in close contact for short periods of time thereby allowing molecular transport to occur in the presence of steep driving force gradients. This can be promoted in gas-liquid systems if the gas is highly dispersed in the liquid and also has the added advantage of increasing interfacial surface areas (since mass transfer *rate* is proportional to surface area). Thus practical mass transfer equipment almost always involves some kind of low-energy contacting device such as spargers, distributors, mechanical agitators, or a column packing. Column packings are the most common means of producing intimate fluid contact in large-scale gas-liquid mass transfer operations, and issues related to the pressure drop (energy loss) and "flooding" characteristics of packed columns have been discussed in Chapter 11.

13-2.a Mathematical Description of Mass Transfer Rates

"Gas absorption" involves the transport of a solute from a gas to a liquid whereas "stripping" involves the transport of a solute from liquid to gas. Note that these two similar operations are classified in terms of what is happening to the liquid; i.e., whether the solute is being absorbed or stripped from the liquid. Commercial applications of these two mass transfer operations are numerous and include the absorption of "acid gases" (e.g., CO_2, H_2S) with aqueous solutions, the absorption of NH_3 and NO_2 by water and the stripping of "sour water" with steam. Although many gas absorption systems actually employ a chemical reagent in the liquid-phase (Diethanolamine for example in the absorption of acid gases), we will concentrate here on nonreactive systems, leaving the subject of simultaneous mass transfer and chemical reactions until later in the chapter. In order to maximize the mass transfer driving force, gas absorber flows are typically countercurrent.[*] Figure 13-2 is a sketch of an absorber and the labeled streams indicate the molar flow rates and mole fractions at the top and bottom of the absorber for both the liquid and gas-phases. These values would typically be known as a result of an overall material balance, since the incoming gas conditions (flow rate and composition), the conditions of the inlet liquid stream and the required degree of separation would all be stated as part of the design problem. Note that the temperatures of each stream are also labeled in the sketch, since the possibility of finite heat transfer rates between gas and liquids could cause temperature differences between the two streams. Usually these differences are small, and so we will assume that, at any point in the column, $T^G = T^L$.

The gas and liquid-phases contact each other in a rather chaotic and tortuous manner throughout the column (see the inset in Figure 13-2). From a practical point of view, we can deal with this complexity by resorting to mass transfer coefficients to account for interfacial mixing and thereby allow us to treat the system as if the gas and liquid flows are uniform and parallel. If so, there will be no macroscopic gradients in the radial direction and the applicable differential volume element at any point in the column is a disk with area, A_z and thickness, δz as shown in Figure 13-3. The disk element is portrayed as consisting of two parts, a gas-phase and a liquid-phase with the interphase transport of A occurring from the gas to the liquid as indicated by the arrow between the two portions of the disk. The first step towards obtaining equations which describe the rate of mass transport in the column is to take gas and liquid material balances for the transferring component, A, over the differential element. Starting with the gas-phase, we have

$$N_{A_{a_z}}{}^G A_z - N_{A_z|z+\delta z}{}^G A_z - N_{AI}\, a_I\, A_z\, \delta z = 0$$

where $N_{A\,I}$ is the interfacial molar flux of A between gas and liquid, $N_{A_z}{}^G$ is the gas-phase molar flux in the z-direction (i.e., up the column) and a_I is the interfacial area per unit volume

---

[*]   However, if a chemical reagent can react rapidly with the solute, then co-current flow might be justifiable.

of packing. Dividing through by A_z and δz and letting δz approach zero in the limit, the differential equation describing the transport of A in the gas-phase is

$$-\frac{d N_{A_z}^{G}}{dz} - N_{A_I}\, a_I = 0$$

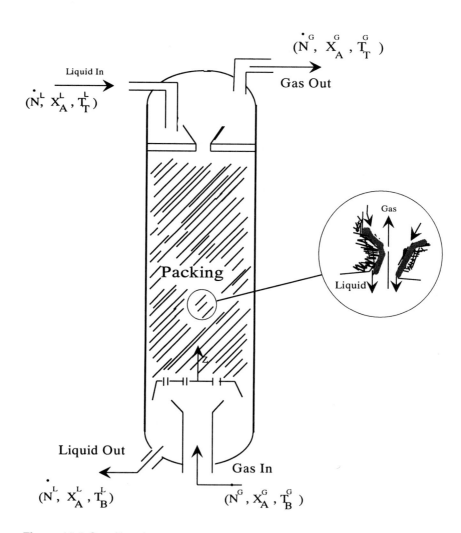

Figure 13-2 Gas Absorbers

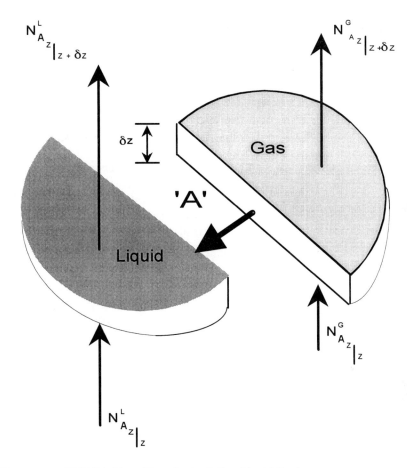

Figure 13-3 DVE For Mass Transfer In A Gas-Liquid System

or, in terms of the mole fraction of A

$$-\frac{d(\,N_z^G\,X_A^G\,)}{dz}-N_{AI}\,a_I = 0 \tag{13.27}$$

Since we are interested in concentration changes of A in the bulk gas and liquid-phases, the overall mass transfer coefficient based on gas-side driving forces is appropriate here and, therefore

$$N_{AI} = K_{OG_x}\,(\,X_A^G-[X_A^G]^*\,) \tag{13.28}$$

where K_{OG_x} is related to the individual mass transfer coefficients by Equation (13.5)

and $[X_A^G]^*$ is defined in conjunction with Equation (13.4).

At this point in the development, it is appropriate to discuss the equilibrium conditions in somewhat more detail. Although we have employed the simple linear relationship given by Equation (13.3),. the equilibrium relationship between gas and liquid can be far more complex. In fact there are entire textbooks devoted to the subject of non-ideal gas-liquid equilibria. With the advent of modern computing it is only necessary to know that these expressions do exist and can be easily introduced into computer models if the need is there.[#] Since the objective here is to introduce the subject of gas absorber calculations we will continue to employ the simple linear equilibrium relationship since it can lead to rather simple mathematics and can also become quite involved if we consider non-isothermal gas absorption. For gas absorption of low solubility gases, the linear relationship of Equation (13.3) is valid and is typically written in the form of <u>Henry's Law</u>, which is

$$P_A = H_A \, X_A^L \tag{13.29}$$

where P_A is the partial pressure of A in the gas-phase and H_A is Henry's Law constant. We can re-write Henry's Law in the form of Equation (13.3) by using Dalton's Law,

$$P_A = X_A^G \, P = H_A \, X_A^L$$

so that the equilibrium parameter, m_A, is equal to $\dfrac{H_A}{P}$.[##] However, for simplicity, we will retain the m parameter in the development that follows. Table 13-1 lists the Henry's Law constants for some representative gas solubilities in water. It should be noted that, even though data are presented for both Cl_2 and SO_2, both gases dissociate upon dissolving in water, and thus Henry's Law is not strictly applicable.

Because of the dependency of $[X_A^G]^*$ on the liquid-phase composition of A, it is also necessary to do a material balance on the liquid-phase entering and leaving the differential volume element. In a similar manner, we then obtain Equation (13.30)

$$-\frac{d(\,N_z^L\,X_A^L\,)}{dz} + N_{A_I} \, a_I = 0 \tag{13.30}$$

[#] Modern process simulation software (e.g., PRO II, Aspen, etc.) all have sophisticated equilibrium correlations for a wide variety of gas-liquid systems.

[##] Note that for an ideal solution following Raoult's Law, $H_A = P_A^*$.

Table 13-1 Henry's Law Constants: Gases in Water [H_A x 10^{-4}, atm/mole fraction]

	0 °C	20 °C	40 °C	60 °C	80 °C
H$_2$	5.81	6.85	7.63	7.63	7.52
O$_2$	2.51	3.88	5.43	6.37	6.94
N$_2$	5.38	7.58	10.0	11.5	-
CO	-	4.9	-	-	-
CO$_2$	-	0.16 (a)	0.25	-	0.35 (b)
COS	0.92	2.19	-	-	-
CH$_4$	2.24	3.76	5.20	6.26	6.82
C$_2$H$_2$	0.07	0.12	-	-	-
C$_2$H$_4$	0.49	0.99	1.62	-	-
C$_2$H$_6$	1.26	2.63	4.23	5.65	6.61
Cl$_2$ *	-	0.006	0.0063	-	-
SO$_2$ *	0.002 (c)	0.0033	0.0065		
Acetone			0.00041	0.00085	0.0019
NH$_3$ *		0.000076		-	-

(a) - 18 °C, (b) - 75 °C (c) - 10 °C, * highly soluble, dissociating gas

Keep in mind that this balance was taken by assuming that the flux of A in both phases was in the positive z-direction, consistent with the conventions discussed in Chapter 2. In this case, N_z^L is an inherently negative value since the liquid is flowing in the negative z-direction (i.e., down the column).

There are two additional factors which must also be considered when describing a gas absorber system: the effect of temperature and the possibility of the transport of liquid (solvent) to gas (the evaporation of water, for example). Temperature is important because the equilibrium parameter m, can be a strong function of temperature. In a gas absorber, the evaporation of the solvent can be important because it is endothermic and will have a direct influence on the temperature.[#] It has already been stated that temperature gradients between the gas and liquid-phases at any point in the column will be neglected, which is equivalent to assuming an instantaneous heat transfer rate (infinite heat transfer coefficient). Consequently there is only one energy flux term entering and leaving the differential volume element; namely, the combined convective energy flux of the two phases, $\Phi \equiv (N_z^G \tilde{c}_P^G + N_z^L \tilde{c}_P^L) T$.

[#] Cooling tower performance is a good example of where this is important and, in fact, the same equations are also applicable.

However, because of heats of solution and heats of evaporation, there will be sources and sinks of energy. Assuming that the heat of solution adds energy to the fluid streams, it becomes a SOURCE term while the evaporation of the solvent, B, is a SINK term. With these considerations, the differential energy balance becomes

$$(\Phi_{|_z} - \Phi_{|z+\delta z}) \, A_z + \left[\, \Delta \tilde{H}_s N_{A_I} - \Delta \tilde{H}_v \, N_{B_I} \, \right] a_I \, A_z \delta z = 0$$

Dividing through by the volume of the differential element, using the definition of Φ, and taking the limit as $\delta z \to 0$, we obtain the differential equation describing the energy balance in the tower

$$-\frac{d}{dz} \left[\, N_z^G \tilde{c}_P^G + N_z^L \tilde{c}_P^L \,) \, T \, \right] + \Delta \tilde{H}_s N_{A_I} - \Delta \tilde{H}_v \, N_{B_I} = 0 \tag{13.31}$$

Note that at this point no assumptions have been made as to the invariance of the fluxes or the heat capacities. To complete the mathematical description of the absorber we still need an expression for the interphase transport of species B, and the differential equations must be expressed in terms of common dependent variables. Typically, species B will be the liquid-phase solvent, $X_B^L \sim 1$, and thus there will be no mass transfer resistance in the liquid and thus, $K_{OG_x} \sim k_{mx}^G$. Then N_{B_I} is simply

$$N_{B_I} = k_{mx}^G \, ([X_B^G]^* - X_B^G) \tag{13.32}$$

where $[X_B^G]^* = m_B \, X_B^L$ and m_B can be calculated from the solvent vapor pressure, $m_B = \dfrac{P_B^*}{P}$. In this case we need only take a material balance for B in the gas-phase.

$$\frac{d(\, N_z \, X_B^G \,)}{dz} - a_I \, N_{B_I} = 0 \tag{13.33}$$

Since we are allowing for the total liquid and gas fluxes to change along the tower length, we need separate equations describing their changes with z. This can be done by writing <u>overall</u> differential mass balances. For the gas molar flux

$$-\frac{dN_z^G}{dz} - N_{A_I} a_I + N_{B_I} a_I = 0 \tag{13.34}$$

and then N_z^L can be calculated at any point in the tower from an overall molar balance, keeping in mind that convention requires the assumption that the direction of all fluxes is in the positive z-direction.[#]

$$N_z^L = - N_z^G + (N_z^L)_B + (N_z^G)_B \qquad (13.35)$$

For most gas absorbers or strippers, the total liquid molar flux is not likely to change very substantially and N_z^L can be taken out of the differential in both Equations (13.30) and (13.31) and set equal to $(N_z^L)_T$. N_z^G can be calculated by solving Equation (13.33) or it can be expressed in terms of the gas mole fractions of A. That is, defining $(N_z^G)'$ as the flux of *solute-free* gas,

$$N_z^G = \frac{(N^G)'}{(1 - X_A^G)} \qquad (13.36)$$

Equation (13.36) can then be substituted into Equations (13.31), (13.32), and (13.34), with $(N_z^G)'$ equal to its value at the bottom of the tower.

In order to determine the height of a tower required for a desired degree of separation, there are six variables which must be solved as a function of distance along the tower: $X_A^G, X_A^L, X_B^G, T, N_z^G, N_z^L$. Table 13-2 gives a summary of these variables along with the 6 equations required for their solution. The interfacial fluxes in these equations, N_{A_I} and N_{B_I} are given by Equations (13.28) and (13.32), respectively.

The differential equations in Table 13-2 can be solved using any number of differential equation solvers (the MATLAB program is used in this text), without making any of the "simplifying" assumptions so often employed in design calculations for mass transfer equipment (see Section 13-1b, below). However countercurrent flow does introduce one complexity to the use of such "solvers." That is, typical ordinary differential equation solvers employ "marching" algorithms which means they start from one point and march to the other; for example, from the bottom of the tower to the top. Depending on the application, some kind of iteration may be necessary since the composition in one of the phases may not be known at either the bottom or the top. For example, if an existing gas absorber is being analyzed for a new set of conditions, the composition in the liquid-phase entering the absorber would be known but the composition at the top would not be known. In this case the exiting liquid composition must be guessed and then the equations are integrated up to $z = Z_T$, the tower

---

[#]   If only one differential equation needs to be solved (see "Simplified Mass Transfer Equations", 13-2.b), it is easier to ignore this convention.

height, and a check is made to see if the calculated liquid composition at the top of the tower agrees with the known value at that point. The iteration procedure then proceeds to change the liquid composition at the bottom of the tower until the calculated concentration at the top is correct. In the typical design problem, the desired separation is known and an overall material balance will yield the compositions of both streams at the top and bottom of the tower. In this case, iterations are not needed and the equations can be solved progressively up the tower until the specified gas concentration at the top of the tower is calculated. The distance, z, at that point is then the required tower height, Z_T.

Table 13-2 Equation Set To Determine Tower Height

Variable	Equation	
X_A^G	$- \dfrac{d(\ N_z^G \ X_A^G \ )}{dz} - N_{A_I} \ a_I = 0$	(13.27)
X_A^L	$- \dfrac{d(\ N_z^L \ X_A^L \ )}{dz} + N_{A_I} \ a_I = 0$	(13.30)
T	$- \dfrac{d}{dz} \left[(\ N_z^G \tilde{c}_P^G + N_z^L \ ctilde_P^L \ ) \ T \right] + [\Delta \tilde{H}_s \, N_{A_I} - \Delta \tilde{H}_v \ N_{B_I}] a_I = 0$	(13.31)
X_B^G	$\dfrac{d(\ N_z^G \ X_B^G \ )}{dz} - a_I \ N_{B_I} = 0$	(13.33)
N_z^G	$- \dfrac{d N_z^G}{dz} - N_{A_I} \ a_I + N_{B_I} \ a_I = 0$	(13.34)
N_z^L	$N_z^L = - \ N_z^G + (\ N_z^L \ )_B + (\ N_z^G \ )_B$	(13.35)

13-2.b Simplified Mass Transfer Equations

The differential equations listed in Table 13-2 allow for variable temperatures, energies associated with phase changes and dissolution and variable gas flow.[#] However, there are many occasions where it is either not necessary to account for these phenomena or they are not significant. In these cases, considerable simplifications can be made and the differential equations can be separated and integrated. If the gas and liquid fluxes are constant, the only equation needed will be Equation (13.27)

$$- \frac{d(\ N_z^G \ X_A^G \ )}{dz} - N_{A_I} \ a_I = 0 \qquad (13.27)$$

[#]   They can also be modified to account for finite heat transfer rates (Section 13-3) and chemical reactions (Section 13-5).

With a constant gas flux, N^G, and substituting for N_{A_I} from Equation (13.28)

$$-N^G \frac{d X_A^G}{dz} - a_I K_{OG_x} (X_A^G - [X_A^G]^*) = 0$$

This equation is separable and can be integrated; i.e.,

$$-\int_{(X_A^G)_B}^{(X_A^G)_T} \frac{d X_A^G}{(X_A^G - [X_A^G]^*)} = \frac{a_I K_{OG_x}}{N^G} \int_0^{Z_T} dz \qquad (13.37)$$

The right-hand side can be integrated and solved for the height of the tower, Z_T

$$Z_T = \frac{N^G}{a_I K_{OG_x}} \int_{(X_A^G)_T}^{(X_A^G)_B} \frac{d X_A^G}{(X_A^G - [X_A^G]^*)} \qquad (13.38)$$

Quite often this "design" equation is written in the form

$$Z_T = H_{OG} (NTU)^G \qquad (13.39)$$

where H_{OG} has units of length and is referred to as the "height of a transfer unit"

$$H_{OG} = \frac{N^G}{a_I K_{OG_x}} \qquad (13.40)$$

The OG subscript simply denotes that it is based on the overall mass transfer coefficient with gas-phase driving forces. It can be related to the gas and liquid-side heights of a transfer unit by Equation (13.41).

$$H_{OG} = H^G + \frac{m_A N^G}{N^L} H^L \qquad (13.41)$$

$(NTU)^G$ is called the "number of transfer units" and is numerically equal to the integral in Equation (13.38). Physically, the number of transfer units is a measure of the degree of separation and is dependent on the magnitude of $(X_A^G)_B - (X_A^G)_T$. The height of a transfer unit is a measure of the difficulty of the separation. A large value of H_{OG} (low mass transfer coefficient) means that, for a given separation [fixed value of $(NTU)^G$], the required tower height will be large.

The calculation of the required tower height depends on the ability to integrate Equation (13.38) and this cannot be done until we relate $[X_A^G]^*$ to X_A^G. Remember that $[X_A^G]^*$ is directly related to the liquid-phase concentration by means of the equilibrium relationship [Equation (13.3)]. Consequently we must first use the equilibrium relationship to express $[X_A^G]^*$ in terms of X_A^L and then get the liquid-phase mole fraction in terms of the gas-phase mole fraction before the integration can be carried out. The latter can be done by taking a material balance over a section of the tower. Figure 13-4 shows such a balance from the bottom of the tower (the top could also be used) to any point in the column. For constant gas and liquid flow rates and no solute in the entering liquid, this material balance gives

$$X_A^L = (X_A^L)_B - \frac{\dot{N}^G}{\dot{N}^L} [(X_A^G)_B - X_A^G]$$
(13.42)

Equation (13.42) gives the liquid mole fraction in terms of the gas mole fraction. It now remains to decide on the appropriate equilibrium relationship. If this is complex, the integral in Equation (13.38) would have to be evaluated numerically. However, if we can use the linear equilibrium relationship of Equation (13.3), then the integral can be evaluated analytically. Substituting for $[X_A^G]^*$ by using Equations (13.3) and (13.42), Equation (13.38) can be written as

$$Z_T = H_{OG} \int_{(X_A^G)_T}^{(X_A^G)_B} \frac{d\,X_A^G}{\phi_1 X_A^G + \phi_2}$$
(13.43)

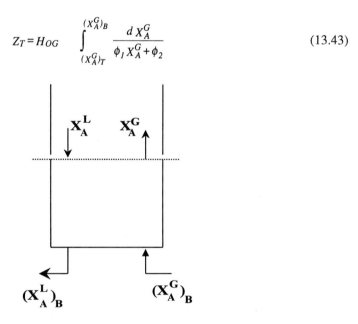

Figure 13-4 Material Balance over Bottom of Tower

where

$$\phi_1 = 1 - \frac{m_A \dot{N}^G}{\dot{N}^L}$$

(13.44)

$$\phi_2 = m_A \left[\frac{\dot{N}^G}{\dot{N}^L} (X_A^G)_B - (X_A^L)_B \right]$$

Integration gives the tower height in terms of the gas concentrations at the top and bottom of the tower

$$Z_T = \frac{H_{OG}}{\phi_1} \ln \left[\frac{\phi_1 (X_A^G)_B + \phi_2}{\phi_1 (X_A^G)_T + \phi_2} \right]$$

(13.45)

or, in terms of Equation (13.39)

$$Z_T = H_{OG} (NTU)^G$$

where

$$(NTU)^G = \frac{1}{\phi_1} \ln \left[\frac{\phi_1 (X_A^G)_B + \phi_2}{\phi_1 (X_A^G)_T + \phi_2} \right]$$

$$H_{OG} = \frac{N^G}{K_{OG_x} a_I}$$

EXAMPLE 13-1: Calculation of Gas Absorber Height

Air containing 5% SO_2 (by volume) at 1 atmosphere pressure is to have 90% of the SO_2 removed by countercurrent absorption in chilled (20 °C) water. The entering air mass velocity is 250 lbm/ft²-hr and the water will enter the tower at a mass velocity of 5,000 lbm/ft²-hr. If the tower is to be packed with 1"-Raschig rings, calculate the required packed tower height.

Solution

On an hourly basis and per square foot of column cross section, the moles entering and leaving the column are equal to $(.95)(250/29) = 8.19$ and the moles of SO_2 entering the column are equal to $(8.19)(.05)/(.95) = 0.43$. Therefore, the mole fraction of SO_2 in the exiting gas streams is:

$$(X_A^G)_T = \frac{0.043}{8.19 + 0.043} = 0.0052$$

Assuming that the gas and liquid flow rates are essentially constant, an overall material balance gives the mole fraction of SO_2 in the exiting liquid

$$(X_A^L)_B = \frac{\dfrac{250}{29}(.05 - .0052)}{\dfrac{5000}{18}} = .00139$$

It is important to insure that this value is not higher than the equilibrium value at the bottom of the tower. From Table 13-1, Henry's Law constant is 33 atm/mole fraction, which is also the value of m_A at 1 atmosphere. If the liquid leaving the tower were in equilibrium with the incoming gas, its mole fraction would be

$$(X_A^L)^*{}_B = \frac{(X_A^G)_B}{m_A} = \frac{0.05}{33} = 0.00152$$

The exiting liquid concentration is very close to equilibrium and this could result in either a very large tower or unstable operation (if concentrations and/or flows fluctuate).

Since SO_2 is a highly soluble gas, it is likely that gas-side mass transfer will be controlling. However this should be checked. Using the correlation given in Equation (13.23) for the liquid-side mass transfer coefficient and with the Schmidt number in the liquid = 550

$$\frac{G_L}{k_{mc}^L a_I \, \rho} = \frac{G_L}{k_{mx}^L a_I \, M} = .011 \left(\frac{G_L}{\mu}\right)^{.22} Sc^{.50}$$

$$k_{mx}^L a_I = \frac{\mu^{.22}}{.011 \, M} \frac{G_L^{.78}}{Sc^{.50}} = \frac{(2.42)^{.22}}{(.011)(18)} \frac{(5000)^{.78}}{(550)^{.50}} = 201 \ \frac{\text{lb-moles}}{\text{ft}^3 \ \text{hr}}$$

The gas-side mass transfer coefficient is calculated from Equation (13.24)

$$\frac{G_G}{M\,k_{mx}^G\,a_I}=1.01\left(\frac{G_G}{G_L}\right)^{.31}$$

$$k_{mx}^G\,a_I=\frac{G_G\,/\,M}{1.01\left(\dfrac{G_G}{G_L}\right)^{.31}}=\frac{250\,/29}{1.01\left(\dfrac{250}{5000}\right)^{.31}}=21.8\quad\frac{\text{lb-moles}}{\text{ft}^3\text{-hr}}$$

So that, from Equation (13.5), the overall mass transfer coefficient is

$$K_{OG_x}\,a_I=\frac{1}{\dfrac{33}{201}+\dfrac{1}{21.8}}=4.76\quad\frac{\text{lb-moles}}{\text{ft}^3\text{-hr}}$$

which, as expected, is very nearly gas-side controlled. Since the gas and liquid flow rates are constant, the equilibrium relationship is linear and the entering liquid is solute-free, all the conditions are met for the use of Equation (13.44), and it can be used to calculate the required packing height. The parameters in Equation (13.44) are:

$$\phi_1=1-\frac{m_A\,N^G}{N^L}=1-\frac{(33)(250)/29}{5000/18}=-.0241$$

$$\phi_2=m_A\left[\frac{N^G}{N^L}(X_A^G)_B-(X_A^L)_B\right]=33\left[\frac{250/29}{5000/18}(.05)-.0014\right]=.0051$$

The packed tower height can then be calculated from Equations (13.45) and (13.39)

$$(NTU)^G=\frac{1}{-.0241}\ln\left[\frac{(-.0241)(.05)+.0051}{(-.0241)(.0052)+.0051}\right]=10.15$$

$$H_{OG}=\frac{250/29}{4.76}=1.81$$

$$Z_T=(10.15)(1.81)=18.4\text{ ft}$$

Another simplified system, one which is amenable to numerical integration, is where the entering gas is *rich* in the absorbing species but the equilibrium relationship is still linear and isothermal conditions can still be assumed. In this case, the quantity of absorbed gas is sufficiently large that there is a considerable change in the overall molar gas flow rate as it proceeds up the tower.[#] Thus it is not valid to assume that gas flow rates are constant and the generalized equations given in Table 13-2 may have to be solved simultaneously. However, if all the other conditions associated with the use of Equations (13.39) and (13.45) are met and $\dot{N}^L$ is constant, it is possible to solve Equation (13.27) in terms of a complex integral.

Using Equation (13.36) to substitute for $\dot{N}^G$ in Equation (13.27)

$$-(N^G)'\frac{d}{dz}\left(\frac{d\ X_A^G}{1-X_A^G}\right)=\frac{(N^G)'}{(1-X_A^G)^2}\frac{dX_A^G}{dz}=N_{AI}\ a_I$$

Defining the interfacial molar flux in terms of K_{OG_x} and re-arranging

$$\int_0^{Z_T}dz=\frac{(N^G)'}{K_{OG_x}a_I}\int_{(X_A^G)_T}^{(X_A^G)_B}\frac{d\ X_A^G}{(1-X_A^G)^2(X_A^G-m_AX_A^L)} \tag{13.46}$$

X_A^L can be expressed in terms of X_A^G by means of the mass balance illustrated in Figure 13-4, keeping in mind that the gas flux is not constant

$$X_A^L=\frac{N_z^G}{N^L}\ X_A^G+(X_A^L)_B-(X_A^G)_B\frac{(N^G)_B}{N^L}$$

which, when substituted into Equation (13.46), results in

$$Z_T=\frac{(N^G)'}{K_{OG_x}a_I}\int_{(X_A^G)_T}^{(X_A^G)_B}\frac{d\ X_A^G}{(1-X_A^G)^2(X_A^G-\gamma_2)+\gamma_1X_A^G(1-X_A^G)} \tag{13.47}$$

where

$$\gamma_1=-\frac{m_A(N^G)'}{N^L}$$

$$\gamma_2=m_A\left[(X_A^L)_B-\frac{(N^G)'}{N^L}\frac{(X_A^G)_B}{1-(X_A^G)_B}\right]$$

The integral in Equation (13.47) can be evaluated by standard numerical methods.

[#]   Unless the liquid molar flow rate is much greater than the gas, it will also vary.

13-2.c Specification of Gas Absorber Diameters

Note that the equations in section 13-2a involve the <u>fluxes</u> of gas and liquid within the tower. However, in the typical design problem, only the gas and liquid flow <u>rates</u> , $\dot{N}^G$ and $\dot{N}^L$, will be stipulated (or calculated). Since the fluxes are equal to the flow rates divided by the cross section of the tower, one of the first steps in the design of an absorber is to specify the diameter of the tower. Because the magnitude of the mass fluxes,[#] G_G , G_L , affect the pressure drop in the tower (see Section 11-2), pressure drop considerations dictate the allowable mass velocities and, hence, the diameter of the tower. Traditionally this was done by deciding on the flooding characteristics of the packed column and how close to flooding we wish to operate. The diameter was then chosen on the basis of flooding correlations [12] to determine the *flooding velocity* (really the mass velocity of the gas, G_G , at the flooding point) of a particular packed column. Typical design constraints were then to operate at some fraction of this flooding velocity; usually at ~ ½ of the flooding velocity. With the stipulation of the total molar flow rate of the gas, the tower diameter is calculated from Equation (13.48)

$$D = \left(\frac{4 \; \dot{N}_G \; M}{\pi \; G_G} \right)^{\frac{1}{2}}$$ (13.48)

where G_G is the value of the gas mass flux at the specified fraction of the flooding value. This approach to the specification of the diameter of a packed tower is illustrated in Example 13-2.

Although the basic approach is still the same, modern design methods rely instead on the specification of an allowable pressure drop per unit column height and then utilize pressure drop correlations for two phase countercurrent flow over packings to decide on the gas mass velocity. According to Robbins [13], gas absorbers are designed to operate in the 0.2 to 0.6 in. H_2O/ft of packing range, with the lower value used if there is a potential for foaming. As discussed in Section 11-2, Robbins also developed a generalized pressure drop correlation for countercurrent gas-liquid flows in packed beds and this forms the basis for deciding on the diameter of a gas absorption tower.

The procedure for selecting the column diameter is to first choose a packing and an operational pressure drop per height of packing (inches of H_2O/ ft. of packing) and then use Robbins' correlation to back out the required value of G_G . This is done by forming the ratio of Robbins' definitions of the liquid and gas "loading factors", Equations (11.36) and (11.37), to give

---

[#]   Frequently referred to as "mass velocities."

$$\frac{G_{L_f}}{G_{G_f}} = \frac{62.4}{\rho_L} \left(\frac{\rho_G}{0.075}\right)^{1/2} (\mu_L)^{.10} \frac{G_L}{G_G}$$ (13.49)

But $\dfrac{G_L}{G_G}$ is known, since $\dot{N}^G$ and $\dot{N}^L$ are known; i.e.,

$$\frac{G_L}{G_G} = \frac{\dot{N}^L}{\dot{N}^G} \frac{M_L}{M_G}$$

This ratio, together with the chosen pressure drop per height of packing, can then be used to enter Figure 11-6 to determine the value of G_{G_f}. Then, using the value of F_P for the chosen packing (see Table 11-3), G_G can be calculated from Equation (11.36)

$$G_G = G_{G_f} \left(\frac{\rho^G}{.075}\right)^{\frac{1}{2}} \left(\frac{20}{F_P}\right)^{\frac{1}{2}}$$ (11.36)

The diameter is then calculated from Equation (13.48), making sure that

$$\frac{D}{d_P} > 6$$

in order to minimize channeling in the packing ("wall effects").

EXAMPLE 13-2: Use of Flooding Correlations to Calculate Tower Diameters

One of the most common methods used in the calculation of tower diameters is to specify that the gas mass velocity must operate at a certain percentage of flooding. The calculations then focus on the use of flooding correlations to predict the gas mass velocity where flooding will occur. Figure 13-5 shows a widely used flooding correlation where the parameter read from the ordinate axis corresponds to the gas mass velocity at flooding. The units in the parameters are in lbm, ft, cp, and hr. Use this figure to determine the diameter of the tower where

$\Phi_y = 2.4 \times 10^{-9} \dfrac{G_G^2 \, a_p \, (\mu_L)^{.20}}{\varepsilon^3 \rho_G \rho_L}$, $\Phi_x = \dfrac{G_L}{G_G} \left(\dfrac{\rho_G}{\rho_L}\right)^{.50}$ if it is to operate at 50% of flooding for a

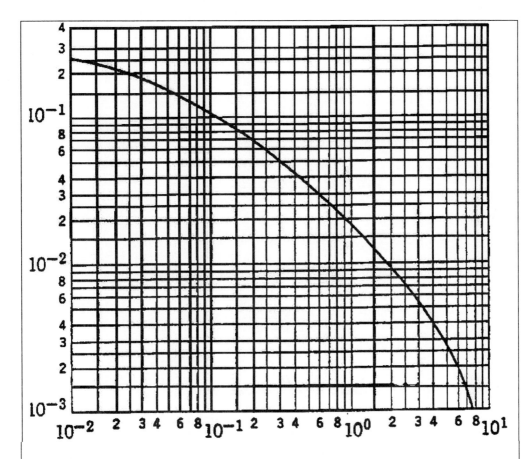

Figure 13-5 Flooding Correlation for Packed Towers, From Foust, *Principles of Unit Operations*, John Wiley & Sons, N.Y., 1960, reprinted by permission of John Wiley & Sons

situation where 12,000 SCFH of gas is to be contacted with 1,400 lbm/hr of liquid in counter current flow in a tower packed with 1"-intalox saddles. Assume the properties of gas and liquid are essentially those of air and water at 25 °C and 1 atm.

<u>Solution</u>

The density of gas and vapor are 0.074 and 62.4 lbm/ft^3, respectively and, from Table 11-3, $a_p = 78$ and $\varepsilon = .77$. The gas mass flow rate is

$$\dot{m}_G = \frac{(12,000)(29)}{359} = 969 \ \text{lbm / hr}$$

and since $\dfrac{G_L}{G_G} = \dfrac{\dot{m}_G}{\dot{m}_L}$, The abscissa of Figure 13-5 is calculated to be

$$\frac{1{,}400}{969}\left(\frac{0.074}{62.4}\right)^{.50} = 0.05$$

The ordinate from Figure 13-5 is read as 0.15, so that the gas mass velocity at flooding is

$$G_G^2 = \frac{0.15}{2.4\mathrm{x}10^{-9}}\ \frac{(.77)^3(0.074)(62.4)}{(78)(1)^{.20}} = 1.68\mathrm{x}10^6$$

$$G_G = 1300\,\mathrm{lbm}/\mathrm{ft}^2\text{-}\mathrm{hr}$$

For a gas mass velocity at 50% of flooding, the diameter is calculated from Equation (13.48)

$$D = \left(\frac{(4)(969)}{(3.14)(650)}\right)^{.50} = 1.38'$$

13-2.d Design Procedure for Gas Absorbers

All of the mathematical equations needed for gas absorber design have been presented in Sections 13-1.a and 13-1.c. Before these equations can be used it is first necessary to choose an appropriate liquid flow rate. To do this we must be careful to insure that the liquid exiting the absorber has a concentration of the solute which is sufficiently below its saturation value. Therefore, the *minimum* liquid flow rate that can be used is that value which causes the exiting liquid to be in equilibrium (saturated) with the entering gas stream. Since equilibrium (the value of m) depends on temperature, the temperature of the liquid at the bottom of the tower must be known.[#] The temperature at the top of the tower is taken to be the entering liquid temperature (specified) and the temperature of the liquid at the bottom of the tower must be calculated from an overall energy balance. Note that this is an iterative process since the bottoms temperature will depend on the liquid flow rate (see Example 13-3). Once the minimum liquid flow rate is calculated, the operating liquid flow rate is specified as either some multiple of the minimum or at a value designed to produce a liquid concentration at some fraction of the saturation value. However, this too can be iterative, since changing the liquid flow rate will also change the temperature.

[#]   This is a consequence of assuming identical gas and liquid temperatures at each point in the column which, in turn, is taken to be the liquid temperature.

Using this value of the solute concentration in the exiting liquid combined with a knowledge of the mole fraction of A which is to be reached at the top of the tower, a material balance calculation determines the solute-free liquid flow rate necessary to achieve that value. Specifically, if the entering liquid is solute-free, then $(\dot{N}_L)'$ can be determined from Equation (13-50)

$$(\dot{N}^L)' \left[\frac{X_A^L}{1 - X_A^L} \right]_B = (\dot{N}^G)' \left[\left(\frac{X_A^G}{1 - X_A^G} \right)_B - \left(\frac{X_A^G}{1 - X_A^G} \right)_T \right] \qquad (13.50)$$

Since mass velocities are needed to calculate the mass transfer coefficient and the equations which determine the packing height are in terms of fluxes, the next step is to calculate the tower diameter using the methods of Section 13-2c. In these calculations, the ratio of the liquid/gas mass velocities is taken to be the ratio of the liquid/gas mass flow rates at the bottom of the tower.

If the entering gas is *lean* (low concentration of A) or the quantity to be absorbed is sufficiently small, then, $N_z^G \sim (N_z^G)'_B, N_z^L \sim (N_z^L)'_T$. If the tower is isothermal and the equilibrium relationship is linear, Equations (13.45) can be used to calculate the height of the packing for a lean gas and Equations (13.36) and (13.47) can be used if the gas is *rich* (i.e., gas and liquid flow rates are not constant). If none of these assumptions apply, then the differential equations listed in Table 13-2 must be solved simultaneously using MATLAB or some other differential equation solver.

Starting with the compositions at the bottom of the column, the design procedure is to solve Equations (13.27) through (13.35) simultaneously until we reach the value of z where the composition of A in the gas-phase is equal to the specified value. This is then the height of the packed tower and, combined with the value of D obtained from Equation (13-48), the column dimensions are now specified. This design procedure is summarized below in Table 13-3 in a step-by-step manner and Example 13-4 is an illustration of how to use the MATLAB program to solve the differential equations pertaining to an ammonia absorption problem.

EXAMPLE 13-3: Minimum Liquid Flow in an Ammonia Gas Absorber

An air stream contains 8 mole % NH_3 and it is desired to reduce the ammonia to 100 ppm by contacting it with water in a gas absorber column. The entering gas flow rate is 2000 kg-mol/hr and is delivered at 25 C and 1 atm pressure. The water is ammonia-free and is available at 25 °C. Assume that the tower is adiabatic and that the NH_3-water equilibrium relationship follows the linear relationship in Equation (13.4) where

$$m_A = 3.09 \times 10^6 \, \exp\!\left[\frac{-4236}{T}\right]$$

Calculate the minimum water flow rate if the entering gas is at 100% humidity. The heat of solution for NH_3–water is - 8,300 cal/mole NH_3.

<u>Solution</u>

Since the entering gas is saturated with water and the exiting gas is also at 25 C, there will be no evaporation of water into the gas stream. Essentially all of the entering ammonia is absorbed by the water or, $(.08)(2,000) = 160$ kg moles/hr . Consequently the overall energy balance, neglecting the effect of the absorbed ammonia on the liquid properties, is simply

$$\dot{N}^L \tilde{c}_p^{\,L} (T_B - 298) = (160)(8,300)$$

An overall ammonia balance, with the exiting ammonia concentration in the liquid calculated from the equilibrium relationship, $(X_A^L)_B = \dfrac{(X_A^G)_B}{m_A}$, is

$$160 = (X_A^L)_B \, \dot{N}^L = \frac{(X_A^G)_B}{m_A} \, \dot{N}^L = \frac{.08 \, \dot{N}^L}{m_A}$$

Solving the above equation for $\dot{N}^L$ and substituting it and the equation for m_A as a function of temperature into the energy balance

$$T_B - 298 = \frac{1.193 \times 10^6}{3.09 \times 10^6 \, \exp\!\left[\dfrac{-4236}{T_B}\right]}$$

which is easily iterated to give a value of 309 $^\circ$K for the temperature at the bottom of the tower. At this temperature, the mole fraction of ammonia in the exiting liquid is computed to be

$$(X_A^L)_B = \frac{.08}{3.09 \times 10^6 \, \exp\!\left[\dfrac{-4236}{309}\right]} = 0.023$$

so that the minimum liquid flow rate is $(\dot{N}^L)_{min} = 160 / 0.023 = 6{,}888$ kg moles/hr

Table 13-3 Design Procedure for Gas Absorbers

Figure 13-2 shows the nomenclature being used. Note that the *total* gas and liquid flow rates ($\dot{N}^G$, $\dot{N}^L$) can vary throughout the column, depending on the quantity of A that is absorbed by the liquid.

Design Specifications: $(\dot{N}^G)_B, (X_A^G)_B, (T^G)_B, (X_A^G)_T$ (or the fraction of A absorbed); and usually $(X_A^L)_T$. Equilibrium data ($m_A = f(X_A^L, T)$).

Design Procedure:

1. Determine the minimum liquid flow rate (*pinch point*); that is, solve for the minimum solute-free liquid flow rate, $(\dot{N}^L)'$ which would have $(X_A^L)_B$ in equilibrium with $(X_A^G)_B$. This is accomplished by using the equilibrium data in conjunction with an overall material balance on A ; e.g., Equation (13.50)

2. Choose $(\dot{N}^L)'$ as some multiple of the minimum value so as to achieve the desired approach to saturation

3. Choose the packing and the allowable pressure drop per inch of packing.

4. Using the maximum flow rate ratio, $\dfrac{\dot{m}^G}{\dot{m}^L} = \dfrac{\dot{N}^G \, M_G}{\dot{N}^L \, M_L}$ (probably at the bottom of the column), calculate $\dfrac{G_{G_f}}{G_{L_f}}$ from Equation (13.49). Use Figure 11-6 and Equation (11.36) to determine G_G

5. Calculate the diameter of the column using Equation (13.48)

6. With knowledge of G_G, G_L calculate the individual mass transfer coefficients and $K_{OG_{x,c}}$ (or $K_{OL_{x,c}}$).

7. Refer to the differential material and energy balances at any point in the column (Table 13-2). If requirements are met, use either Equations (13.44) or (13.47) and solve for Z_T. If tower is not isothermal and/or the equilibrium relationship is not linear, solve the equations in Table 13-2 simultaneously using a differential equation solver. See MATLAB Program in Example 13-4.

EXAMPLE 13-4: Design of an Ammonia Gas Absorber

An air stream contains 17 mole % NH_3 and it is desired to reduce the ammonia content by 95% by contacting it with water in a column packed with -in Raschig Rings. The gas flow rate is 100 lb-moles/hr and is delivered at 35 °C and 10 psig at 10% relative humidity. The water is ammonia-free and is available at 20 °C. Assume that the tower is adiabatic and that the NH_3-water equilibrium relationship follows the linear relationship in Equation (13.3) where

$$m_A = 3.09 \times 10^6 \exp\left[\frac{-4236}{T}\right]$$

The water flow rate is to be chosen so that the ammonia concentration in the exiting liquid is 50% of saturation and the packed tower is to operate at a pressure drop of 0.5 inches H_2O/ft of packing. Determine the packed column diameter and height to achieve the desired ammonia absorption.

<u>Solution</u>

The vapor pressure of water is calculated from the Antoine Equation, $P_B^* = 10^\alpha$, $\alpha = B_1 - \dfrac{B_2}{T + B_3}$, and the following physio-chemical properties apply: $\hat{C}_P^G = 0.24$ BTU/lbm-F, $\hat{C}_P^L = 1.0$ BTU/lbm-F, $\Delta \tilde{H}_S = 1.3 \times 10^4$ BTU/lb-mole NH_3, $\mu_L = 0.70\ cp$, $(Sc)_L = 570$

At saturation, the bottoms liquid mole fraction will be

$$(X_A^L)_B^* = \frac{0.17}{m_A} = \frac{0.17}{3.09 \times 10^6 \exp\left[\dfrac{-4236}{308}\right]} = 0.052$$

At 50% of this value, an overall material balance will provide the liquid flow rate at the bottom of the tower

$$(X_A^L)_B (\dot{N}^L)_B = (.95)(0.17)(100) = 16.15 \ \text{lb-moles } NH_3/\text{hr}$$
$$(X_A^L)_B = (0.5)(.052) = 0.026$$
$$(\dot{N}^L)_B = \frac{16.15}{0.026} = 621 \ \text{lb-moles/hr}$$

The gas and liquid flow rates at the top of the tower will be 84.35 and 641.35 lb-moles/hr, respectively and the mole fraction of ammonia in the gas at the top of the tower = 0.01

The tower diameter is determined by using Robbins' correlation at a pressure drop of 0.5 inches H_2O/ft. Using Equation (13.49) for conditions at the bottom of the tower where the average molecular weight of the gas is 27,

$$\frac{G_L}{G_G} = \frac{(621)(18)}{(100)(27)} = 4.13$$

$$\frac{G_{Lf}}{G_{Gf}} = \frac{62.4}{62.4}\left(\frac{075}{.075}\right)^{1/2}(.70)^{.10}\ 4.13 = 3.99$$

Using this and the design criterion of 0.5" H_2O/ft of packing, the value of G_{G_f} is read from Figure 11-6 as 2000 and this is used in Equation (11.36) to calculate G_G. Then D is calculated from Equation (13.48)

$$G_G = 2000\,(1.0)^{1/2}\,(20/155)^{1/2} = 718\ \text{lbm}/\,ft^2\text{-}hr$$

$$D = \left[\frac{(4)(100)(27)}{(3.14)(718)}\right]^{1/2} = 2.1'9,\ \text{and}\ G_L = (718)(4.13) = 2965\ \text{lbm}/\,ft^2\text{-}hr$$

Since 1" Raschig rings are to be used, the Sherwood-Holloway correlations, Equations (13.23) and (13.24) are applicable for both the liquid and gas coefficients. Since $m_A = 3.29$ at 35 °C and Sc = 570,

$$k_{mx}^L\,a_I = \frac{(2994)^{.78}[(.70)(2.42)]^{.22}}{(.011)(18)(570)^{1/2}} = 122\ \text{lb-moles}/\,ft^3\text{-}hr$$

$$k_{mx}^G\,a_I = \frac{\dfrac{718}{27}(4.13)^{.32}}{1.01} = 41.6\ \text{lb-moles}/\,ft^3\text{-}hr$$

$$K_{OG_x}\,a_I = \frac{1}{\dfrac{1}{41.6}+\dfrac{3.29}{122}} = 19.6\ \text{lb-moles}/\,ft^3\text{-}hr$$

Because the tower will not be isothermal, we will need to solve the set of equations listed in Table 13-2 and integrate them from the bottom of the tower to the value of z where $(X_A^G)_T = 0.01$. The inset, below is a listing of these equations in the MATLAB format, where the initial conditions (i.e., at the bottom of the tower) are:

$$(N_z^G)_B = 718/27 = 26.6, (N_z^L)_B = -2965/18 = -164.7$$

$$(N_z^G X_A^G)_B = 4.522, \quad (N_z^L X_A^L)_B = -4.28, \quad T_B = 35,$$

$$(N_z^G X_B^G)_B = (26.6)(.10)(\frac{P_B^*}{P_{TOT}}) = 0.146$$

The vapor pressure of water at 35 $^{\circ}$C, P_B^*, is calculated from the Antoine Equation to be 0.055 atm.

The MATLAB solution to these equations is illustrated in the plot of the temperature and X_{AG} as a function of distance up the tower.

MATLAB Program For Example 13-4

<u>M-File</u>

```
function dy=absorber(z,y)
dy=zeros(6,1);
XAGB=.17;XALB=.026;XAGT=.01;NGB=26.6;
NLB=-164.7;DHV=55.6;DHS=1.3E4;CPG=7.0;
CPL=18;PTOT=1.0;TB=35;KOGAI=19;
A1=1.6E9;A2=6276;B1=8.10765;B2=1750.286;
B3=235;NXAGB=4.522;NXALB=-4.28;
NXBGB=.146;

XAG=y(1)/y(5);
y(6) = XAG;
NL=-y(5)+NGB+NLB;
MA=A1*exp(-A2/(y(3)+273));
XAL=y(2)/NL;
XAGS=MA*XAL;
NAI=KOGAI*(XAG-XAGS);
ALPHA=B1-B2/(B3+y(3));
PBST=10^ALPHA;
XBGS=PBST/(PTOT*760);
XBG=y(4)/y(5);
NBI=KOGAI*(XBGS-XBG);

dy(1)=-NAI;
dy(2)=NAI;
dy(3)=(1/1.8)*(-DHV*NBI+DHS*NAI)/(y(5)*CPG+NL*CPL);
dy(4)=NBI;
```

```
dy(5)=NBI-NAI;
dy(6)=(1/y(5))*(dy(1) - y(6)*dy(5));
```

<u>COMMAND File</u>

```
clear;
y0(1) = 4.522;
y0(2) = -4.28;
y0(3) = 35;
y0(4) = .146;
y0(5) = 26.6;
y0(6) =  .17;
[z,y]=ode45('absorber',[0 5],[y0]);
plotyy(z,y(:,3),z,y(:,6))
ylabel('Temperature (F)')
 ylabel('XAG')
 xlabel('z (ft)')
 title('{Mole Fraction and Temperature in Gas Absorber}')
```

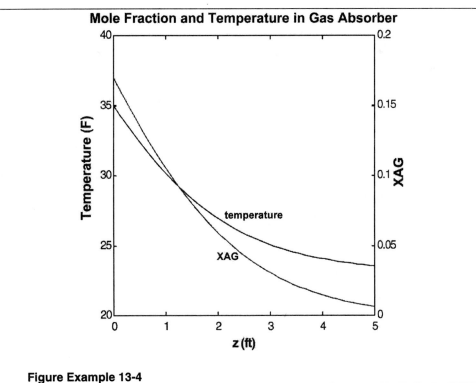

Figure Example 13-4

13-3 MASS AND HEAT TRANSFER IN COOLING TOWERS

Humidification operations involve the simultaneous transport of both mass (water) and energy. While any type of drying operation will fall under this classification, we will only discuss one particular application — cooling towers. The objective here is simple — to evaporate a fraction of a warm water stream so that the energy required for evaporation is extracted from the water, thus cooling the stream. As might be expected, this is a typical operation in any plant where cooling water is used since it allows for the recycle of cooling water back to the process units which require cooling. The alternative would be to use "single-pass" cooling water and that can be expensive since cooling waters generally require controls on their corrosion and scaling potential.

As far as the equations describing mass and heat transfer in cooling towers is concerned, they are very similar to those used to describe gas absorption. That is, they both involve gas-liquid systems, and they both experience mass transfer between the phases. There are some differences however. In a cooling tower there is no need for a packing material and the mass transfer is only a single component being transferred from the liquid to the gas (water). Both of these factors are a simplification over gas absorbers. Recall that in gas absorbers we generally assume that heat transfer between the gas and liquid-phases is instantaneous. A complicating factor in cooling towers is that there is a finite heat transfer rate and thus we need to account for interfacial energy transport as well as mass transport. Therefore we need to examine this aspect of cooling tower behavior before attempting to write the appropriate differential balances.

Since cooling towers will always involve the air-water system, the nomenclature here can be greatly simplified. The subscripts, w and DA will refer to water and dry air (i.e., water-free air) and X_w will always refer to the mole fraction of water in the air. Since humidity is a common means of specifying water concentrations in air, it is appropriate to review these definitions and to relate them to mole fractions. *Humidity* is defined as the mass of water vapor in the air divided by the mass of dry air, or

$$H \equiv \frac{m_w}{m_{DA}} = \frac{(P_w M_w) / (R_g T)}{(P_{DA} M_{DA}) / (R_g T)}$$

$$H = \frac{P_w M_w}{M_{DA}(P - P_w)}$$

Thus, the mole fraction of water in the air can be calculated from

$$X_w = \frac{\dfrac{H}{M_w}}{\dfrac{1}{M_{DA}} + \dfrac{H}{M_w}}$$

(13.51)

Relative Humidity is defined as the percent of saturation in the air, or

$$H_R = \frac{X_w P}{P_w^*}, \ or, \ X_w = \frac{H_R P_w^*}{P}$$

(13.52)

Another useful concept is that of the *wet bulb temperature*. This is the steady-state temperature that is reached by the water when dry air is passed over water. Since it is related to the humidity of the air passing over the water, it is often used as a measure of humidity. However it is used here to calculate the steady-state temperature at the air-water interface during simultaneous heat and mass transfer. Physically the experiment can be described as a drop of warm water that is suspended in air and which does not change shape or volume during the experiment. If the air is not saturated, water will evaporate into the air and will take the required energy needed for evaporation from the sensible heat of the water. Eventually the water will cool until its temperature is lower than the air temperature and then the air will transport energy to the water until the rate of energy transport to the liquid equals the rate of energy leaving the liquid due to evaporation. The temperature of the water at this point is the wet bulb temperature.

Figure 13-6 illustrates this same experiment at the air-water interface. At steady-state the

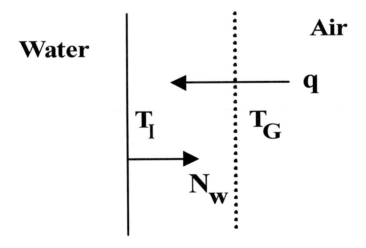

Figure 13-6 Air-Water Interface during Wet Bulb Experiment

energy and molar fluxes through the gas-side phoney film are

$$q = h(T_G - T_I)$$

$$N_w = k_{mx}^G(X_{w_I} - X_w)$$

where T_I is the wet bulb temperature.

If there are no temperature gradients on the liquid side of the interface, then

$$q = \Delta \tilde{H} \, N_w$$

$$h(T_G - T_I) = \Delta \tilde{H}_v \, k_{mx}^G(X_{w_I} - X_w)$$

and the difference between the gas and wet bulb temperatures is

$$T_G - T_I = \frac{\Delta \tilde{H}_v \, k_{mx}^G}{h}(X_{w_I} - X_w) \tag{13.53}$$

Notice that the wet bulb temperature, T_I, also appears on the right-hand side of Equation (13.53), since X_{w_I} is a function of T_I (as a result of its dependence on the water vapor pressure).

The heat and mass transfer coefficients can be related in terms of the usual correlations. For heat transfer we have

$$Nu = a \, Re^m \, Pr^n \tag{13.54}$$

Dividing both sides by the product of the Reynolds and Prandtl numbers, Equation (13.54) can be written as

$$\frac{h}{\hat{c}_P^G \, G_G} = a \, Re^{m'} \, Pr^{n'} \tag{13.55}$$

In a similar manner, the Sherwood number can also be written in terms of the Reynolds and Schmidt numbers

$$Sh = a\, Re^m\, Sc^n$$

(13.56)

$$\frac{Sh}{Re\, Sc} = \frac{M_{DA}\, k_{mx}^G}{G_G} = a\, Re^{m'}\, Sc^{n'}$$

The ratio of the heat and mass transfer coefficients is, from Equations (13.55) and (13.56)

$$\frac{h}{M_{DA}\, k_{mx}^G} = \hat{c}_P^G\, Le^{n'}, \;\; or$$

(13.57)

$$\frac{h}{k_{mx}^G} = M_{DA}\, \hat{c}_s^G \sim \tilde{c}_P^G$$

where the Lewis number is taken to be approximately 1 and $\tilde{c}_s^G$ is the <u>Humid Heat</u>[*] which is approximately equal to the gas molar heat capacity. Equation (13.57) is known as the <u>Lewis Relation</u> and is used to relate the heat and mass transfer coefficients to one another in humidification operations.

In order to determine the height of the cooling tower necessary to achieve the desired outlet water temperature, differential energy and material balances, very similar to those in Table 13-2, must be solved. In this case however, we need separate energy balances on the water and air phases and, since the water phase is only a single component, only a material balance for water in the gas-phase is needed. Since gas and liquid flows will be reasonably constant in the tower,[**] Equation (13.27) can be re-written as

$$-N^G \frac{d\, X_w}{dz} + k_{mx}^G\, a_I\, (\, X_{wI} - X_w\,) = 0$$

(13.58)

Because temperatures in the air and water phases will differ, separate energy balances are needed for both phases and we must also account for interphase energy transport between the two phases. Therefore, the analogous equations to Equation (13.31) are

<u>Liquid</u>
$$-G_L\, \hat{c}_{PL} \frac{d\, T_L}{dz} + h\, a_I\, (T_G - T_L) - \Delta \tilde{H}_v\, k_{mx}^G\, a_I\, (X_{wI} - X_w) = 0$$

(13.59)

---

[*]   This is defined as $\dfrac{h}{k_{mx}^G} = M_{DA}\, \hat{c}_S = M_{DA}\,[\hat{c}_{DA} + H\hat{c}_{P_w}]$.

[**] Most of the water change is due to *drift*, the loss of entrained liquid droplets to the atmosphere.

Gas

$$-G_G \hat{c}_{PG} \frac{dT_G}{dz} - h \, a_I (T_G - T_L) = 0 \qquad (13.60)$$

In the typical design problem, the liquid flow rate and temperature at the bottom of the tower is calculated from design specifications (see below) and thus Equations (13.58) through (13.60) can be solved simultaneously using a differential equation solver such as MATLAB. The only additional parameters that need to be specified are the heat and mass transfer coefficients and the tower diameter. If the cooling tower is to utilize a packing, the methods of Section 13-2 can be employed to calculate the required diameter (or cross section). If the cooling tower is to have an empty cross section, then there are empirical methods to decide on the "plan area" of the tower. Since the Lewis Relation is usually valid in these applications, there is a unique relationship between the heat and mass transfer coefficients and so only one must be specified. Because mass transfer data are more plentiful, the gas-side mass transfer coefficient is usually specified or calculated.

In a cooling tower application, the liquid flow rate is known and it is the gas flow rate which must be calculated. This is often done by stipulating that the air flow rate must be some multiple of the *minimum* air flow rate. The minimum air flow rate is that rate which results in the exiting air being in equilibrium with the incoming water. One way of deciding on the exit water temperature is to stipulate how close it will get to the wet bulb temperature of the entering air. A "rule-of-thumb" is that the "wet bulb approach temperature," $(T_L)_B - T_I$, should be about 3 - 5 ^{0}C.

While the approach used here is to set up the differential equations and solve them numerically without making too many assumptions, there are other, more traditional methods of analyzing cooling tower problems. One simplification, due to Merkel [14], utilizes "enthalpy driving forces" to combine Equations (13.58) through (13.60) into an integral which, after simplifying assumptions, can be solved numerically. The development is as follows. The Lewis relation in terms of the humid heat, Equation (13.57), is used to express the heat transfer coefficient in terms of the mass transfer coefficient so that Equation (13.59) becomes

$$-G_L \hat{c}_{PL} \frac{dT_L}{dz} + k_{mx}^G \, a_I \left[\tilde{c}_S \, (T_G - T_L) - \Delta \tilde{H}_v \, (X_{wI} - X_w) \right] = 0 \qquad (13.61)$$

If this equation is now written in terms of humidity where the reference humidity is the dry air at some temperature, T_0 (usually chosen as 0 C), then, assuming that $T_I \sim T_L$, it can be shown that

$$\tilde{c}_S \, (T_G - T_L) + \Delta \tilde{H}_v (X_{wI} - X_w) = \tilde{H}_I - \tilde{H}_G$$

so that Equation (13.61) can be separated and expressed in the integral form

$$Z_T = \frac{G_L \hat{c}_{PL}}{k_{mx}^G a_I} \int_{T_B}^{T_T} \frac{dT_L}{\tilde{H}_I - \tilde{H}_G} \tag{13.62}$$

Equation (13.62) can then be integrated numerically, using values of enthalpy as a function of temperature and humidity for the air-water system.

13-4 GAS ABSORPTION WITH CHEMICAL REACTIONS

It was pointed out earlier that the absorption of low solubility gases tend to be limited by liquid side mass transfer rates. This is particularly true when it is necessary to remove "acid gases," such as CO_2 and H_2S, from gas streams. In these situations, a reactive agent is often added to the liquid-phase so that the liquid-phase concentration of the absorbing species remains far below its saturation value. For acid gases, inorganic and organic bases such as caustic, hot carbonates, and organic amines are all used for this purpose. For example, diethanolamine (DEA) reacts with CO_2 in the following overall reaction

$$2RNH_2 + CO_2 + H_2O \leftrightarrow (RNH_2)_2 H_2CO_3$$

and hydrogen sulfide can be reacted with sodium carbonate

$$Na_2CO_3 + H_2S \leftrightarrow NaHCO_3 + NaHS$$

The extent to which these reactions affect the overall mass transfer rates is dependent on the magnitude of the chemical reaction rates and the degree of reversibility of the reactions. The effects of chemical reaction on gas absorption are often presented in terms of overall mass transfer coefficients which are dependent on conversions or pH. While these coefficients are useful for specific applications, they are difficult to apply in a general manner since these "psuedo" mass transfer coefficients will be a function of chemical reaction rates and equilibrium constants as well as gas-liqiuid equilibria and convective mixing. The effect of chemical reactions on mass transfer in gas absorbers is presented here in an effort to demonstrate just how they affect the rates of mass transfer and to then show how they can be incorporated into the usual definitions of the mass transfer coefficient.

The desired effect of a chemical reaction is to consume the solute as soon as it enters the liquid-phase so that the liquid solute concentration remains close to zero and the driving force for mass transfer is kept high. The ideal situation would be an irreversible, instantaneous reaction (a reaction rate much faster than the diffusion rate) which would lower the solute concentration to zero at the reaction site. The two example reactions listed above are both reversible and even if they were instantaneous, the degree to which the "free" solute concentration in the liquid can be lowered will be a function of the equilibrium constant for the reaction. The more usual case is where the reaction rate is on the same order as the diffusion rate and the two rate processes must be analyzed simultaneously; much in the same way as illustrated in Example 10-2. In this section of the chapter we will first consider the effects of reversible, instantaneous reactions, then remove the reversibility and finally incorporate the approach used in Chapter 10 to examine the effects of finite reaction rates.

Figure 13-7 shows the progress of gas absorption in two-film theory as the dissolved solute, A, first encounters the liquid-phase reactant, B. Initially at time, t_1, A does not have to diffuse very far until it meets reactant B and reacts instantaneously to form the equilibrium concentration, $(C_A^L)_{eq}$. For simplicity Figure 13-7 shows the equilibrium concentration of B and A to be numerically equal. As time progresses, to t_2, A must diffuse further until it meets with B. The diffusion path for B is now shorter than it was at t_1. Eventually, the diffusion rates of A and B are opposite and equal (for a 1:1 stoichiometric ratio) at steady-state (t_∞), and the reaction point stabilizes at δ_R. Therefore, even though the reaction rate is instantaneous, the real effect of the reaction is to shorten the diffusion distance for A. To quantify this effect we can set the diffusion rate of A equal to the diffusion rate of B (all concentrations are in the liquid-phase, so the superscript, L, is dropped for clarity).

$$\frac{D_A^L}{\delta_R}(C_{AI} - C_{A_{eq}}) = \frac{D_B^L}{\delta_L - \delta_R}(C_B - C_{B_{eq}})$$

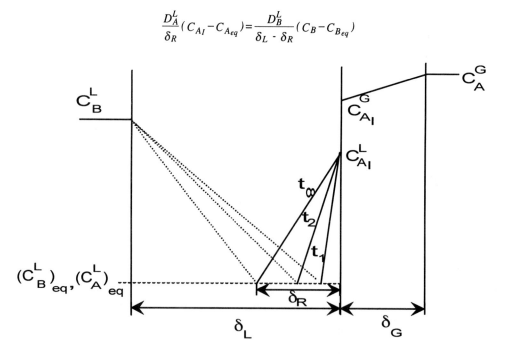

Figure 13-7 Instantaneous Reversible Reaction in a Liquid Film

Solving this equation for δ_R,

$$\delta_R = \frac{\dfrac{\delta_L}{D_B^L \Delta C_B}}{\dfrac{1}{D_A^L \Delta C_A} + \dfrac{1}{D_B^L \Delta C_B}} \tag{13.63}$$

where $\Delta C_{A,B}$ refers to the diffusional driving forces for A and B. The steady-state flux A is given by

$$N_A = \frac{D_A^L}{\delta_R}(C_{AI} - C_{Aeq})$$

Substituting for δ_R from Equation (13.63), the flux becomes

$$N_A = \frac{D_A^L \Delta C_A}{\delta_L}\left[1 + \frac{D_B^L \Delta C_B}{D_A^L \Delta C_A}\right] \tag{13.64}$$

Since $N_A = k_{mc}^L \Delta C_A$, a comparison with Equation (13.64) gives the "psuedo" mass transfer coefficient in the presence of an instantaneous, reversible reaction

$$k_{mc}^L = (k_{mc}^L)_0\left[1 + \frac{D_B^L \Delta C_B}{D_A^L \Delta C_A}\right] \tag{13.65}$$

Where $(k_{mc}^L)_0$ is the mass transfer coefficient in the absence of a chemical reaction and the term in parenthesis is called the "enhancement factor", ξ.

The equilibrium concentrations can be expressed in terms of the equilibrium conversion of B. If the reaction is of the type, A + B = AB, and $C_{Beq} = (1 - \chi_e)C_B$, then the enhancement factor can be expressed as

$$\xi = 1 + \frac{D_B^L}{D_A^L}\frac{\chi_e(1 - \chi_e)K_{eq}C_B}{(1 - \chi_e)C_{AI}K_{eq} - \chi_e} \tag{13.66}$$

It now becomes apparent why the mass transfer coefficients in chemically reacting gas absorbers are often given as a function of conversion (χ_e) and pH (K_{eq}).

In the particular case of an instantaneous <u>irreversible</u> reaction, $\Delta C_B \sim C_B^L$ and $\Delta C_A \sim C_{AI}^L$, so that the enhancement factor simplifies to

$$\xi = 1 + \frac{D_B^L C_B^L}{D_A^L C_{AI}^L} \tag{13.67}$$

and N_A becomes, from Equation (13.64)

$$N_A = \frac{D_A^L}{\delta_L} \left[C_{AI}^L + \frac{D_B^L}{D_A^L} C_B \right] \tag{13.68}$$

At steady-state, N_A can also be calculated from the gas-side of the film as

$$N_A = k_{mc}^G (C_A^G - C_{AI}^G) \tag{13.69}$$

Since $C_{AI}^L = \dfrac{C_{AI}^G}{m'}$, Equation (13.68) can be written in terms of the bulk gas concentration by eliminating C_{AI}^L from Equations (13.68) and (13.69)

$$N_A = \frac{\dfrac{C_A^G}{m'} + \dfrac{D_B^L C_B^L}{D_A^L}}{\dfrac{\delta_L}{D_A^L} + \dfrac{1}{m' k_{mc}^G}} \tag{13.70}$$

The denominator in Equation (13.70) is the sum of the resistances in the liquid and gas-phases and the numerator is the sum of two driving forces which are related to the gas-liquid equilibrium and the reaction of A with B.

The development above assumed that the reactions were instantaneous. If the reaction rates are on the same order of magnitude as the diffusion rates, then the concentration profiles for A in the liquid-side phoney film will be as shown in Figure 13-8. The profiles illustrate the situation for both first-order and second-order reaction rates.

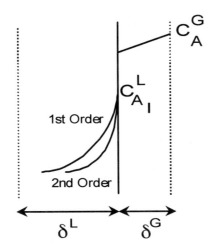

Figure 13-8 Finite Reaction Rates in Liquid Film

Since the reactions are bimolecular, they would normally be expected to be second order ($r = k_r C_A C_B$), but the reactant is often in large excess, so C_B is approximately constant and can be lumped together with the actual rate constant, k_r. This makes the reaction a "psuedo" first order reaction; i.e., $r = k_{r'} C_A$, where $k_{r'} = k_r C_B$. Note that the profiles in Figure 13-8 show the concentration going to zero within the film. This applies to a situation where the reaction rate is still fast, but not instantaneous. If the reaction rate were considerably slower, then the concentration of A in the bulk liquid may not be zero and unless it were significantly lower than the saturation value in the liquid, there would be no point in employing the reactant.

The mathematics describing the simultaneous occurrence of a first order reaction and diffusion has already been described in Section 10-3. There it was shown that the enhancement factor can be calculated from

$$\xi = \beta \frac{C_{AI}}{C_{AI} - \overline{C}_A} \coth \beta$$

where, β, the Hatta Number, is

$$\beta = \left(\frac{k_{r'} (\delta^L)^2}{D_A^L} \right)^{\frac{1}{2}} \tag{13.71}$$

If the reaction is sufficiently fast ($k_r \ll D_A^L$), then $C_A \sim 0$ and $\coth \beta \sim 1$ so that the enhancement factor is $\xi \sim \beta$. That is,

$$\xi = \frac{(k_{r'} D_A^L)^{1/2}}{(k_{mc})_0} \tag{13.72}$$

and since this is a "psuedo" first-order reaction, the k_r value to be used in Equation (13.72) is actually the product of the rate constant and the prevailing concentration of the reactant, B, or $k_r C_B$. Thus, in this case, the enhancement factor is a function of the mass transfer coefficient in the absence of a chemical reaction as well as the reaction rate constant and diffusivity.

13-5 FLUID-SOLID SYSTEMS (ADSORPTION)

Most molecules have some affinity for adsorbing onto solid surfaces and this usually involves weak inter-molecular forces between the adsorbing molecule (the *adsorbate*) and the surface (*adsorbant*). These forces, similar to Van Der Waals forces, usually occur at lower temperatures, and adsorption is classified as *physi-sorption*. However some molecules have a particularly strong affinity for some specific solids and, when this occurs there is potential for separation of these molecules from those which do not adsorb very readily on that particular solid. In some cases adsorption of a molecule on a specific solid only occurs at higher temperatures and this typically involves weak covalent bonds between the molecule and a "site" on the surface. This type of adsorption is referred to as *activated adsorption*, or *chemisorption*. Most, if not all, heterogeneous catalysis phenomena involve some form of chemisorption of reactants. The attraction of adsorption as a separation process is that it is potentially a low energy process, and it is often capable of efficient separation, even with very low concentrations of the adsorbate in the fluid phase. The disadvantage of adsorption separation is that it is inherently an unsteady-state process since the surface continues to adsorb until saturation is reached and then adsorption will cease.

When adsorption is used to remove molecules with fluid phase concentrations of 10% or more, it is termed, "bulk separation" and when the concentrations are less than about 1%, it is classified as a "purification" process. One of the most common purification processes is in tertiary water treatment where activated carbon adsorbants, due to their extremely high surface area,[#] are extremely effective at removing trace quantities of organics. An example of bulk separation is Pressure Swing Adsorption (PSA) which separates oxygen and nitrogen and has become a commercial success. Table 13-4, taken from Keller [15], lists the four most common adsorption processes in terms of their estimated annual sales, along with their characteristics,

---

[#]   Surface areas on the order of 2000 m²/g have been achieved.

applications and disadvantages. Activated carbon dominates, because of stricter environmental regulations which necessitates the removal of trace organics from aqueous and gaseous streams. However, zeolites, which separate on the basis of molecular size[##], has sales in excess of $100 million as a consequence of its use in drying and air separation processes. As can be seen, silica gel and activated alumina have similar properties and are competitors for the same market. Not listed in this table are some of the newer adsorption processes which are finding increased use in the biotechnology industry because of their selective separation properties and the fact that large quantities of separated product are not usually required.

Table 13-4 Major Commercial Adsorbants

Type	Annual Sales	Characteristics	Use	Disadvantages
Activated Carbon	$380 MM	Hydrophobic, favors organics over water	Removal of organic pollutants	Difficult to regenerate
Zeolites	$100 MM	Hydrophillic, polar, regular channels	Air separation, dehydration	Low total capacity
Silica Gel	$27 MM	High capacity, hydrophillic	Drying gas streams	Trace removal not effective
Activated Alumina	$26 MM	High capacity, hydrophillic	Drying gas streams	Trace removal not effective

Some of the comments on the disadvantages of each of the adsorbants listed in Table 13-4 point to some process considerations when choosing adsorption for a separation application. For example, the capacity of an adsorbant for a particular molecule will affect how often the adsorbant must be regenerated. Regeneration is an important process consideration and because it is usually an exothermic process, higher temperatures do not favor adsorption and this is advantageous for regeneration purposes. That is, as long as the surrounding fluid contains little or no adsorbate, raising the temperature of the adsorbant will drive off the adsorbate. This method of regeneration is called "Temperature Swing Adsorption"(TSA). There are two other common methods for regeneration, "Pressure Swing Adsorption"(PSA) and "Displacement Purge Adsorption" (DPA). In PSA, the adsorption is usually accomplished at elevated pressures and the bed is regenerated by lowering the pressure in the bed. This has an advantage over TSA in that the pressure can be lowered more rapidly, allowing for faster regeneration. In DPA the adsorbant is first saturated with an adsorbate that is easily displaced by the desired adsorbate. Upon regeneration, the bed conditions are changed to favor displacement of the desired adsorbate by the original adsorbate, which continually recycles from the adsorber to the desorber. The decision to regenerate an adsorber bed is based on the "breakthrough" point. The breakthrough point is the maximum allowable concentration of adsorbate in the gas exiting the

---

[##] They are often referred to as "molecular sieves."

adsorber and is illustrated in Figure 13-9 where the exit adsorbate concentration is plotted as a function of time. In the case of a "sharp" separation, the adsorbate is not detected in the exit gas until the entire bed is close to saturation and the exit concentration climbs quickly to the inlet concentration. In a diffuse separation, the exit concentration climbs slowly and, at the breakthrough concentration, C_{BT}, the bed must be regenerated. At the breakthrough concentration, the bed with a diffuse separation has removed much less adsorbate than in the case of a sharp separation as indicated by the area between the inlet and outlet concentration curves. The time at which the breakthrough concentration is reached is designated the "cycle" time, τ_{cycle}, and the bed is often characterized by the "number of bed volumes" processed at that point, which is calculated by

$$N_{BV} = \frac{\tau_{cycle} \dot{V}}{V_{Bed}}$$

Adsorbents are typically chosen on the basis of their selectivity for a particular adsorbate, their saturation capacity and the equilibrium relationship between the concentration of an adsorbate in the fluid and its adsorbed concentration on the solid. This latter relationship is called an *adsorption isotherm*; so-called because the equilibrium data are taken at a constant temperature. Figure 13-10 shows three radically different isotherms where the adsorbed concentration, $[C_A^{ad}]^*$, at equilibrium is plotted as a function of the concentration of the adsorbate in the fluid.

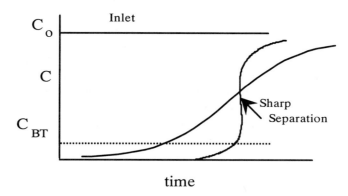

Figure 13-9 Exit Concentration in an Adsorber Bed

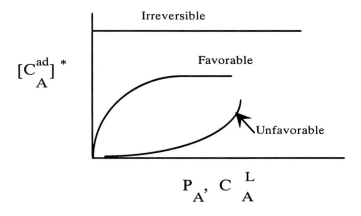

Figure 13-10 Adsorption Isotherms

With an irreversible isotherm, the adsorbed concentration is independent of the adsorbate concentration in the fluid and the other two isotherms are classified as either "favorable" or unfavorable. The reason that the concave up isotherm is unfavorable is because it requires rather high fluid concentrations before any appreciable adsorption will occur. The favorable isotherm, on the other hand, has high adsorbed concentrations at low fluid concentrations.

A prime example of a favorable isotherm is the Langmuir isotherm and since it is so prevalent, it is worthwhile to derive the mathematics associated with it. The assumptions involved are rather restrictive but for a variety of reasons, it even seems to describe many situations which violate the assumptions. The basic assumption behind the Langmuir isotherm is that the surface is homogeneous and only one molecule adsorbs on one surface site. The last requirement is equivalent to assuming a monolayer coverage of the adsorbate on the adsorbent. With these assumptions the rate of adsorption is taken to be proportional to both the number of unoccupied sites and the concentration of the adsorbate in the surrounding fluid. Since there are no interactions between adsorbed molecules, the desorption rate is taken to be proportional to the number of adsorbed molecules. Mathematically, these two rates can be expressed as

$$r_a = k_a P_A (\overline{C}_s - \overline{C}_A)$$

$$r_d = k_d \overline{C}_A$$

where $\overline{C}_s$ is the maximum number of adsorption sites available on a particular adsorbent; i.e., when the adsorbent is completely saturated with adsorbate, $\overline{C}_A = \overline{C}_s$. At equilibrium the adsorption rate will equal the desorption rate and these equations can be solved for $\overline{C}_A$ to give its value at equilibrium, $\overline{C}_A^*$

$$\overline{C}_A^* = \frac{\overline{C}_s K_A P_A}{1 + K_A P_A} \tag{13.73}$$

Note that the isotherm is linear at low partial pressures of A and $\overline{C}_A^*$ approaches $\overline{C}_s$ at high partial pressures. K_A is the adsorption equilibrium constant ($K_A = \dfrac{k_a}{k_d}$) and since adsorption is usually exothermic, it decreases exponentially with temperature. Since both k_a and k_d are rate constants, their variation with temperature would be similar to that shown in Figure 13-11. At low temperatures k_a is higher than k_d but k_d is more sensitive to temperature (has a higher activation energy) and it becomes greater at high temperatures. Adsorption would take place at the lower temperatures and desorption (regeneration) will take place at the higher temperatures. One modification to the Langmuir model is the Fruendlich model which attempts to account for heterogeneous sites on the surface. The assumption here is that there is a maxwellian distribution of sites which is characterized by the binding energy of the adsorbate to the site. Expressing K_A as a function of temperature and noting that the fraction of adsorbed sites, θ, is $\dfrac{\overline{C}_A}{\overline{C}_s}$, Equation (13.73) can be written for a particular site of energy, E, as

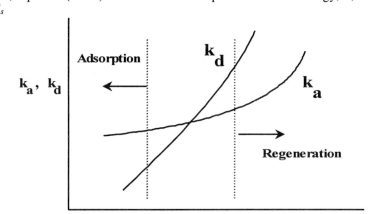

TEMPERATURE

Figure 13-11 Adsorption Rate Constants as a Function of Temperature

$$\theta(E) = \frac{A_o P_A \, \mathrm{Exp}\left[\dfrac{E}{R_g T}\right]}{1 + A_o P_A \, \mathrm{Exp}\left[\dfrac{E}{R_g T}\right]}$$

The fraction of all occupied sites is found by integrating θ (E) with respect to E over all sites, using the Maxwellian distribution, P(E), as a weighting function

$$\theta = \frac{\int_0^\infty \theta(E)\ P(E)dE}{\int_0^\infty P(E)dE}$$

The integration yields the Fruendlich isotherm

$$\theta = K_{A'}\ P_A^m \tag{13.74}$$

where $K_{A'}$ and m are parameters obtained from fitting the adsorption equilibrium data. Note that this equation predicts that the fraction of adsorbed sites increases without limit and so it is commonly employed in the Langmuir form, or

$$\theta = \frac{K_{A'}\ P_A^m}{1 + K_{A'}\ P_A^m} \tag{13.75}$$

The above equations describe the situation at equilibrium but the size of an adsorber will depend on the rates of adsorption. The mathematical description of adsorption rates in an adsorber bed is rather complex since it is an inherently unsteady-state process. In any case, the approach is the same as in the previous sections of this chapter; an analysis of the interfacial transport at any point in the bed, followed by differential material balances within the bed. Assuming adsorption of a gas, the sketch in Figure 13-12 shows the rate steps involved in getting species A from the bulk gas to an adsorption site. Step 1 is the convective mass transport of A from the bulk gas to the solid surface (through the phoney film) and Step 2 is the adsorption rate of A on to the surface. Since most solid adsorbents will have large surface areas, adsorption on the outer surface area of the adsorbent will be small compared to the adsorption on the surfaces of the pores. Consequently Step 2 can be ignored and Step 3 is in series with convective mass transport. This step involves simultaneous pore diffusion and adsorption, very similar to diffusion in a catalyst pore which was analyzed in detail in Section 4-3. However, in this case the mathematics are not as tractable due to the fact that the adsorption reaction is reversible. Another characteristic of solid adsorbents is that the rates of adsorption are usually very fast, so that equilibrium is established instantaneously. In this case the rate of adsorption is controlled by the rate of mass transport which involves both convective gas-solid mass transport as well as pore diffusion. The rate may then be expressed in terms of an overall volumetric mass transfer coefficient, as

$$r_{ad} = K_{OGc}\ a_l\ (C_A^G - \overline{C}_A^*) \tag{13.76}$$

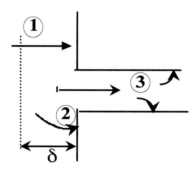

Figure 13-12 Rate Steps in Adsorption

For gas adsorption in a packed bed, $a_I = a_p(1 - \varepsilon)$, where ε is the void fraction in the bed. For spherical particles, $a_p = \dfrac{6}{d_p}$. Although pore diffusion can be complex,[#] a reasonable estimate to the combined convective and pore diffusion resistance can be calculated from Equation (13.77)

$$K_{OGC} = \frac{1}{\dfrac{1}{k_{mc}^G} + \dfrac{d_p}{10\,D_{eff}}} \tag{13.77}$$

The same difficulty as existed with overall gas-liquid mass transfer coefficients also exists in Equation (13.76), the units of the gas-phase and adsorbed concentrations are incompatible. However, if adsorption is fast, the adsorbed concentration of A is in equilibrium with the gas-phase concentration and Equation (13.76) can be written as

$$r_{ad} = K_{OGC}\,a_I\,(C_A^G - [C_A^G]^*) \tag{13.78}$$

where $[C_A^G]^*$ is the equilibrium concentration of the gas corresponding to the adsorbed concentration and is obtained from the adsorption isotherm expression.

Figure 13-13 is a sketch of an adsorber bed with an entering superficial velocity (velocity in the absence of the packing), v_0, and concentration, C_{A0}^G. Since accumulation of A only occurs in the void space of the bed, the differential material balance for species A in the bed is

---

[#]   For example, the migration of adsorbed molecules along the surface ("surface" diffusion) can be a significant factor.

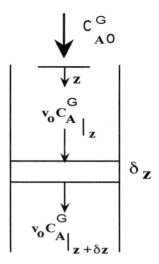

Figure 13-13 Adsorber Differential Material Balance

$$[v_0 C_{A|_z}^G - v_0 C_{A|_{z+\delta}}^G] A_x \, \delta t - r_{ad} A_x \delta z \, \delta t = [C_{A|_{t+\delta t}}^G - C_{A|_t}^G] \varepsilon A_x \, \delta z$$

and this leads to the differential equation

$$-v_0 \frac{\partial C_A^G}{\partial z} - K_{OG_c} a_I (C_A^G - [C_A^G]^*) = \varepsilon \frac{\partial C_A^G}{\partial t} \tag{13.79}$$

If the Langmuir isotherm applies, $[C_A^G]^*$ is obtained from Equation (13.73) and is a function of the adsorbate concentration, $\overline{C}_A$, which in turn, varies with time and position in the bed. The needed equation to calculate $\overline{C}_A$ comes from a differential material balance for A in the solid bed, which is

$$[\overline{C}_{A|_{t+\delta t}} - \overline{C}_{A|_t}] \rho_p (1 - \varepsilon) A_x \, \delta z = K_{OG_c} a_I (C_A^G - [C_A^G]^*) A_x \, \delta z \delta t \tag{13.80}$$

$$\frac{\partial \overline{C}_A}{\partial t} = \frac{K_{OG_c} a_I}{\rho_p (1 - \varepsilon)} (C_A^G - [C_A^G]^*)$$

The solution to Equations (13.79) and (13.80) will depend on the particular isotherm which applies to the situation but, in general, the required time, τ, to reach a certain degree of bed saturation, depends on the following two dimensionless parameters

$$t^* = \frac{v_0 C_{Ao}^G \tau}{\rho_p (1 - \varepsilon) Z_T \overline{C}_s} \tag{13.81}$$

and

$$N_{ad} = \frac{K_{OGc} a_I Z_T}{v_0} \tag{13.82}$$

where t^* is a dimensionless time, τ is the breakthrough time and N_{ad} is a dimensionless adsorption rate parameter, sometimes referred to as the number of transfer units. If adsorption takes place as an ideal plug flow operation, then the sharp separation shown in Figure 13-9 will be a perfect step change and the ideal time required to saturate the entire bed will be

$$\tau_{Id} = \frac{Z_T \rho_p (1 - \varepsilon) \overline{C}_s}{v_0 C_{Ao}^G} \tag{13.83}$$

and thus, comparing Equations(13.81) and (13.83), $t^* = \dfrac{\tau}{\tau_{Id}}$.

The mathematics can be simplified further if we now assume that the adsorption is irreversible and that we can ignore the accumulation of A in the gas. Then Equation (13.79) becomes

$$-v_0 \frac{d C_A^G}{dz} - K_{OGc} a_I C_A^G = 0$$

which can be integrated over the height of the tower to give

$$-\ln \left[\frac{C_A^G}{C_{Ao}^G} \right] = N_{ad} = \frac{K_{OGc} a_I Z_T}{v_0} \tag{13.84}$$

Equation (13.84) can be used to predict the *minimum* bed length. That is, it can be solved for Z_T, given the breakthrough criterion, $C_A^G = C_{BT}$. In practice, a commercial adsorber bed is designed to operate with a multiple number of the minimum bed lengths with the number dictated by the desired regeneration (or disposal) time. Thus, the real interest is in predicting the time for breakthrough in a bed length much larger than the minimum.

For a reasonably sharp separation, only a rather narrow portion of the bed will be undergoing active adsorption. The gas concentration in the saturated portion of the bed will be equal to the inlet concentration and it will be essentially zero in regions below the active portion of the bed. This is illustrated in Figure 13-14, where for simplicity, the bed is shown as being horizontal and Z_1 is the zone of active adsorption. If t_1 is the time it takes to saturate zone, Z_1, then the quantity of A fed to the zone over that time period is $v_0 C_{A_o}^G A_x t_1$. The quantity of A retained by the zone over that same period will be $\overline{C}_s \rho_p (1 - \varepsilon) A_x Z_1$. In the limit, we can apply this balance to a zone of length, δZ_1, over time, δt_1. Then, if the two terms are set equal to each other and solved for $\dfrac{\delta Z_1}{\delta t_1}$, the velocity at which the zone progresses through the bed, u_z is just the limit as $\dfrac{\delta Z_1}{\delta t_1}$ approaches zero, or

$$u_z = \frac{v_0 \, C_{A_o}^G}{\overline{C}_s \, \rho_p (1 - \varepsilon)} \tag{13.85}$$

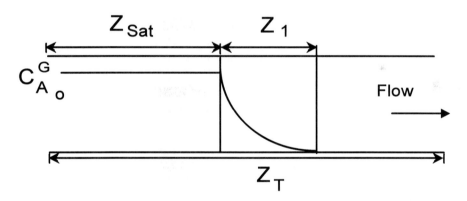

Figure 13-14 Adsorber with Irreversible Adsorption

Equation (13.84) only applies to zone, Z_1, so when it is applied to the entire bed, it has to be rewritten as

$$- \ln \left[\frac{C_A^G}{C_{A_o}^G} \right] = \frac{K_{OGC} \, a_1}{v_0} (Z_T - Z_{Sat}) \tag{13.86}$$

If τ_1 is the time required to saturate the first Z_1 zone (i.e., at the entrance to the bed), then the saturated length of the bed can be calculated by the product of u_z and the time since the zone started moving, or

$$Z_{Sat} = u_z \, (\tau - \tau_1) \tag{13.87}$$

and τ_1 is just the quantity retained by the bed in zone Z_1 divided by the mass transfer rate of A to the bed, or

$$\tau_1 = \frac{\overline{C}_s \rho_p (1 - \varepsilon) A_x Z_1}{K_{OGC} \, a_I \, C_{A_o}^G \, A_x Z_1} = \frac{\overline{C}_s \rho_p (1 - \varepsilon)}{K_{OGC} \, a_I \, C_{A_o}^G} \tag{13.88}$$

If Z_{Sat}, u_z, and τ_1 from Equations (13.87), (13.85) and (13.88) are substituted into Equation (13.86), the breakthrough concentration is predicted by

$$\ln \left[\frac{C_A^G}{C_{A_o}^G} \right] = N_{ad}(t^* - 1) - 1 \tag{13.89}$$

Equation (13.89) predicts that the breakthrough concentration increases exponentially with $N_{ad}(t^* - 1)$ and the exit gas concentration will equal the inlet when $(N_{ad}t^* - 1) = 1$. In reality the breakthrough curves are sigmoid shaped and are heavily influenced by pore diffusion when the bed approaches saturation.

13-6 MEMBRANE SEPARATION

The final application of mass transfer operations to be considered in this chapter is membrane separation. It is significantly different than the other mass transfer operations in that the separation is not limited by equilibrium considerations. In membrane separation, the transported component is removed from one fluid to another by moving across the membrane at a rate that is higher than other components. After traversing the membrane, it is physically removed and thus the membrane serves only as a barrier between the two fluids. The key to successful membrane separation is the degree to which the membrane is "semi-permeable" to a particular component. There are many different types of membranes and they are characterized by the mechanism which determines the separation as well as by the temperatures at which they operate. Table 13-5 lists some of the major membrane separators in terms of the fluid phases to which they are most often applied.

TABLE 13-5
Membrane Classification

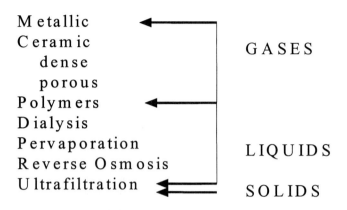

Typically, membrane separation of gases is conducted at higher temperatures and so, metallic and ceramic membranes are employed. An example of a metallic membrane is palladium, which is used for hydrogen separation. The mechanism here is that of hydrogen dissociation, followed by solid state diffusion of protons through the palladium crystal structure, with recombination of the protons to molecular hydrogen at the other side of the membrane. Because the dissociation of hydrogen to protons requires high temperatures (> 400 ^{0}C) and solid state diffusion is "activated"; i.e., is exponentially dependent on temperature, this separation is carried out at high temperatures. An example of dense ceramic membrane separation is zirconia (stabilized by yttria) where oxygen dissociates on the surface and the ions are transported across the membrane. Again this requires high temperatures (~ 1000 ^{0}C), although there is now a class of "perovskite" ceramics which will give even higher oxygen transport fluxes at temperatures of about 800 ^{0}C.[#] Gas separation can also be carried out by using porous ceramics, which depend on differing diffusivities in the pores of the membrane. Polymeric membranes, on the other hand, are restricted to lower temperatures and are usually employed in liquid systems although they are also used for some low temperature gas separation. Dialysis takes advantage of different solute concentrations in aqueous phases on either side of the membrane and the most common application is blood purification in individuals with severe kidney malfunction. Pervaporation is a membrane separation that actually involves a phase change within the membrane. A liquid on the high pressure side of the membrane is evaporated within the membrane itself by imposing low pressures on the other side. Reverse osmosis relies on osmotic pressure for the separation. Here the pressure on the dilute <u>solvent</u> phase is raised above the osmotic pressure and the solvent then permeates from the low concentration to the high concentration side of the membrane. The most common application here is in the

---

[#]   An example of such a perovskite is $La_{0.6}Sr_{0.4}Co_{0.2}Fe_{0.8}O_{3-\delta}$, where δ represents oxygen "vacancies" in the crystalline structure as a result of Sr and Fe substitution.

desalination of brackish water. Finally, ultrafiltration separates on the basis of pores being smaller than the size of the species being separated. Pore sizes of these membranes can range from 1 μm to 1 nm, thus they can be used for separating particulates or molecules with high molecular weights, such as proteins.

The remaining section of this chapter will deal with gas separation in both porous ceramic membranes and polymeric membranes since the intention is only to introduce the methodology of mathematically analyzing membrane separators. Using the same approach as in the other sections of this chapter, attention is first focused on interfacial transport across the membrane and then on describing separation in a membrane separator. Separation in porous ceramic membranes is dependent on different rates of diffusion through the membrane and, in gases, this usually means the separation depends on having very different molecular sized molecules. Two types of diffusion can occur in small pore membranes, molecular diffusion and Knudsen diffusion. Molecular diffusion was discussed in Chapter 4 and methods to predict molecular diffusivities were covered in Chapter 6. Whereas molecular diffusion comes about by molecule-molecule collisions, Knudsen diffusion occurs as the result of collisions of the molecules with the walls of the pores and both mechanisms are depicted in Figure 13-15. In general, Knudsen diffusion becomes important when the mean free path of colliding molecules is equal or greater than the characteristic dimension of the system. For a small cylindrical pore, this criterion can be written as $\lambda \geq r_p$, where r_p is the pore radius. The pore radius in Figure 13-15 is shown as being on the same size as the mean free path of the molecules and thus, some of the molecules are colliding with other molecules and others are colliding with the walls of the pore. The Knudsen diffusivity depends on the pore radius and the temperature and molecular weight of the molecule; specifically,

$$D_{K_A} = 9700 \, r_p \left(\frac{T}{M_A} \right)^{1/2} \tag{13.90}$$

where the pore radius is in cm, the temperature in K and the diffusivity has units of cm²/s. Knudsen diffusivities are calculated using an average pore radius and this is obtained from surface area measurements of the membrane, called BET area, and the void volume of the

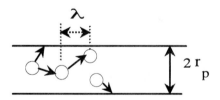

Figure 13-15 Molecular and Knudsen Diffusion in a Pore

membrane. The former is based on measuring the quantity of N_2 which is adsorbed, per mass of material, onto the surface at liquid nitrogen temperatures. The BET area, $\hat{S}$, is then calculated by using a value for the surface area per adsorbed nitrogen molecule. The void volume per mass is calculated from the density and porosity of the material

$$\hat{V} = \frac{\varepsilon}{\rho_p}$$

The average pore radius is calculated by assuming that there are N cylindrical pores of the same length and radius,

$$\frac{\hat{S}}{\hat{V}} = \frac{N\pi(2r_P)L}{N\pi r_p^2 L}$$

(13.91)

$$r_p = \frac{2\hat{V}}{\hat{S}}$$

Even though these two processes take place simultaneously, the combined effect is usually treated as a geometric mean, or

$$\overline{D} = \frac{1}{\dfrac{1 - \phi X_A}{D_{AB}} + \dfrac{1}{D_{KA}}}$$

(13.92)

Note that the molecular diffusivity in Equation (13.92) is corrected for bulk diffusion, using the parameter, ϕ (see Chapter 4). Of course the pores are not straight and cylindrical and therefore, a correction is needed for the "tortuosity" of the pores to get the "effective" diffusivity, D_{eff_A}, and this is usually estimated by

$$D_{eff_A} = \varepsilon^2 \overline{D}_A$$

(13.93)

where ε is the porosity of the membrane.

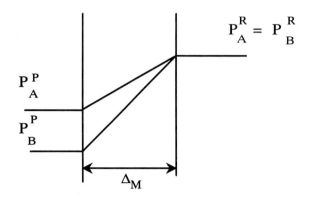

Figure 13-16 Pressure Driving Forces in a Binary System

To illustrate the effect of differing diffusivities and total pressures on separation in a porous membrane, Figure 13-16 shows the partial pressure profile across such a membrane for a binary system. Note that the side of the membrane which has a higher ratio of A/B is called the "permeate" and the other side is referred to as the "retentate." The molecular flux of A across the membrane is constant and is given by

$$J_A = -D_{eff\,A}\frac{dC_A}{dz} = D_{eff\,A}\,R_g T\,\frac{(P_A^R - P_A^P)}{\Delta_M}$$

(13.94)

In a binary system, the mole fraction of A in the permeate is simply

$$X_A^P = \frac{J_A}{J_A + J_B} = \frac{D_A \Delta P_A}{D_A \Delta P_A + D_B \Delta P_B}$$

(13.95)

where the driving forces are as defined in Equation (13.94) and $D_{A,B}$ are taken to be the effective diffusivities. Then the ratio of the mole fractions of A and B in the permeate is

$$\frac{X_A^P}{1 - X_A^P} = \frac{D_A (P_A^R - P_A^P)}{D_B (P_B^R - P_B^P)}$$

(13.96)

If the partial pressures are written in terms of the total pressures on the permeate and retentate sides of the membrane, the above equation can also be written as

$$\frac{X_A^P}{1 - X_A^P} = \frac{D_A}{D_B}\frac{[X_A^R - \gamma X_A^P]}{[X_B^R - (1 - \gamma X_A^P)]}$$

where $\gamma = \dfrac{P^P}{P^R}$. This equation is a quadratic equation which can be solved for X_A^P. For example, with an equimolar composition on the retentate side and $\gamma = .5$, the mole fraction of A in the permeate is only 0.59, not a large increase over 0.5 in the retentate. Thus, in order to get a large separation, it would be necessary to "stage" a number of the membranes in series.

While the mathematical description of polymeric membranes is similar, the major difference is that the gases first dissolve in the "skin" of an asymmetric membrane and diffusion then takes place through the skin. Thus, the separation is enhanced by two effects: the difference between the solubilities of the species as well as the difference in their diffusivities. This situation is illustrated in Figure 13-17.

The diffusion flux is written in terms of the gradient of the dissolved concentrations and they, in turn, must be expressed in terms of the partial pressures. The latter is done by means of the solubilities of A and B (S_A and S_B); i.e.,

$$C_{AI}^P = S_A P_A^P \ , \ C_{AI}^R = S_A P_A^R$$

So that the diffusion flux of A is

$$J_A = \frac{S_A D_A}{\Delta M} [P_A^R - P_A^P] \tag{13.97}$$

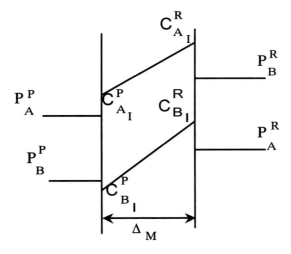

Figure 13-17 Driving Forces in a Polymeric Membrane

The term, $S_A D_A$, is the "permeability coefficient" of A in the membrane and the units are usually expressed in terms of "barrers," where 1 barrer is equivalent to 10^{-10} scc-cm/ cm^2-s-mm Hg. More often, the permeability coefficient is combined with the membrane thickness so that Equation (13.97) is written as

$$J_A = P_{M_A} (P_A^R - P_A^P)$$ (13.98)

where $P_{M_A} = \dfrac{S_A D_A}{\Delta_M}$, is the "permeability" of Air the membrane. Another useful term in polymeric membranes is the "membrane selectivity," α, which is defined as

$$\alpha = \frac{P_{M_A}}{P_{M_B}} = \frac{S_A D_A}{S_B D_B}$$ (13.99)

For a polymeric membrane, the mole fraction of A in the permeate of a binary system is calculated in the same way as in Equation (13.95) and, again, results in a quadratic equation. This equation can be written in terms of the selectivity and the total pressure ratio by combining Equations (13.95), (13.98) and (13.99) to obtain

$$\gamma (\alpha - 1)(X_A^P)^2 - [1 + (\alpha - 1)(X_A^R + \gamma)]X_A^P + \alpha X_A^R = 0$$ (13.100)

There is one property of Equation (13.100) which should be pointed out. That is, if the total pressure ratio is 1, there is no separation and $X_A^P = X_A^R$. This can be easily verified by substituting $\gamma = 1$ and $X_A^P = X_A^R$ into equation (13.100) and showing that the left-hand side of the equation does indeed equate to zero.

The next step in the analysis of a polymeric membrane system is to describe the separation in an actual separator. The retentate and permeate flow patterns can take on a variety of configurations. Two of the most common configurations involving only retentate and permeate streams (no sweep gases) are shown in Figure 13-18. In the first configuration, the permeate stream is "dead-headed" so that it is only removed at the retentate feed end of the separator. This is a true countercurrent flow operation. In the other configuration, the permeate is removed at either end of the separator, creating a mixed, parallel-countercurrent flow pattern. While the basic mathematical analysis of the two systems are the same, they do differ in the boundary conditions that are used. In addition to the permeate composition achieved in a membrane separator, another important factor is the yield of the desired component. The latter is usually defined in terms of the "stage cut" of the separator, which is defined as the fraction of feed recovered in the permeate, which in the nomenclature of Figure 13-18, would be

$$Stage\,Cut = \frac{\dot{N}_{ex}^{P}}{\dot{N}_{f}^{R}} \qquad (13.101)$$

The procedure for designing a countercurrent membrane separator depends on how the problem is stipulated. Usually the desired purity of either the retentate or the permeate stream is specified and the stage cut is a calculated value. If stage cuts higher than the calculated value are required, then a number of staged membrane separators may be necessary. The basic equations describing the operation of a membrane separator arise from differential material balances on the retentate and permeate streams. For mathematical purposes, either of the configurations in Figure 13-18 can be pictured as two concentric ducts as shown in Figure 13-19. Even though they are pictured as cylinders, they can be of any geometry. The permeate duct is described in terms of a perimeter, P_w, so that the surface area of the differential volume element is $P_w \delta z = \delta A_s$. For a binary system the differential material balance for A in the retentate stream is

Countercurrent

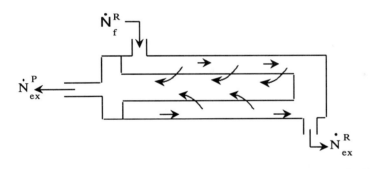

Parallel - Countercurrent

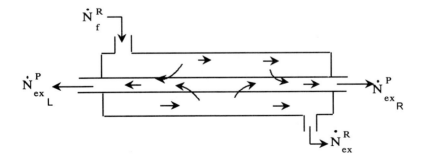

Figure 13-18 Membrane Separator Flow Systems

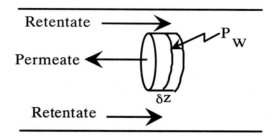

Figure 13-19 Differential Element in a Membrane Separator

$$(\dot{N}^R X_A^R)|_z - (\dot{N}^R X_A^R)|_{z + \delta z} - J_A P_W \delta z = 0$$

(13.102)

$$-\frac{d}{dz}(\dot{N}^R X_A^R) - J_A P_W = 0$$

The balance for B in the retentate is

$$-\frac{d}{dz}[\dot{N}^R (1 - X_A^R)] - J_B P_W = 0$$ (13.103)

Similar balances for A and B in the permeate stream result in

$$-\frac{d}{dz}(\dot{N}^P X_A^P) + J_A P_W = 0$$ (13.104)

and

$$-\frac{d}{dz}[\dot{N}^P (1 - X_A^P)] + J_B P_W = 0$$ (13.105)

The auxiliary equations for J_A and J_B complete the description

$$J_A = P_{M_A}(P^R X_A^R - P^P X_A^P)$$
$$J_B = P_{M_B}(P^R X_B^R - P^P X_B^P)$$

(13.106)

These equations can be solved using MATLAB once the boundary conditions at $z = 0$ are specified. At this point we have values for the feed conditions, $\dot{N}^R{}_f$, $(X_A^R)_f$ and the specified exit concentration in the retentate stream, $(X_A^R)_{ex}$. Three variables remain to be specified: the exit flow rates of the permeate and retentate streams, $(\dot{N}^P)_{ex}$, $(\dot{N}^R)_{ex}$, and the composition of A in the exiting permeate stream, $(X_A^P)_{ex}$.

The latter can be calculated by solving the quadratic equation in Equation (13.100) and then the two exiting flow rates can be calculated from overall material balances

$$(\dot{N}^R)_f = (\dot{N}^P)_{ex} + (\dot{N}^R)_{ex}$$

$$(13.107)$$

$$(\dot{N}^R)_f (X_A^R)_f = (\dot{N}^P)_{ex} (X_A^P)_{ex} + (\dot{N}^R)_{ex} (X_A^R)_{ex}$$

With the boundary conditions now specified, the design problem is solved by integrating Equations (13.102) through (13.105), using MATLAB, until the specified concentration in the exiting retentate stream is reached. The required surface area of the membrane is then the product of P_w and the value of z at that point.

EXAMPLE 13-5: Calculation of Membrane Surface Area

The concentration of oxygen in a 8% oxygen-nitrogen gas is to be reduced to 0.3% by using a membrane with an O_2-N_2 selectivity of 6.4 and an oxygen permeability of 0.34 SCF/ft²-h-atm. A counter current flow membrane is to be used with a pressure ratio of 0.067. If the retentate pressure is 250 psia and the gas feed rate is 500 SCFM,

(a) Calculate the exiting permeate concentration and the stage cut

(b) Specify the numerical values for the boundary conditions at z= 0 (the permeate exit location) corresponding to Equations (13.102)-(13.105)

(c) Write a MATLAB program to calculate the required membrane area, using Equations (13.102)-(13.105)

Solution

(a) The relationship between the compositions of the entering retentate feed and the exiting permeate feed is given by Equation (13.100), where $\gamma = 15$, $\alpha = 6.4$ and, at this location, $X_A^R = (X_{A_f})^R = 0.08$. The quadratic equation is then

$$0.067\,(6.4\,-\,1)(X_A^P)_{ex}^2-[1\,+\,(6.4\,-\,1)(0.08\,+\,0.067)](X_A^P)_{ex}+(6.4)(0.08)=0$$

$$(X_A^P)_{ex}=\frac{1.792\pm1.572}{0.72}$$

Therefore, choosing the negative root gives an oxygen composition in the exiting permeate of 0.304.

The stage cut is dependent on the exiting permeate flow rate and it is obtained from the overall material balances given in Equations (13.107). In hourly time units, the retentate feed rate is $\dot{N}_f^R = (500)(60)/359 = 83.56\,\text{lb-moles/hr}$. The material balance equations are then

$$83.56=\dot{N}_{ex}^P+\dot{N}_{ex}^R$$

$$(83.56)(0.08)=(.304)(\dot{N}_{ex}^P)+(0.003)(\dot{N}_{ex}^P)$$

with the result that $\dot{N}_{ex}^P = 21.3\,\text{lb-moles/hr}$ which gives a stage cut $= 21.3/83.56 = 0.256$.

(b)The boundary conditions are now calculated as:

$$\dot{N}_{ex}^R=83.56,\ \dot{N}_{ex}^P=-21.3$$

$$X_{A_f}^R=0.08,\ X_{A_{ex}}^P=0.304$$

$$X_{B_f}^R=0.92,\ X_{B_{ex}}^P=0.696$$

The following MATLAB Program can be used to solve the system of equations to determine the required surface area. Note, that in this case the independent variable is the cumulatove membrane surface area. For example, Equation (13.104) is written as

$$-\frac{d}{dA}[\dot{N}^P(1\,-\,X_A^P)]+J_B=0$$

where $dA = P_W\ dz$

However, in order to properly pose the problem, it must be expressed in terms of the three truly dependent variables, which are chosen as X_A^P, X_A^R, and $\dot{N}^P$. With this choice, and recognizing that

$$\dot{N}^R = 83.56 - \dot{N}^P$$

the three equations to be solved are Equations (13.102), (13.104) and (13.105). The MATLAB program identifies these variables as:

$$y(1) = X_A{}^P$$
$$y(2) = X_A{}^R$$
$$y(3) = \dot{N}^P$$

The Table below lists the MATLAB output. As can be seen, the oxygen mole fraction in the retentate $(X_A{}^P)$, reaches a minimum value of 0.0395, far above the specified value of 0.003. This is because, at that point in the membrane, $P_A{}^R = P_A{}^P$, and no further separation is possible. To achieve the goal of $X_A{}^R = 0.003$, will require that the membrane be "staged". That is, the retentate stream exiting this unit (at about $X_A{}^R = 0.04$) will become the feed for a subsequent stage and so on. Note that the membrane area for this first unit is about 1.3 ft². because the driving forces will be smaller , the areas of subsequent stages will become progressively larger.

MATLAB Output - Example 13-5

A(ft²)	$X_A{}^P$	$X_A{}^R$	N^P
0	0.304	0.08	- 21.3
0.075	0.3044	0.0756	-20.07
0.15	0.3056	0.0714	-18.88
0.30	0.3112	0.064	-16.61
0.45	0.3219	0.0576	-14.49
0.60	0.3394	0.0522	-12.51
0.75	0.3662	0.0477	-10.68
0.9	0.4062	0.0441	-8.99
1.01	0.4485	0.0420	-7.83
1.2	0.5533	0.0399	-6.13
1.275	0.6117	0.0396	-5.53
1.313	0.6454	0.0395	-5.26

Matlab Program: Example 13-5

<u>M-File</u>

```
function dy=membrane(A,y)
dy=zeros(3,1);

NRF=83.56;ALFA=6.4;PMA=.34;GAMMA=15;PR=250;

PMB=PMA/ALFA;
PP=PR/GAMMA;
JA=PMA*(PR*y(2)-PP*y(1));
JB=PMB*(PR*(1-y(2))-PP*(1-y(1)));
```

$% \; y(1) = X_A^P;$
$% \; y(2) = X_A^R;$
$% \; y(3) = N_A^P;$

```
dy(1)=(1/(y(3)*(1+1/y(1))))*(-JB+((1-y(1))/y(1))*JA);
dy(2)=(-1/(NRF-y(3)))*(y(2)*(1/y(1))*(JA-y(3)*dy(1))+JA);
dy(3)=(1/y(1))*(JA-y(3)*dy(1));
```

<u>Command File</u>

```
y0(1) = .304;
y0(2) = .08;
y0(3) = -21.3;

[A,y]=ode45('membrane', [0 1.3],y0);
res = [A,y];
```

REFERENCES

[1] Wankat, P.C., *Rate-Controlled Separations*, Elsevier, Barking, UK (1990).

[2] Sherwood & Pigford, 1975 - see McCabe, W.L., Smith, J.C. & P. Harriot, *Unit Operations of Chemical Engineering*, 5th ed., Mcgraw-Hill, N.Y., 1993, p. 663.

[3] Ranz, W.E. and W.R. Marshall, Chem. Eng. Progr., 48, p. 141, 173 (1952).

[4] Kuni, D. and O. Levenspiel, Krueger Pub. Co., Huntington, N.Y.,(1977).

[5] Chu, Kalil and Wetteroth, Chem. Eng. Progr., 49, p. 141 (1953).

[6] Gilliland E.R. and T.K. Sherwood, I&EC, <u>26</u>, p. 516 (1934)

[7] Vivian, J.E. and Peacemen, AIChE J., <u>2</u>, p. 437 (1956)

[8] Onda, K., Takeuchi, H. J. and O. Yoshio, ChE Japan, <u>1</u>, p. 56 (1968)

[9] Shullman, H.L., AIChE J., <u>1</u>, p. 253 (1955)

[10] Sherwood, T.K. and F.A.L. Hollaway, Trans. Inst. Chem. Engrs., <u>36</u>, p. 39 (1940)

[11] Trybal, R.E., *Mass Transfer Operations*, 3rd ed., (1980), p. 206

[12] McCabe, W.L., Smith, J.C. & P. Harriot, *Unit Operations of Chemical Engineering*, 5th ed., (1993), p. 693

[13] Robbins, L.A., Chem. Eng. Progr., <u>87</u>, No. 5, p. 87, (1991)

[14] Merkel, Ver. Deut. Ing., Forschungsarb., No. 275, Berlin (1925)

[15] Keller, G.E., Chem. Eng. Progr., p. 56, October, 1995

PROBLEMS

13-1 Using a similar approach as was used in Section 13-1,

   (a) Derive $(K_{OL})_c$. Assume that the equilibrium relationship is $C_A^G = m' \, C_A^L$

   (b) Using the result from (a) and Equation 13.5, obtain a relationship between $(K_{OL})_c$ and $(K_{OG})_x$.
   HINT: relate m' and m .

13-2 Compare the predicted value of k_{mx} for packed beds of 60 mesh spherical particles, using Equations 13.14 and 13.15 and the following correlation, proposed by Sherwood and Pigford

   $Sh_p = 1.17 \, Re^{.585} \, Sc^{1/3}$

   which is purported to be valid for conditions of moderate void fractions (ε ~ .40-.50) and $10 < Re_p < 2,500$. Make this comparison for:

   (a) H_2 - in - air at 200 $^\circ$C and 1 atm pressure and for $Re_p = 30$ and 1000

   (b) NH_3 -in - water at 20 $^\circ$C for $Re_p = 30$ and 1000

13-3   Schulman et al. [AIChE J., 1 , p. 253 (1955)] obtained independent measurements of a_i for 1 in-Raschig Rings. At gas and liquid mass velocities of 300 and 750 lbm/ft^2 - h, respectively, they measured a_i to be 10 ft^2 / ft^3. Compare this value with that predicted by Onda et al. (Equation (13.19) for ceramic rings. The data were taken in an ammonia-water gas absorber system. Assume a water temperature of 35 $^{\circ}$C.

13-4   Weissman and Bonilla obtained the following correlation for a_i data using data from the evaporation of water over 1-in Raschig rings

$$\frac{a_i}{a_p} = .44 \ G^{.31} \ L^{.07}$$

where a_p is the specific area for the dry packing and the correlation corresponds to $540 < L < 2,600$ lb/ft^2 - h. Compare this correlation with Onda et al's correlation, Equation (13.19) for the three sets of (G,L) values: (300, 540), (750,750), (750, 3000) and at 50 $^{\circ}$C.

13-5   Brittany and Woodburn [AIChE J., 12, p. 544, 1966] used CO_2 and N_2 in H_2O to obtain the following correlation for the liquid-side mass transfer coefficient

$$k_{mx}^L \, a_l = 1.107 \ G_L^{.43}$$

with $5500 < G_L < 9500$, $4 < G_G < 10$ lbm/ft^2-h and where G_L has units of lbm/ft^2-hr and $k_{mx}^L \, a_l$ has units of lbm/ft^3-hr.

Compare this prediction for the volumetric mass transfer coefficient for CO_2 - in water at 20 $^{\circ}$C, G_L = 8000, G_G = 10 lbm/ft^2-hr, for 1" ceramic Raschig rings to that predicted by :

(a) Onda's correlations

(b) The correlation of Sherwood and Holloway

13-6   Program the problem illustrated in Example 13-1 on a spreadsheet and determine the effect of changing $(X_A^G)_B$, $(X_A^G)_T$, G_G , G_L . Write this program so that it directly calculates H$_{OG}$ and (NTU)^G . In making these changes, be careful that you do not violate thermodynamics and produce a bottoms liquid concentration that is higher than saturation.

(a) Present some of these results by holding all other parameters at the conditions given in Example 13-1 and plotting: $(NTU)^G$ versus the ratio $\dfrac{G_G}{G_L}$ with parameters of $(X_A^G)_B$, and H_{OG} versus G_L with parameters of G_G.

(b) Plot Z_T versus the temperature of the tower, over the range 10 - 40 $^{\circ}$C for $G_L = 10,000$ lbm/ft^2-hr

13-7

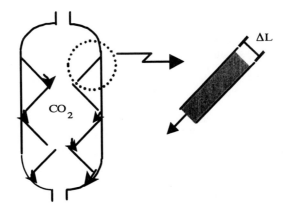

A beverage is to be carbonated by passing it at 20 C over a series of inclined planes, as shown in the sketch. Under normal operating conditions the vat contains pure CO_2 at 2 atmospheres pressure, the incoming beverage flows at 2 barrels/plane and contains 0.3 lbs CO_2/barrel and the exit CO_2 concentration is 0.5 lbs/barrel. What will be the exit CO_2 concentration if the beverage flow is doubled? Assume that the liquid on the planes is in turbulent flow and the increased flow does not affect the applicable mass transfer coefficient. The concentration of CO_2 at the gas-liquid interface is saturated and Henry's Law applies; i.e., $C_A^* = \dfrac{P_{Ab}}{H}$ (where $H = 1.93\ \dfrac{atm\ -\ barrel}{lb_m}$). In other words, the concentration gradient in the gas can be neglected.

13-8 Air at 1 atm and containing 2 % NH_3, is to be reduced to 500 ppm by scrubbing with ammonia-free water at 25 C. Determine the height of 1" Berl saddles required if the mass velocities of gas and liquid are each 300 lbm/ft^2-hr. The Henry's Law constant for this system is 0.746 atm/mole fraction.

13-9   Starting with the definition of the overall mass transfer coefficient, Equation(13.5), derive Equation (13.41)

13-10 Derive Equations (13.46) and (13.47), which are used to calculate the packed tower height for "rich" gases under isothermal conditions and for a linear equilibrium relationship.

13-11 250 lbm/ft^2-hr of a gas consisting of 24% SO_2-in-air and delivered at 2 atm pressure and 40 C is to be reduced to a concentration of less than 0.4 % by scrubbing with water. The SO_2 concentration in the liquid bottoms is to be restricted to 50% of saturation. Assuming that Henry's Law applies and that the tower will be operated isothermally, calculate the height of 3/4 inch Intalox saddles required to achieve the separation.

13-12 Whitney and Vivian (Chem. Engr. Progress, 45, p. 323, 1949) ran countercurrent experiments in a SO_2-water system in order to measure values of $K_{OLC} a_I$. The equipment consisted of a 2 ft high column packed with 1" Raschig rings at various gas and liquid flow rates and at 1 atm pressure between 50-90 $^{\circ}$F. The following data were taken in one experiment: G_L = 3000, G_G = 179 lbm/ft^2-hr, respectively, $(P_{SO2})_B$ = 0.15 atm, $(P_{SO2})_T$ = 0.09 atm, T = 68 $^{\circ}$F. Assuming that Henry's Law applies, that the system is isothermal and that the gas and liquid flow rates are constant, calculate the value of $K_{OLC} a_I$ for this experiment. The entering water in these experiments was free of SO_2.

13-13 A tower packed with 1" Raschig Rings is to be designed to treat 35,000 SCFH of a gas containing 3% NH_3 so that its exit concentration will be reduced to 50 ppm. The entering liquid contains 0.5 mole % NH_3 and its properties are those of water. If the liquid flow rate is to be 1.5 times the minimum, calculate the diameter of the tower using:

(a) Flooding criteria; gas mass velocity to be ½ the flooding value.

(b) Robbins' method; pressure drop to be 0.5 inches water

13-14 Use MATLAB to solve Example 13-3 for the height of the packed tower required to achieve an ammonia concentration in the exit gas of 0.016.

13-15 Redo 13-11 using MATLAB, assuming an adiabatic tower. The heat of solution of SO_2 can be assumed to be 30,000 BTU/Lb-mole and the relative humidity of the entering air is 30%.

13-16 Calculate the wet bulb temperature at a point in a cooling tower where the air temperature is 45 $^{\circ}$C and its relative humidity is 55%. Does the use of the humid heat rather than the gas heat capacity, change this result in any significant manner?

13-17 Warm cooling water is to be recycled after cooling it from 110 $^{\circ}$F to a temperature which is 8 $^{\circ}$F greater than the wet bulb temperature of the entering air in a forced draft cooling tower. The water mass velocity is 1,750 lbm/ft^2-hr and the ambient air is at 90 F and 20% relative humidity. The gas flow rate is to be twice the minimum and the gas-side volumetric mass transfer coefficient is 5.0 lb-moles/ft^3-hr.

(a) Calculate the required gas mass velocity and the exit water temperature.

(b) Write an ACSL program to solve Equations (13.58)-(13.60) for the required height of the tower.

13-18 Sherwood and Pigford ("Absorption and Extraction", 2nd ed., McGraw-Hill, N.Y., 1952) report that the mass transfer coefficient, $K_{OG_x} a_I$, for CO_2 absorption in water at 77 $^{\circ}$F, increased by a factor of 8 when 3N DEA was added and its conversion to the carbonate was 50%. The data were taken with 1" Raschig rings and G_L = 2500 and G_G = 300 lbm/ft^2 -hr.

(a) Assuming that the gas was pure CO_2 and gas-phase resistance is negligible, compare this increase with the enhancement factor for an instantaneous, irreversible reaction. The diffusivity of DEA-in-H_2O is taken from Hikita et al., J. Chem. Eng. Data, 25, p. 324 (1980) and is 4.3 x 10^{-6} cm^2/s for a 3 N solution.

(b) At high concentrations of DEA, its reaction with CO_2 can be considered a "psuedo" first order reaction; i.e., the apparent rate constant is $C_B k_r$. The second order rate constant can be calculated from

$$\log k_r = 7.443 - \frac{2275.5}{T(K)}$$

with k_r in units of m^3/ mole-s. Calculate the enhancement factor for this psuedo first order reaction rate and compare it to the value calculated in (a) by assuming an instantaneous irreversible reaction. Use Onda's correlations, Equations (13.19) and (13.21), to calculate the mass transfer coefficient in the absence of a chemical reaction.

13-19 Determine which equation(s) in Table 13-2 would have to be modified (and/or added) to account for a second order, irreversible reaction in the liquid-phase and then show the modified set of equation(s) which would need to be solved. Assume that the entering concentration of the reactant is $(C_B)_T$ and that the reaction rate constant, k_r, varies with temperature according to an Arrenhius expression.

13-20 The following adsorption isotherm data were taken for the adsorption of SO_2 on activated charcoal at 273 K. In addition, separate experiments, conducted at a constant pressure of 180 mm Hg and at temperatures of 302 °K and 351 °K gave the following results for the quantity of SO_2 adsorbed on the charcoal: 0.25 g/g at 302 °K and 0.10 g/g at 351 °K.

P (mm Hg)	$\overline{C}_{SO_2}$ (g/g adsorbent)
73	0.214
180	0.471
309	0.726
540	1.089
882	1.494

(a) Determine the Langmuir isotherm parameters at 273 °K.

(b) Determine the Fruendlich Isotherm parameters at 273 K

(c) Assuming the Langmuir isotherm applies, what will be the adsorbed SO_2 concentration in equilibrium with P_{SO2} = 325 mm Hg and T = 325 K?

13-21 Quite often large surface area adsorbents are impregnated with compounds which react with trace quantities of contaminants in gas or liquid streams and, upon saturation, the saturated bed is sent to hazardous waste storage or reclamation. This type of process is only practical when the contaminant concentrations are very low and the adsorbent capacity is high. Assume that the chemists have developed a sulfur-impregnated activated carbon adsorbent that is capable of removing trace cadmium vapors from gases by reacting with the sulfur to form CdS. The reaction is very rapid and irreversible. The following specifications apply:

Adsorbent particles: spherical, 4x6 mesh, density= 0.5 g/cm³, D_{eff} = 0.15 D_{Cd-N2}

Bed: $\Delta P = 1$ in H_2O/ft, $\varepsilon = 0.42$, T = 25 C, $\overline{C}_S = .15$

Feed gas: N_2 at 5000 SCFH, Cd concentration = 30 ppt

If the cycle time is to be 1 year and the Cd concentration cannot exceed 0.1 ppt, we wish to examine the feasibility of this process by specifying:

(a) The bed diameter

(b) The minimum bed length

(c) The "ideal" bed length

(d) The bed length, using Equation (13.89)

13-22 A methane reformer produces a gas with a H_2 :CO_2 ratio of 4:1. It is proposed to use a porous ceramic membrane to separate the hydrogen from the CO_2. Calculate composition of the H2:CO2 mixture on the permeate side of the membrane if the pressure ratio is 20 and the temperature of the system is 750 C. The membrane is an alumina with a porous surface area of 600 m^2/g and a porosity of 0.5.

13-23 An air-separation membrane has an oxygen selectivity of 7.0 and $P_{MO2} = .31$ SCFH/ft²-atm. If the separator is to be operated in a countercurrent manner with a pressure ratio of 10 and the N_2 concentration in the exiting retentate phase is to be 94%, calculate the required surface area using MATLAB. The air is fed to the separator at a flow rate of 380 SCFM and a pressure of 180 psia

APPENDIX A

Generalized Equations of Change

MOMENTUM, ENERGY AND SPECIES CONSERVATION EQUATIONS IN CARTESIAN, CYLINDRICAL, AND SPHERICAL COORDINATES

The approach taken throughout this textbook has been to use differential conservation balances to derive differential equations which describe the transport of energy, mass and momentum for a particular physical problem. In other words, the mathematical model is applied individually to each physical situation. Another type of approach is to derive a general mathematical model and to then simplify it according to a particular physical problem. Other than the difficulty of obtaining a truly general model, this latter approach has a certain appeal to it. From an educational point of view, the approach in the text is probably more valuable at the introductory level. However, when the physical problems become complex, particularly when dealing with problems in non-cartesian coordinate systems, it is far simpler to utilize the general equations.

This Appendix presents the general "equations of change" (mass, momentum and energy) in the three most common coordinate systems, but for the specific case of Newtonian fluids with constant fluid properties; the so called *Navier-Stokes Equations*.[#] In order to illustrate this

---

[#]   Refer to any of the Graduate level text books on transport phenomena for a more general approach.

approach, we first derive the Momentum and Energy Equations in cartesian coordinates; similar to that which was done in Chapter 8 for total mass conservation (the Continuity Equation). Once the results are obtained in cartesian coordinates, coordinate transformations can be employed to obtain the results for cylindrical and spherical coordinates and these are listed for the Continuity, Momentum, and Energy Equations in Tables A-1 through A-3. The derivation of a general species conservation equation is left as an exercise (see Problems at the end of Chapter 8) but the results are given here in Table A-4.

Derivation of The Momentum Equation

Figures A-1 and A-2 show a three-dimensional differential volume element, around which a momentum balance is taken. Since momentum is a vector quantity, this balance must be taken for each of the three components. For simplicity, only the x-component is shown in the figures and only this component will be derived in detail. Figure A-1 illustrates the *convective* x-momentum components entering and leaving the DVE and Figure A-2 depicts the surface forces[#] acting on the DVE. To be perfectly general we would have to allow for any type of surface and body forces but instead, only the shear forces resulting from molecular momentum transport, the pressure forces and the gravitational body force will be considered.

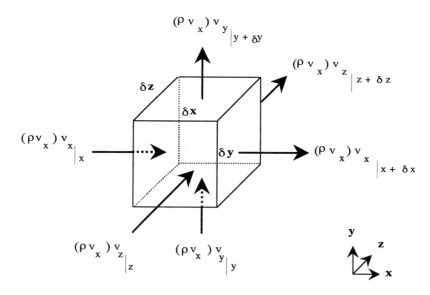

Figure A-1 X-Component Of Momentum: Convective Fluxes

[#] In this derivation the molecular momentum fluxes are represented as shear stresses.

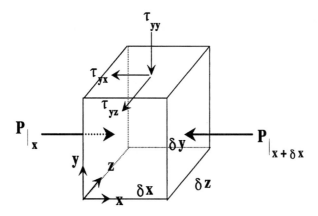

Figure A-2 X-Component Of Momentum: Surface Forces

With respect to the convective momentum fluxes, note that we are only dealing with the volumetric concentration of x-momentum, ρv_x, which can be transported in all three directions. With respect to the shear stresses, they are shown in Figure A-2 as acting on the xz-plane at $y + \delta y$. Consequently, consistent with convention, they all act in the <u>negative</u> direction of the coordinate system at this surface. Substituting into the general conservation equation (Equation 2.20), the x-component of the momentum balance becomes

$$\left[\rho\, v_x\, v_{x|x} - \rho\, v_x\, v_{x|x+\delta x}\right]\delta y\,\delta z\,\delta t + \left[\rho\, v_x\, v_{y|y} - \rho\, v_x\, v_{y|y+\delta y}\right]\delta x\,\delta z\,\delta t$$

$$+\left[\rho\, v_x\, v_{z|z} - \rho\, v_x\, v_{z|z+\delta z}\right]\delta x\,\delta y\,\delta t + \left[\tau_{xx|x} - \tau_{xx|x+\delta x}\right]\delta y\,\delta z\,\delta t$$

$$+\left[\tau_{yx|y} - \tau_{yx|y+\delta y}\right]\delta x\,\delta z\,\delta t + \left[\tau_{zx|z} - \tau_{zx|z+\delta z}\right]\delta x\,\delta y\,\delta t + \left[P_{|x} - P_{|x+\delta x}\right]\delta y\,\delta z\,\delta t$$

$$+(\rho\, g_x)\delta x\,\delta y\,\delta z\,\delta t = \left[\rho\, v_{x|t+\delta t} - \rho\, v_{x|t}\right]\delta x\,\delta y\,\delta z$$

Dividing through by $\delta x\,\delta y\,\delta z\,\delta t$ and taking the limit as each approaches zero, the x-component of the momentum equation can be written as

$$\frac{\partial}{\partial t}(\rho v_x) = -\frac{\partial}{\partial x}(\rho v_x v_x) - \frac{\partial}{\partial y}(\rho v_x v_y) - \frac{\partial}{\partial z}(\rho v_x v_z)$$

$$-\frac{\partial \tau_{xx}}{\partial x} - \frac{\partial \tau_{yx}}{\partial y} - \frac{\partial \tau_{zx}}{\partial z} - \frac{\partial P}{\partial x} + \rho g_x$$

(A.1)

For Newtonian fluids, the shear stresses are proportional to the velocity gradients and, in order to illustrate their complexity, these relationships are given for rectangular coordinates in Table A-2d. They can be found for all three coordinate systems in Reference [1]. For the particular example of constant density and viscosity, it can be shown [2] that the x-component of the momentum balance in terms of velocities for a Newtonian fluid takes the form

$$\rho\left[\frac{\partial v_x}{\partial t} + v_x\frac{\partial v_x}{\partial x} + v_y\frac{\partial v_x}{\partial y} + v_z\frac{\partial v_x}{\partial z}\right] = -\frac{\partial P}{\partial x} + \mu\nabla^2 v_x + \rho g_x$$

where $\nabla^2 = \frac{\partial^2}{\partial x^2} + \frac{\partial^2}{\partial y^2} + \frac{\partial^2}{\partial z^2}$ is called the *Laplacian Operator*.

The y- and z-components of the momentum balance are derived in a similar fashion and the results are given in Table A-2 for rectangular, cylindrical and spherical coordinates. The three components can be vectorially combined and the momentum balance, in terms of shear stresses becomes[#]

$$\rho\frac{D\vec{v}}{Dt} = -\nabla P - [\nabla\cdot\vec{\tau}] + \rho\vec{g}$$

(A.2)

where $\frac{D}{Dt}() = \frac{\partial}{\partial t}() + v_x\frac{\partial}{\partial x}() + v_y\frac{\partial}{\partial y}() + v_z\frac{\partial}{\partial z}()$ is known as the *substantial time derivative* and has the physical significance of changes with time as an observer moves with the fluid motion. For the particular case of a Newtonian fluid with constant density and viscosity, the vector form of the Momentum Equation in terms of velocity gradients, becomes

$$\rho\frac{D\vec{v}}{Dt} = -\nabla P + \mu\nabla^2\vec{v} + \rho\vec{g}$$

(A.3)

[#]   To obtain the form of Equation A.2, the Continuity Equation (Equation 8.2) is combined with the vector sum of all three components of the Momentum Equation.

Derivation of the Energy Equation

A generalized form of the Energy Equation can also be derived, again with the proviso that not all forms of energy SOUCES/SINKS will be included. Because there are many forms of energy, there are also many forms of the Energy Equation. Due to its wide applicability, here we will deal only with the Thermal Energy Equation, the reader is referred to any of the Graduate level textbooks on Transport Phenomena (see, for example [3]) for a discussion and presentation of other forms. In addition to generalizing the derivation to include time and spatial dimensions, comparisons will also be made to the First Law of Thermodynamics for an open system. This is done so that the student can relate the approaches taken here with the more familiar thermodynamic energy conservation principle.

Applying the general conservation equation, Equation 2.20, in terms of energy, $E^{\#}$, we can write it in a slightly different form than was done in Chapters 2 and 3, so that

$$[\text{ACCUMULATION - (IN - OUT)}_{\text{CONVECTIVE}}] = [\text{IN - OUT}]_{\text{MOLECULAR}} + [\text{SOURCES - SINKS}] \tag{A.4}$$

If we now recognize that

$$[\text{ACCUMULATION - (IN - OUT)}_{\text{CONVECTIVE}}] = \text{Increase of } E \text{ in the system} = \Delta E$$

and

$$[\text{IN - OUT}]_{\text{MOLECULAR}} = \text{Net Inflow of Heat} = \delta Q$$

and

$$[\text{SOURCES - SINKS}] = - \{\text{Total work done by system on surroundings}\} = \delta W$$

With this we can see that Equation A.4 is another way of writing the first law of thermodynamics; that is,

$$\Delta E = \delta Q - \delta W \tag{A.5}$$

Of course we must keep in mind that, in this case, the system is the differential volume element. Note also that, as defined here, work done by the system on the surroundings is inherently positive.

$^{\#}$ In general the energy would include internal, kinetic and potential energy although, here, the potential energy term is considered as work done against the gravitational force; i.e., a SINK.

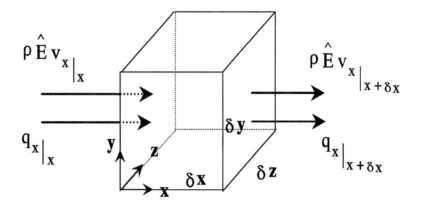

Figure A-3 Energy Fluxes: X-Direction

Figure A-3 shows both the convective and molecular fluxes of internal energy being transported in the x-direction. Although not shown, there are also convective and molecular fluxes of internal energy in the y and z-directions ($\rho v_y \hat{U}$, $\rho v_z \hat{U}$ and q_y, q_z, respectively). So, over some increment of time, δt ,

$$\delta E = [(\rho \hat{U})_{|t + \delta t} - (\rho \hat{E})_{|t}]\delta x \delta y \delta z - \left[(\rho v_x \hat{E})_{|x} - (\rho v_x \hat{E})_{|x + \delta x}\right]\delta y \delta z \delta t$$

$$- \left[(\rho v_y \hat{E})_{|y} - (\rho v_y \hat{E})_{|y + \delta y}\right]\delta x \delta z \delta t - \left[(\rho v_z \hat{E})_{|z} - (\rho v_z \hat{E})_{|z + \delta z}\right]\delta x \delta z \delta t$$

(A.6)

and

$$\delta Q = \left[q_{x\,|\,x} - q_{x\,|\,x + \delta x}\right]\delta y \delta z \delta t + \left[q_{y\,|\,y} - q_{y\,|\,y + \delta y}\right]\delta x \delta z \delta$$

$$+ \left[q_{z\,|\,z} - q_{z\,|\,z + \delta z}\right]\delta x \delta y \delta t t$$

(A.7)

Various types of energy sources and sinks were discussed in Chapter 3 but here we will only consider those sources and sinks resulting from work done by or on the system as a result of forces. Since work is simply the *Scalar Dot Product* of the force and position vectors, the <u>rate</u> of work can be computed from the dot product of the force and velocity vectors. As was done in the derivation of the Momentum Equation, only the three most common forces will be considered: pressure and shear surface forces and the gravitational body force. Figure A-4 shows these forces acting on the differential volume element.[#] Consistent with convention, the

---

[#]   For clarity, only the shear forces on the x-faces (yz plane) and the pressure forces on the y-face (xz plane) are shown.

surface forces act in the positive direction on the leading faces of the DVE and are negative on the opposite faces. The velocity components are already assumed to be in the positive direction and, in this derivation, the direction of the gravitational components is arbitrarily chosen to be in the negative direction.

Dealing first with the gravitational force, it can be seen from Figure A-4a that the fluid entering the DVE opposes gravity and thus there is work done by the surroundings on the system (the DVE). Since the δW term in Equation (A.5) represents the work done by the system on the surroundings,

$$\delta W_{grav} = -\Big[\rho g_x v_x + \rho g_y v_y + \rho g_z v_z \Big] \delta x \, \delta y \, \delta z \, \delta t \tag{A.8}$$

Turning now to the surface forces, note from Figure A-4b that the fluid moves <u>against</u> the pressure force at $y + \delta y$ and thus, this is work done by the system on the surroundings. Correspondingly, the pressure force at y is work done by the surroundings on the system. For the particular pressure force shown in Figure A-4b, the *net* work done by the system on the surroundings is

$$(\delta W_{Press})_{y\text{-}face} = [\,(P \ \delta x \delta z)\, v_{y|_y \, + \, \delta y} - (P \ \delta x \delta z)\, v_{y|_y}\,] \delta t \tag{A.9}$$

Similarly, the net work done by the system on the surroundings due to the shear forces at the leading x-face of the DVE is[#]

$$\delta W_{shear|_x} = -\Big[(\tau_{xx} \, \delta y \delta z)\, v_x + (\tau_{xy} \, \delta y \, \delta z)\, v_y + (\tau_{xz} \, \delta y \, \delta z)\, v_z \Big]_{|_x} \, \delta t \tag{A.10}$$

When the results of Equations A.6 - A.10 are put into the energy conservation equation (Equation A.5) and the limits are allowed to approach zero, we obtain

---

[#] Note that the work is negative at this face since all three components of velocity as well as the shear forces are positive.

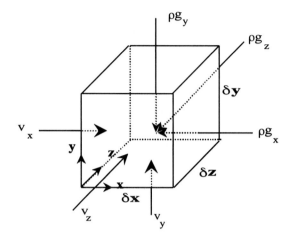

Gravitational Forces

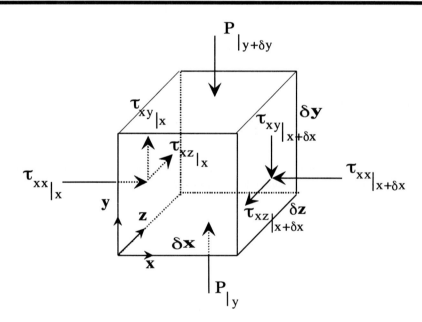

Surface Forces

Figure A-4 Work Resulting from Forces

$$\frac{\partial}{\partial t}(\rho\hat{E}) = -\left[\frac{\partial}{\partial x}(\rho\hat{E}\,v_x) + \frac{\partial}{\partial y}(\rho\hat{E}\,v_y) + \frac{\partial}{\partial z}(\rho\hat{E}\,v_z)\right]$$

$$-\left[\frac{\partial q_x}{\partial x} + \frac{\partial q_y}{\partial y} + \frac{\partial q_z}{\partial z}\right] + [\rho g_x v_x + \rho g_y v_y + \rho g_z v_z]$$

$$-\left[\frac{\partial}{\partial x}(P\,v_x) + \frac{\partial}{\partial y}(P\,v_y) + \frac{\partial}{\partial z}(P\,v_z)\right] - \left[\frac{\partial}{\partial x}\left(\tau_{xx}v_x + \tau_{xy}v_y + \tau_{xz}v_z\right)\right.$$

$$\left. + \frac{\partial}{\partial y}\left(\tau_{yx}v_x + \tau_{yy}v_y + \tau_{yz}v_z\right) + \frac{\partial}{\partial z}\left(\tau_{zx}v_x + \tau_{zy}v_y + \tau_{zz}v_z\right)\right]$$

or, in vector notation

$$\frac{\partial}{\partial t}(\rho\hat{E}) + \nabla\cdot(\rho\hat{E}\,\vec{v}) = -\nabla\cdot\vec{q} - \nabla\cdot(P\,\vec{v}) + \rho\vec{g}\cdot\vec{v} - \nabla\cdot(\bar{\tau}\cdot\vec{v}) \qquad (A.12)$$

Rather than work with an equation in terms of $\hat{E}$, it is advantageous to express the energy equation in terms of the physical properties of the system and the temperature. To do this we can express $\hat{E}$ in terms of internal and kinetic energy and the internal energy in terms of enthalpy. Then thermodynamic principles can be used to relate enthalpy in terms of specific heats and an *equation of state*. Finally, the Mechanical Energy Equation is subtracted to derive a form of the Thermal Energy Equation.

First the Continuity Equation (Equation 8.2) is combined with the basic thermodynamic relationship between the different forms of energy

$$\hat{E} = \hat{U} + \frac{1}{2}v^2 = \hat{H} - \frac{P}{\rho} + \frac{1}{2}v^2$$

so that Equation A.12 becomes[#]

$$\rho\frac{D}{Dt}\left(\hat{H} + \frac{1}{v^2}\right) = \frac{\partial P}{\partial t} - \nabla\cdot\vec{q} + \rho\vec{g}\cdot\vec{v} - \nabla\cdot(\bar{\tau}\cdot\vec{v}) \qquad (A.13)$$

To obtain a form of the thermal energy equation, the contributions of mechanical energy are removed from Equation A.13. A relationship for the mechanical energy can be obtained by

---

[#] See [4] for a thorough explanation of the vector-tensor operations appearing in Equations A.13-15.

forming the scalar dot product of the velocity vector with the Momentum Equation, Equation A.2, which is

$$\rho \frac{D}{Dt}\left(\frac{v^2}{2}\right) = -\vec{v}\cdot\Delta\vec{P} + \rho\vec{v}\cdot\vec{g} - \vec{v}\cdot[\nabla\cdot\vec{\tau}] \tag{A.14}$$

If Equation A.14 is subtracted from A.13, we have

$$\rho\frac{D\hat{H}}{Dt} = \frac{DP}{Dt} - [\vec{\tau} : \nabla\vec{v}] - \nabla\cdot\vec{q} \tag{A.15}$$

where $\vec{\tau} : \nabla\cdot\vec{v} \equiv \nabla\cdot[\vec{\tau}\cdot\vec{v}] - \vec{v}\cdot[\nabla\cdot\vec{\tau}]$

The next step is to express A.15 in terms of temperature and the fluid properties. From thermodynamics we have

$$\hat{H} = \left(\frac{\partial\hat{H}}{\partial T}\right)_P dT + \left(\frac{\partial\hat{H}}{\partial P}\right)_T dP = \hat{c}_P\, dT + \left[\frac{1}{\rho} - T\left(\frac{\partial\frac{1}{\rho}}{\partial T}\right)_P\right] dP$$

which can be differentiated <u>substantially</u> with respect to time and substituted into Equation A.15 with the result

$$\rho\,\hat{c}_P\frac{DT}{Dt} = \left[\frac{\partial \ln\left(\frac{1}{\rho}\right)}{\partial \ln T}\right]_P \frac{DP}{Dt} - \nabla\vec{q} - [\vec{\tau} : \nabla\vec{v}] \tag{A.16}$$

The term on the left hand side of A.16 represents the net increase in energy due to accumulation and/or convective inflow. The terms on the right hand side represent energy contributions due to compressibility, molecular transport and viscous dissipation (i.e., friction), respectively. The term in front of the substantial derivative of pressure can be computed from an appropriate Equation of State and is equal to 1 for an ideal gas and equal to 0 for fluids where the density is constant or is independent of temperature.

For the particular case of a Newtonian fluid with constant $\rho, \mu \text{ and } k$, Equation A.16 becomes

$$\rho \, \hat{c}_P \frac{DT}{Dt} = k \, \nabla^2 T + \mu \, \Phi_v \qquad (A.17)$$

where Φ_v is the *viscous dissipation* (usually negligible except in cases of small channel widths or high speed gas flows). Expanded forms of Equation A.17 are given in each of the coordinate systems in Table A.3.

CONSERVATION OF SPECIES EQUATION

The derivation of a generalized conservation of species equation can be done in a similar manner. While we will only consider molecular diffusion, for the most general case, accounting must be done for all of the mechanisms by which molecular transport can take place (eg., mechanical and thermal driving forces) and this is treated in detail by Bird, Stewart and Lightfoot [5]. The general source/sink terms are related to chemical reaction rates and these are accounted for in terms of a generalized, homogeneous reaction rate term, R_A, which will be positive or negative depending on whether the species is being created or consumed by a chemical reaction. In this case the vector form of the conservation equation for species A is relatively simple; viz.,

$$\frac{\partial C_A}{\partial t} + (\nabla \cdot \vec{N}_A) = \sum_j (R_A)_j \qquad (A.18)$$

In Equation A.18, N_A includes both molecular and convective fluxes (see discussion in Chapter 4) and j is the number of reactions in which species A is participating. For a binary system and the case of constant diffusivity and total concentration, Equation A.18 takes the form

$$\frac{\partial C_A}{\partial t} + (\vec{v} \cdot \nabla C_A) = D_{AB} \nabla^2 C_A + \sum_j (R_A)_j \qquad (A.19)$$

Table A.4 lists the individual terms of Equation A.19 in all three coordinate systems.

REFERENCES

[1] Bird, R.B., Stewart, W.E., and E.N. Lightfoot, *Transport Phenomena*, John Wiley & Sons, N.Y. (1960) p.88-90

[2] Bird, R.B., Stewart, W.E., and E.N. Lightfoot, *op. cit*, p.80

[3] Bird, R.B., Stewart, W.E., and E.N. Lightfoot, *op. cit*, p.322-324

[4] Bird, R.B., Stewart, W.E., and E.N. Lightfoot, *op. cit*, p.726-731

[5] Bird, R.B., Stewart, W.E., and E.N. Lightfoot, *op. cit*, p.567

Table A-1 The Continuity Equation

<u>Rectangular Coordinates</u>

$$\frac{\partial \rho}{\partial t} + \frac{\partial}{\partial x}(\rho v_x) + \frac{\partial}{\partial y}(\rho v_y) + \frac{\partial}{\partial z}(\rho v_z) = 0 \qquad \text{[A-1].1}$$

<u>Cylindrical Coordinates</u>

$$\frac{\partial \rho}{\partial t} + \frac{1}{r}\frac{\partial}{\partial r}(\rho r v_r) + \frac{1}{r}\frac{\partial}{\partial \theta}(\rho v_\theta) + \frac{\partial}{\partial z}(\rho v_z) = 0 \qquad \text{[A-1].2}$$

<u>Spherical Coordinates</u>

$$\frac{\partial \rho}{\partial t} + \frac{1}{r^2}\frac{\partial}{\partial r}(\rho r^2 v_r) + \frac{1}{r\sin\theta}\frac{\partial}{\partial \theta}(\rho v_\theta \sin\theta) + \frac{1}{r\sin\phi}\frac{\partial}{\partial \phi}(\rho v_\phi) = 0 \qquad \text{[A-1].3}$$

Table A-2.a Momentum Equation – Rectangular Coordinates [Newtonian Fluid, Constant ρ, μ]

X-Component

$$\rho\left[\frac{\partial v_x}{\partial t} + v_x\frac{\partial v_x}{\partial x} + v_y\frac{\partial v_x}{\partial y} + v_z\frac{\partial v_x}{\partial z}\right] = -\frac{\partial P}{\partial x} + \rho\, g_x + \mu\left[\frac{\partial^2 v_x}{\partial x^2} + \frac{\partial^2 v_x}{\partial y^2} + \frac{\partial^2 v_x}{\partial z^2}\right] \qquad \text{[A-2a].1}$$

Y-Component

$$\rho\left[\frac{\partial v_y}{\partial t} + v_x\frac{\partial v_y}{\partial x} + v_y\frac{\partial v_y}{\partial y} + v_z\frac{\partial v_y}{\partial z}\right] = -\frac{\partial P}{\partial y} + \rho\, g_y + \mu\left[\frac{\partial^2 v_y}{\partial x^2} + \frac{\partial^2 v_y}{\partial y^2} + \frac{\partial^2 v_y}{\partial z^2}\right] \qquad \text{[A-2a].2}$$

Z-Component

$$\rho\left[\frac{\partial v_z}{\partial t} + v_x\frac{\partial v_z}{\partial x} + v_y\frac{\partial v_z}{\partial y} + v_z\frac{\partial v_z}{\partial z}\right] = -\frac{\partial P}{\partial z} + \rho\, g_z + \mu\left[\frac{\partial^2 v_z}{\partial x^2} + \frac{\partial^2 v_z}{\partial y^2} + \frac{\partial^2 v_z}{\partial z^2}\right] \qquad \text{[A-2a].3}$$

Table A-2.b Momentum Equation – Cylindrical Coordinates [Newtonian Fluid, Constant ρ, μ]

Z-Component

$$\rho\left[\frac{\partial v_z}{\partial t} + v_r\frac{\partial v_z}{\partial r} + \frac{v_\theta}{r}\frac{\partial v_z}{\partial \theta} + v_z\frac{\partial v_z}{\partial z}\right] =$$

$$-\frac{\partial P}{\partial z} + \rho g_z + \mu\left[\frac{1}{r}\frac{\partial}{\partial r}\left(r\frac{\partial v_z}{\partial r}\right) + \frac{1}{r^2}\frac{\partial^2 v_z}{\partial \theta^2} + \frac{\partial^2 v_z}{\partial z^2}\right]$$

[A.2b].1

r-Component

$$\rho\left[\frac{\partial v_r}{\partial t} + v_r\frac{\partial v_r}{\partial r} + \frac{v_\theta}{r}\frac{\partial v_r}{\partial \theta} - \frac{v_\theta^2}{r} + v_z\frac{\partial v_r}{\partial z}\right] =$$

$$-\frac{\partial P}{\partial r} + \rho g_r + \mu\left[\frac{\partial}{\partial r}\left(\frac{1}{r}\frac{\partial}{\partial r}(r\,v_r)\right) + \frac{1}{r^2}\frac{\partial^2 v_r}{\partial \theta^2} - \frac{2}{r^2}\frac{\partial v_\theta}{\partial \theta} + \frac{\partial^2 v_r}{\partial z^2}\right]$$

[A.2b].2

Θ-Component

$$\rho\left[\frac{\partial v_\theta}{\partial t} + v_r\frac{\partial v_\theta}{\partial r} + \frac{v_\theta}{r}\frac{\partial v_\theta}{\partial \theta} + \frac{v_r v_\theta}{r} + v_z\frac{\partial v_\theta}{\partial z}\right] =$$

$$-\frac{1}{r}\frac{\partial P}{\partial \theta} + \rho g_\theta + \mu\left[\frac{\partial}{\partial r}\left(\frac{1}{r}\frac{\partial}{\partial r}(r\,v_\theta)\right) + \frac{1}{r^2}\frac{\partial^2 v_\theta}{\partial \theta^2} + \frac{2}{r^2}\frac{\partial v_r}{\partial \theta} + \frac{\partial^2 v_\theta}{\partial z^2}\right]$$

[A.2b].3

Table A-2.c Momentum Equation – Spherical Coordinates [Newtonian Fluid, Constant ρ, μ]

r-Component

$$\rho\left[\frac{\partial v_r}{\partial t}+\frac{v_r\partial v_r}{\partial r}+\frac{v_\theta}{r}\frac{\partial v_r}{\partial\theta}+\frac{v_\phi}{r\sin\theta}\frac{\partial v_r}{\partial\phi}-\frac{v_\theta^2+v_\phi^2}{r}\right]=$$

$$-\frac{\partial P}{\partial r}+\rho g_r+\mu\left[\nabla^2 v_r-\frac{2}{r^2}\left(v_r+\frac{\partial v_\theta}{\partial\theta}+v_\theta\cos\theta\right)+\frac{1}{\sin\theta}\frac{\partial v_\phi}{\partial\phi}\right]$$

[A.2c].1

Θ-Component

$$\rho\left[\frac{\partial v_\theta}{\partial t}+v_r\frac{\partial v_\theta}{\partial r}+\frac{v_\theta}{r}\frac{\partial v_\theta}{\partial\theta}+\frac{v_\phi}{r\sin\theta}\frac{\partial v_\theta}{\partial\phi}+\frac{v_r v_\theta}{r}-\frac{v_\phi^2\cot\theta}{r}\right]=$$

$$+\mu\left[\nabla^2 v_\theta-\frac{2}{r^2}\left(\frac{\partial v_r}{\partial\theta}-\frac{v_\theta}{2\sin^2\theta}-\cos\frac{\theta}{\sin^2}\theta\frac{\partial v_\phi}{\partial\phi}\right)\right]-\frac{1}{r}\frac{\partial P}{\partial\theta}+\rho g_\theta$$

[A.2c].2

φ-Component

$$\rho\left[\frac{\partial v_\phi}{\partial t}+v_r\frac{\partial v_\phi}{\partial r}+\frac{v_\theta}{r}\frac{\partial v_\phi}{\partial\theta}+\frac{v_\phi}{r\sin\theta}\frac{\partial v_\phi}{\partial\phi}+\frac{v_r v_\phi}{r}+\frac{v_\phi v_\theta\cot\theta}{r}\right]=$$

$$+\mu\left[\nabla^2 v_\phi-\frac{v_\phi}{r^2\sin^2\theta}+\frac{2}{r^2\sin\theta}\frac{\partial v_r}{\partial\phi}+\frac{2\cos\theta}{r^2\sin^2\theta}\frac{\partial v_\theta}{\partial\phi}\right]-\frac{1}{r\sin\theta}\frac{\partial P}{\partial\phi}+\rho g_\phi$$

[A.2c].3

Note: $\nabla^2=\dfrac{1}{r^2}\dfrac{\partial}{\partial r}\left(r^2\dfrac{\partial}{\partial r}\right)+\dfrac{1}{r^2\sin\theta}\dfrac{\partial}{\partial\theta}\left(\sin\theta\dfrac{\partial}{\partial\theta}\right)+\dfrac{1}{r^2\sin^2\theta}\left(\dfrac{\partial^2}{\partial\phi^2}\right)$

Table A-2.d Relationship Between τ_{ig} and Velocity Gradients for a Newtonian Fluid
[Rectangular Coordinates]

$$\tau_{xx} = -\mu\left[2\frac{\partial v_x}{\partial x} - \frac{2}{3}\nabla\cdot\vec{v}\right]$$

$$\tau_{yy} = -\mu\left[2\frac{\partial v_y}{\partial y} - \frac{2}{3}\nabla\cdot\vec{v}\right]$$

$$\tau_{zz} = -\mu\left[2\frac{\partial v_z}{\partial z} - \frac{2}{3}\nabla\cdot\vec{v}\right]$$

$$\tau_{xy} = \tau_{yx} = -\mu\left[\frac{\partial v_x}{\partial y} + \frac{\partial v_y}{\partial x}\right]$$

$$\tau_{xz} = \tau_{zx} = -\mu\left[\frac{\partial v_x}{\partial z} + \frac{\partial v_z}{\partial x}\right]$$

$$\tau_{yz} = \tau_{zy} = -\mu\left[\frac{\partial v_y}{\partial z} + \frac{\partial v_z}{\partial y}\right]$$

Table A-3 Thermal Energy Equation [Newtonian Fluid, Constant ρ, μ, k]

Rectangular Coordinates

$$\rho\, c_P \left[\frac{\partial T}{\partial t} + v_x \frac{\partial T}{\partial x} + v_y \frac{\partial T}{\partial y} + v_z \frac{\partial T}{\partial z} \right] =$$

$$k\left[\frac{\partial^2 T}{\partial x^2} + \frac{\partial^2 T}{\partial y^2} + \frac{\partial^2 T}{\partial z^2} \right] + 2\mu \left[\left(\frac{\partial v_x}{\partial x} \right)^2 + \left(\frac{\partial v_y}{\partial y} \right)^2 + \left(\frac{\partial v_z}{\partial z} \right)^2 \right] \qquad \text{[A-3].1}$$

$$+ \mu \left[\left(\frac{\partial v_x}{\partial y} + \frac{\partial v_y}{\partial x} \right)^2 + \left(\frac{\partial v_x}{\partial z} + \frac{\partial v_z}{\partial x} \right)^2 + \left(\frac{\partial v_y}{\partial z} + \frac{\partial v_z}{\partial y} \right)^2 \right]$$

Cylindrical Coordinates

$$\rho\, \hat{c}_P \left[\frac{\partial T}{\partial t} + v_r \frac{\partial T}{\partial r} + \frac{v_\theta}{r} \frac{\partial T}{\partial \theta} + v_z \frac{\partial T}{\partial z} \right] = k\left[\frac{1}{r} \frac{\partial}{\partial r} \left(r \frac{\partial T}{\partial r} \right) + \frac{1}{r^2} \frac{\partial^2 T}{\partial \theta^2} + \frac{\partial^2 T}{\partial z^2} \right]$$

$$+ 2\mu \left[\left(\frac{\partial v_r}{\partial r} \right)^2 + \frac{1}{r^2} \left(\frac{\partial v_\theta}{\partial \theta} + v_r \right)^2 + \left(\frac{\partial v_z}{\partial z} \right)^2 \right] \qquad \text{[A-3].2}$$

$$+ \mu \left[\left(\frac{\partial v_\theta}{\partial z} + \frac{1}{r} \frac{\partial v_z}{\partial \theta} \right)^2 + \left(\frac{\partial v_z}{\partial r} + \frac{\partial v_r}{\partial z} \right)^2 + \left(\frac{1}{r} \frac{\partial v_r}{\partial \theta} + r \frac{\partial}{\partial r} \left(\frac{v_\theta}{r} \right) \right)^2 \right]$$

Spherical Coordinates

$$
\rho\,\hat{c}_P\left[\frac{\partial T}{\partial t}+v_r\frac{\partial T}{\partial r}+\frac{v_\theta}{r}\frac{\partial T}{\partial \theta}+\frac{v_\phi}{r\sin\theta}\frac{\partial T}{\partial \phi}\right]=
$$

$$
k\left[\frac{1}{r^2}\frac{\partial}{\partial r}\left(r^2\frac{\partial T}{\partial r}\right)+\frac{1}{r^2\sin\theta}\frac{\partial}{\partial \theta}\left(\sin\theta\frac{\partial T}{\partial \theta}\right)+\frac{1}{r^2\sin^2\theta}\frac{\partial^2 T}{\partial \phi^2}\right]
$$

$$
+2\mu\left[\left(\frac{\partial v_r}{\partial r}\right)^2+\left(\frac{1}{r}\frac{\partial v_\theta}{\partial \theta}+\frac{v_r}{r}\right)^2+\left(\frac{1}{r\sin\theta}\frac{\partial v_\phi}{\partial \phi}+\frac{v_r}{r}+\frac{v_\theta\cot\theta}{r}\right)^2\right]
$$

$$
+\mu\left[\left(r\frac{\partial}{\partial r}\left(\frac{v_\theta}{r}\right)+\frac{1}{r}\frac{\partial v_r}{\partial \theta}\right)^2+\left(\frac{1}{r\sin\theta}\frac{\partial v_r}{\partial \phi}+r\frac{\partial}{\partial r}\left(\frac{v_\phi}{r}\right)\right)^2\right]
$$

$$
+\mu\left(\frac{\sin\theta}{r}\frac{\partial}{\partial \theta}\left(\frac{v_\phi}{\sin\theta}\right)+\frac{1}{r\sin\theta}\frac{\partial v_\theta}{\partial \phi}\right)^2
$$

[A-3].3

Table A-4 Species Conservation Equation [Constant C, D_{AB}]

Rectangular Coordinates

$$\frac{\partial C_A}{\partial t} + v_x \frac{\partial C_A}{\partial x} + v_y \frac{\partial C_A}{\partial y} + v_z \frac{\partial C_A}{\partial z} = D_{AB}\left[\frac{\partial^2 C_A}{\partial x^2} + \frac{\partial^2 C_A}{\partial y^2} + \frac{\partial^2 C_A}{\partial z^2}\right] + R_A \qquad \text{[A-4].1}$$

Cylindrical Coordinates

$$\frac{\partial C_A}{\partial t} + v_r \frac{\partial C_A}{\partial r} + \frac{v_\theta}{r} \frac{\partial C_A}{\partial \theta} + v_z \frac{\partial C_A}{\partial z} =$$

$$D_{AB}\left[\frac{1}{r}\frac{\partial}{\partial r}\left(r\frac{\partial C_A}{\partial r}\right) + \frac{1}{r^2}\frac{\partial^2 C_A}{\partial \theta^2} + \frac{\partial^2 C_A}{\partial z^2}\right] + R_A \qquad \text{[A-4].2}$$

Spherical Coordinates

$$\frac{\partial C_A}{\partial t} + v_r \frac{\partial C_A}{\partial r} + \frac{v_\theta}{r} \frac{\partial C_A}{\partial \theta} + \frac{v_\phi}{r\sin\theta} \frac{\partial C_A}{\partial \phi} =$$

$$D_{AB}\left[\frac{1}{r^2}\frac{\partial}{\partial r}\left(r^2\frac{\partial C_A}{\partial r}\right) + \frac{1}{r^2\sin\theta}\frac{\partial}{\partial \theta}\left(\sin\theta \frac{\partial C_A}{\partial \theta}\right) + \frac{1}{r^2\sin^2\theta}\frac{\partial^2 C_A}{\partial \phi^2}\right] + R_A \qquad \text{[A-4].3}$$

APPENDIX B

Using MATLAB ODE

MATLAB is a generalized mathematical software program with many features but here we discuss only those features associated with the solution of dynamic simulation problems (differential equations). As with any software program, there are a few "rules" and "codes" which must be followed. The structure of MATLAB consists of two separate files, the "M-file" and an execution file which we will call the "Command-file". Simply put, the M-file is the file where you identify the equations you want to be solved and the Command-file is the file that establishes the initial conditions, identifies the solution methodology and dictates the particular output that is desired. Because these two files must be linked, it is important to establish a "file path" to the location where the M-file is saved; that is, a path that allows the Command-file to "find" the M-file upon which it will operate. So, it is first necessary to save the M-file and then use the "path" menu in the Command-file to tell the Command-file where to find the M-file.

THE M-FILE

There are a few general statements to be made about the M-file in MATLAB. First of all, the software is set up to use matrix algebra and therefore it expects most things to be in a matrix form. For our purposes, this means the dependent variables will be a vector (a one-column matrix). Unlike some other differential equation solvers (ACSL, for example), the differential equation solver in MATLAB is <u>not</u> "free formatted." That is, execution of the statements within each block is the usual top-to-bottom, logical sequence you would find in a FORTRAN program. The program also distinguishes between separate parameters and mathematical expressions by means of a semi-colon. Therefore, a set of constant parameters can be inputted as a string of values as long as they are separated from one another by a semi-colon. Since the program is set up in a matrix format, you must be careful to distinguish between parentheses and brackets and use them in EXACTLY the manner prescribed by MATLAB.

The first statement in the M-file must be a defining "function" statement. This creates the function that the Command-File will work on and also identifies the file name under which it will be saved. The form of this statement for a dependent variable, y and an independent variable, t, is as follows

function dy = {file name} (t, y)

Since the dependent variable, y, can be a vector, it is necessary to tell MATLAB how many components are in the vector. The most convenient method of doing this is to use a "zeros" statement immediately following the function statement. For example, the zeros statement,

dy = zeros(3,1)

tells MATLAB that the dy variable is an array consisting of one column (a vector) and 3 rows (3 components). Later in the M-file, you must provide defining equations for the vector, dy, (in this case, dy(1), dy(2) and dy(3)).

The mathematical operations in MATLAB are done in essentially the same manner as in a Fortran program. You can also use logic statements within the M-file. For example, the series of statements

```
if T(1)<825
   ra=0;
else
   ra=1.8e-6;
end
```

calculates a reaction rate in a different manner, depending on whether the temperature is less than or greater than 825. You can also set up repetitive calculations. For example

```
for i = 1:3
p(i) = y(i)/(R*y(4))
end
```

calculates the partial pressures for three components, given the volumetric concentrations of the three [y(1), y(2) and y(3)] and the temperature, y(4).

Now let's discuss how to go about defining a set of differential equations which need to be solved. First of all, the differential equation solver applies only to initial value problems; i.e., differential equations where all of the boundary conditions are known at one boundary. In this sense, the numerical solution "marches" from one end of the system to the other. If you are dealing with a "split boundary value" problem (boundary conditions stipulated at either of the system), then you must "guess" at the boundary condition at the start of the integration (t = 0, z = 0, etc.) and then solve the differential equation to predict the value at the other boundary. If it doesn't match, the known value at that point, then you must guess again and iterate until a satisfactory match is obtained.

The next important point is that the differential equations must be posed in a certain manner, with the highest order term on the left-hand side of the equation. For a set of differential equations with dependent variables, Y_i , and an independent variable, t , Equation [B.1] shows the required form of the differential equations.

$$\frac{dY_1}{dt} = f_1(Y_1, Y_2, Y_3, \dots Y_n, t)$$

$$\frac{dY_2}{dt} = f_2(Y_1, Y_2, Y_3, \dots Y_n, t)$$

(B.1)

$$-$$
$$-$$
$$-$$

$$\frac{dY_n}{dt} = f_n(Y_1, Y_2, Y_3, \dots Y_n, t)$$

which have initial conditions: at t = 0, $Y_i = Y0_i$, i = 1, ...n . The vector form of these equations is apparent and we could define a vector, Y(n), which would consist of the entire equation set of dependent variables. In this case we need to keep track of what the vector components refer to in the actual physical situation. For example, Y(1) could be the concentration of A, Y(2), the concentration of B, Y(3), the temperature, etc.

If a differential equation is of order higher than 1, it is necessary to express the differential equation with the highest order differential on the left hand side of the equation and to turn it into a first order equation. The latter is done by defining a new dependent variable which is defined in terms of the differential of the previous dependent variable. In effect, this also adds one new differential equation to the set each time this is done. For example, suppose in Equation [B.1], that Y_1, is equal to $\frac{dY_2}{dt}$. The first equation in the set will provide the solution for Y_1 (that is, for $\frac{dY_2}{dt}$) and then Y_2 can be found by solving the additional differential equation,

$$\frac{dY_2}{dt} = Y_1$$

It is easier to see this in an example. Suppose we have the following set of three differential equations to solve:

$$\frac{d^2C_A}{dz^2} = f_1(C_A, C_B, T)$$

$$\frac{dC_B}{dz} = f_2(C_A, C_B, T)$$

$$\frac{dT}{dz} = f_3(C_A, C_B, T)$$

If we now let $Y_1 = \frac{dC_A}{dz}$, then the dependent "vector" variable, Y, will have four components, Y_1, $Y_2 = C_A$, $Y_3 = C_B$, and $Y_4 = T$. The set of differential equations which must be solved is then

$$\frac{dY_1}{dz} = f_1(C_A, C_B, T)$$

$$\frac{dY_2}{dz} = Y_1$$

$$\frac{dY_3}{dz} = f_2(C_A, C_B, T)$$

$$\frac{dY_4}{dz} = f_3(C_A, C_B, T)$$

which must be solved subject to the known "initial" conditions (at z = 0)

$$(Y_1)_0 = \left(\frac{dC_A}{dz}\right)_0 , \ (Y_2)_0 = (C_A)_0$$

$$(Y_3)_0 = (C_B)_0 , \ (Y_4)_0 = T_0$$

THE COMMAND-FILE

Within the command file you can establish the initial conditions, solve the set of differential equations and manipulate the data. This can be done "interactively"; i.e., as you go along, or you can write a separate M-file, save it in the same directory as the function M-file and "run" it directly from this M-file (from the Tools Bar). Doing it this way allows you to run multiple problems without having to re-format the output each time. Keep in mind, though, that when you first open MATLAB, you will be in the Command-file and you still need to establish the file path as described earlier. The Example at the end of the Appendix is an illustration of how you can set up the "execute" statements, including the initial conditions, in a second M-file. To solve a set of differential equations where the independent variable is z, and the dependent variable is a vector, T, the following statement is used

[z, T] = ode45('WSU', [0 30.], [T0])

Here, ode45, is the name of the differential equation solver you want to use for the solution (discussed below), WSU is the name of the function you defined in the M-file (remember to tell the Command-file where to find it), [0 30] is the range of the independent variable (z, in this case) over which you wish to have the solution, and [T0] is the previously established set of initial values for the dependent variable, T.

Notice that there is <u>NOT</u> a semi-colon after this statement. Leaving out the semi-colon tells MATLAB to immediately solve the differential equation and pass the output (tabular) to the Command-file. Once the output has been passed, you can manipulate it in a number of ways. First of all, the default output consists of a column of the independent data (from 0 to 30)[*], followed by an equal sized column of the dependent variable. If the dependent variable consists of a vector with three components, the output will be three parallel columns. To display the data in a more convenient format, you can use the statement

res = [z,T]

which will yield 4 parallel columns consisting of the independent variable and the three dependent variables. The data can also be plotted, using the simple command:

plot(z,T)

Using this statement will plot the dependent variable, T (all 3 components) on the y-axis and z on the x-axis. The plot can be formatted in a number of different ways to give a title, labels on the x and y-axes and text within the body of the plot. The thickness and styles of the plots themselves can also be changed to distinguish one dependent variable from another. The

---

[*] Note the space between 0 and 30

example problem, below illustrates a few of these formats but the MATLAB manual (or the on-line HELP feature should be consulted for more information).

THE DIFFERENTIAL EQUATION SOLVERS

As with most differential equation solvers, there are a number of different numerical methods that can be used. Essentially, they are split between "non-stiff" and "stiff" differential equations. A stiff differential equation is one that has a large and rapidly varying derivative. That is, the derivative changes rapidly and is therefore susceptible to relatively large errors in the numerical analysis. There are also choices within these two broad categories and these choices are based on the type of errors that can be tolerated. The choices are:

NON-STIFF SOLVERS	ode45, ode23, ode113
STIFF SOLVERS	ode15s, ode23s, ode23t, ode23t

The numbers associated with the names indicates the "order" of the numerical method, the "s" indicates a stiff algorithm and the "t" indicates an implementation of the trapezoidal rule. The best bet is to initially try ode45, which utilizes a 4-5th order Runge-Kutta algorithm. If this takes an unduly long time, then try one of the stiff solvers. However, for the type of transport phenomena problems that are dealt with in this text, stiff solvers should never have to be used. If you find that you are having difficulty, the most likely reason is that you have not posed the problem correctly.

Example B-1

In order to illustrate some of these points, let's consider an unsteady state problem where a chemical reaction is taking place in a vessel of volume, V. The reaction is

$$A \rightarrow 2R + S$$

and the uniform reaction rate (moles/vol-t) is given by $r = k\,C_A$ where k is the reaction rate constant. Assume also that the temperature in the vessel starts off at a temperature, T_0, and is raised linearly with time according to $T = T_0 + a\,t$. The reaction rate is an exponential function of temperature; specifically

$$k = A_0 \exp\left(-\frac{E}{R_g T} \right)$$

A differential mass balance can be taken on each of the three species and, for species, A, it is

$$(C_{A_{|t+\delta t}} - C_{A_{|t}}) V = -rV\delta t$$

Taking the limit as $\delta t \to o$, the differential equation describing the disappearance of 'A' is

$$\frac{dC_A}{dt} = -r$$

which must be solved along with the equations

$$r = kC_A$$

$$k = A_0 \exp\left(-\frac{E}{R_g T} \right)$$

$$T = T_0 + at$$

Similar material balances can be taken on the other two species, and the resulting differential equations are

$$\frac{dC_R}{dt} = 2r$$

$$\frac{dC_S}{dt} = r$$

The MATLAB program to solve this set of equations is shown below and the output plot of the time-varying concentrations of all three species is shown in the accompanying plot.

MATLAB Program Concentrations in a Batch Reactor

<u>M-File</u>

```
function dC = rxn(t,C)
dC = zeros(3,1);
a = 1.5; A0 = 2.35e9; T0 = 298;
```

```
    T = T0 + a*t;
    k = A0*exp(-7550./T);
    r = k*C(1);
    dC(1) = - r;
    dC(2) = 2*r;
    dC(3) = r;


COMMAND-File

    clear;
    % Initial Conditions
        C0(1)=1.0;
        C0(2)=0.0;
        C0(3)=0.0;
    % Solve the differential equations and obtain tabular output in the form of time and
then all 3 concentrations
        [t,C] = ode45('rxn', [0 20.], [C0]);
    % To obtain tabular output in 4 parallel columns (t, CA, CR, CS)
        res=[t,C];
    % To plot the output with a labeled plot
        plot(t,C,'linewidth',2);
        ylabel('Concentrations','fontsize',14);
        xlabel('Time  -  min','fontsize',14);
        text(15, 1.5,'\bf{CR}','fontsize', 14);
        text(0.2, 1.1,'\bf{CA}','fontsize', 14);
        text(17, 0.9,'\bf{CS}','fontsize', 14);
        title('\bf{Concentrations in a Batch Reactor}','fontsize',16);
```

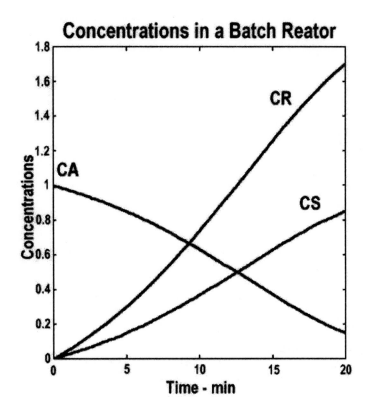

APPENDIX C

Lennard Jones Parameters and Collision Integrals

Table C-1 Lennard-Jones Parameters

Gas		Molecular Weight	$\sigma \times 10^2$ (nm)	$\dfrac{\varepsilon}{K}$ (°K)
H_2	Hydrogen	2.016	29.15	38.0
Monatomic Gases				
He	Helium	4.003	25.76	10.2
Ne	Neon	20.183	27.89	35.7
Ar	Argon	39.944	34.18	124
Kr	Krypton	83.8	36.1	190
Xe	Xenon	131.3	40.55	229
Hg	Mercury	200.59	29.69	750
Polyatomic Gases				
Air		28.97	36.17	97
N_2	Nitrogen	28.02	36.81	91.5
O_2	Oxygen	32	34.33	113

Gas		Molecular Weight	σ x 10^2 (nm)	$\dfrac{\varepsilon}{K}$ (°K)
CO	Carbon Monoxide	28.01	35.9	110
CO$_2$	Carbon Dioxide	44.01	39.96	190
NO	Nitric Oxide	30.01	34.7	119
N$_2$O	Nitrous Oxide	44.02	38.79	220
SO$_2$	Sulfur Dioxide	64.07	42.9	252
H$_2$O	Water	18.016	26.41	809
F$_2$	Fluorine	38	36.53	112
Cl$_2$	Chlorine	70.91	41.15	357
Br$_2$	Bromine	159.83	42.68	520
I$_2$	Iodine	253.82	49.82	550
HF	Hydrogen Fluoride	20.01	31.48	330
HCl	Hydrogen Chloride	36.47	33.39	345
HBr	Hydrogen Bromide	80.91	33.53	449
HI	Hydrogen Iodide	127.91	42.11	289
HCN	Hydrogen Cyanide	27.03	36.3	569
H2$_s$	Hydrogen Sulfide	34.08	36.23	301
NH$_3$	Ammonia	17.03	29	558
SiF$_4$	Silicon Fluoride	104.08	48.8	172
SiH$_4$	Silane	32.12	40.84	208
UF$_6$	Uranium Hexafluoride	314	59.67	237
CF$_4$	Carbon Tetrafluoride	88	46.62	134
CCl$_4$	Carbon Tetrachloride	153.84	58.81	327
COS	Carbonyl Sulfide	60.08	41.3	335
CS$_2$	Carbon Disulfide	76.14	44.38	488
Hydrocarbons				
CH$_4$	Methane	16.04	38.22	137
C$_2$H$_2$	Acetylene	26.04	42.21	185
C$_2$H$_4$	Ethylene	28.05	42.32	205
C$_2$H$_6$	Ethane	30.07	44.18	230
C$_3$H$_6$	Propylene	42.08	46.78	299

Gas		Molecular Weight	$\sigma \times 10^2$ (nm)	$\dfrac{\varepsilon}{K}$ (°K)
C_3H_8	Propane	44.09	50.61	254
$n\text{-}C_4H_{10}$	Butane	58.12	46.87	531
$i\text{-}C_4H_{10}$	iso-Butane	58.12	53.41	313
$n\text{-}C_5H_{12}$	Pentane	72.15	57.69	345
$n\text{-}C_6H_{14}$	Hexane	86.17	59.09	413
$n\text{-}C7H_{16}$	Heptane	100.2	88.8	282
$n\text{-}C_8H_{18}$	Octane	114.22	74.51	320
$n\text{-}C_9H_{20}$	Nonane	128.25	84.48	240
C_6H_{12}	Cyclohexane	84.16	60.93	324
C_6H_6	Benzene	78.11	52.7	440
$CH_3CO_2CH_3$	Acetone	58.08	46	560
CH_3OH	Methanol	32.04	36.26	482
C_2H_5OH	Ethanol	46.07	45.3	363
C_3H_7OH	Propanol	60.09	45.49	577
CH_3Cl	Methyl Chloride	50.49	33.75	855
CH_2Cl_2	Methylene Chloride	84.94	47.59	406
$CHCl_3$	Chloroform	119.39	54.3	327

Table C-2 Collision Integrals as a Function of Temperature

$\dfrac{K}{\varepsilon}T$	$\Omega_{\mu,k}$	Ω_D	$\dfrac{K}{\varepsilon}T$	$\Omega_{\mu,k}$	Ω_D	$\dfrac{K}{\varepsilon}T$	$\Omega_{\mu,k}$	Ω_D
0.30	2.785	2.662	1.60	1.279	1.167	3.80	0.9811	0.8942
0.35	2.628	2.476	1.65	1.264	1.153	3.90	0.9755	0.8888
0.40	2.492	2.318	1.70	1.248	1.140	4.00	0.9700	0.8836
0.45	2.368	2.184	1.75	1.234	1.128	4.10	0.9649	0.8788
0.50	2.257	2.066	1.80	1.221	1.116	4.20	0.9600	0.8740
0.55	2.156	1.966	1.85	1.209	1.105	4.30	0.9553	0.8694
0.60	2.065	1.877	1.90	1.197	1.094	4.40	0.9507	0.8652
0.65	1.982	1.798	1.95	1.186	1.084	4.50	0.9464	0.8610
0.70	1.908	1.729	2.00	1.175	1.075	4.60	0.9422	0.8568
0.75	1.841	1.667	2.10	1.156	1.057	4.70	0.9382	0.8530
0.80	1.780	1.612	2.20	1.138	1.041	4.80	0.9343	0.8492
0.85	1.725	1.562	2.30	1.122	1.026	4.90	0.9305	0.8456
0.90	1.675	1.517	2.40	1.107	1.012	5.00	0.9269	0.8422
0.95	1.629	1.476	2.50	1.093	0.9996	6.00	0.8963	0.8124
1.00	1.587	1.439	2.60	1.081	0.9878	7.00	0.8727	0.7896
1.05	1.549	1.406	2.70	1.069	0.9770	8.00	0.8538	0.7712
1.10	1.514	1.375	2.80	1.058	0.9672	9.00	0.8379	0.7556
1.15	1.482	1.346	2.90	1.048	0.9576	10.0	0.8242	0.7424
1.20	1.452	1.320	3.00	1.039	0.9490	20.0	0.7432	0.6640
1.25	1.424	1.296	3.10	1.030	0.9406	30.0	0.7005	0.6232
1.30	1.399	1.273	3.20	1.022	0.9328	40.0	0.6718	0.5960
1.35	1.375	1.253	3.30	1.014	0.9256	50.0	0.6504	0.5756
1.40	1.353	1.233	3.40	1.007	0.9186	60.0	0.6335	0.5596
1.45	1.333	1.215	3.50	0.9999	0.9120	70.0	0.6194	0.5464
1.50	1.314	1.198	3.60	0.9932	0.9058	80.0	0,6076	0.5352
1.55	1.296	1.182	3.70	0.9870	0.8998	100.	0.5882	0.5170

APPENDIX D

The Error Function

The error function, $erf\,(\eta)$, is defined as

$$erf\,(\eta) \equiv \frac{2}{\sqrt{\pi}} \int_0^\eta e^{-\phi^2}\, d\phi$$

where ϕ is a "dummy" variable. This integral can be evaluated numerically as a function of η and specific values are given in Table D-1. In transport phenomena the error function is usually encountered as solutions to partial differential equations which can be transformed into ordinary differential equations by means of "similarity" transforms (sometimes called *combination of variables*). For example, the solutions for the velocity, temperature and concentration profiles in one-dimensional, unsteady state molecular transport into an infinite media are all of the form

$$\Phi = 1 - erf\,(\eta) = erfc\,(\eta)$$

where Φ and η are defined in the table, below, for each situation.

Transport Of	Φ	η
Momentum	$\dfrac{v}{v_0}$	$\dfrac{y}{\sqrt{4vt}}$
Energy	$\dfrac{T}{T_0}$	$\dfrac{y}{\sqrt{4\alpha t}}$
Mass Species	$\dfrac{C_A}{C_{A0}}$	$\dfrac{y}{\sqrt{4 D_{AB} t}}$

where y is the distance from the boundary at which the velocity, temperature and concentrations are v_0, T_0, and C_{A0}, respectively. The fluxes at the boundary can also be calculated from the appropriate phenomenological law; for example, in energy transport

$$q_{y\,|\,y=0} = -k \left(\frac{\partial T}{\partial y}\right)_{|\,y=0} = -k\,\frac{\partial \eta}{\partial y}\left(\frac{\partial T}{\partial \eta}\right)_{|\,\eta=0} = \frac{k T_0}{\sqrt{\pi \alpha t}}$$

Table D-1 Values Of The Error Function

η	erf(η)	η	erf(η)
0.	0.	0.80	0.7421
0.05	0.0564	0.85	0.7707
0.10	0.1125	0.90	0.7969
0.15	0.1680	0.95	0.8209
0.20	0.2227	1.0	0.8427
0.25	0.2763	1.1	0.8802
0.30	0.3286	1.2	0.9103
0.35	0.3794	1.3	0.9340
0.40	0.4284	1.4	0.9522
0.45	0.4755	1.5	0.9661
0.50	0.5205	1.6	0.9763
0.55	0.5633	1.7	0.9838
0.60	0.6039	1.8	0.9891
0.65	0.6420	1.9	0.9928
0.70	0.6778	2.0	0.9953
0.75	0.7112	2.1	0.9970

APPENDIX E

Viscosity and Thermal Conductivity Data

VISCOSITY DATA FOR GASES

The viscosity for selected gases as a function of temperature can be found by using the nomograph on the following page (Figure E-1). The nomograph is entered with a set of coordinates, (X, Y), to locate a point on the grid. The coordinate sets are given for a number of gases, in Table E-1, below. A straight edge is then placed on this point and on the desired temperature (left-hand side of the figure). The intersection of the straight edge with the column of viscosity values (right-hand side of the figure), gives the value of the viscosity of that gas (in centipoise) at that temperature.

Table E-1 Coordinates for use with Figure E-1

GAS	X	Y	GAS	X	Y
Acetic Acid	7.7	14.3	Fluorine	7.3	23.8
Acetone	8.9	13.0	Helium	10.9	20.5
Acetylene	9.8	14.9	Hexane	8.6	11.8
Air	11.0	20.0	Hydrogen	11.2	12.4
Ammonia	8.4	16.0	Hydrogen Chloride	8.8	18.7
Argon	10.5	22.4	Hydrogen Sulfide	8.6	18.0
Benzene	8.5	13.2	Methane	9.9	15.5
Butene	9.2	13.7	Methyl Alcohol	8.5	15.6
Butylene	8.9	13.0	Nitric Oxide	10.9	20.5
Carbon Dioxide	9.5	18.7	Nitrogen	10.6	20.0
Carbon Monoxide	11.0	20.0	Nitrous Oxide	8.8	19.0
Chlorine	9.0	18.4	Oxygen	11.0	21.3
Chloroform	8.9	15.7	Pentane	7.0	12.8
Cyclohexane	9.2	12.0	Propane	9.7	12.9
Ethane	9.1	14.5	Propyl Alcohol	8.4	13.4
Ethyl Acetate	8.5	13.2	Propylene	9.0	13.8
Ethyl Alcohol	9.2	14.2	Sulfur Dioxide	9.6	17.0
Ethyl Ether	8.9	13.0	Toluene	8.6	12.4
Ethylene	9.5	15.1	Water	8.0	16.0

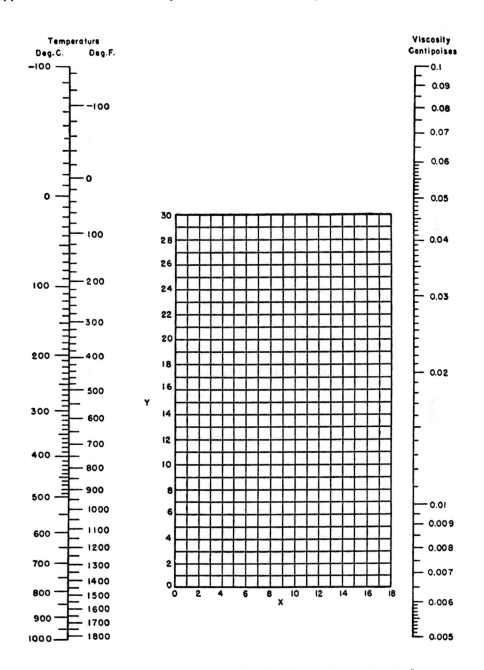

Figure E-1 Gas Viscosity Nomograph, *Chemical Engineer's Handbook*, 5th ed, McGraw-Hill Co., N.Y., 1973, reproduced with permission of McGraw-Hill Companies

VISCOSITY DATA FOR LIQUIDS

The viscosity for selected liquids as a function of temperature can be found by using the nomograph on the following page (Figure E-2). The coordinate sets are given for a number of liquids, in Table E-2, below.

Table E-1 Coordinates For Use With Figure E-2

Liquid	X	Y	Liquid	X	Y
Acetic Acid, 100%	12.1	14.2	Glycerol, 50%	6.9	19.6
Acetic Acid, 70%	9.5	17.0	Heptane	14.1	8.4
Acetone, 100%	7.9	15.0	Hexane	14.7	7.0
Acetone, 35%	14.4	7.4	Hydrochloric Acid, 31.5%	13.0	16.6
Ammonia, 100%	12.6	2.0	Isobutyl Alcohol	7.1	18.0
Ammonia, 26%	10.1	13.9	Kerosene	10.2	16.9
Aniline	8.1	18.7	Methanol, 100%	12.4	10.5
Benzene	12.5	10.9	Methanol, 90%	12.3	11.8
Brine, NaCl, 25%	10.2	16.6	Methanol, 40%	7.8	15.5
Butyl Alcohol	8.6	17.2	Methyl ethyl ketone	13.9	8.6
Carbon Dioxide	11.6	0.3	Naphthalene	7.9	18.1
Carbon Tetrachloride	12.7	13.1	Nitric Acid, 60%	10.8	17.0
Ethyl Acetate	13.7	9.1	Pentane	14.9	5.2
Ethyl Alcohol, 100%	10.5	13.8	Propyl Alcohol	9.1	16.5
Ethyl Alcohol, 95%	9.8	14.3	Sodium Hydroxide, 50%	3.2	25.8
Ethyl Alcohol, 40%	6.5	16.6	Sulfuric Acid, 100%	8.0	25.1
Ethyl Benzene	13.2	11.5	Sulfuric Acid, 60%	10.2	21.3
Ethylene Glycol	6.0	23.6	Toluene	13.7	10.4
Formic Acid	10.7	15.8	Vinyl Acetate	14.0	8.8
Glycerol, 100%	2.0	30.0	Water	10.2	13.0

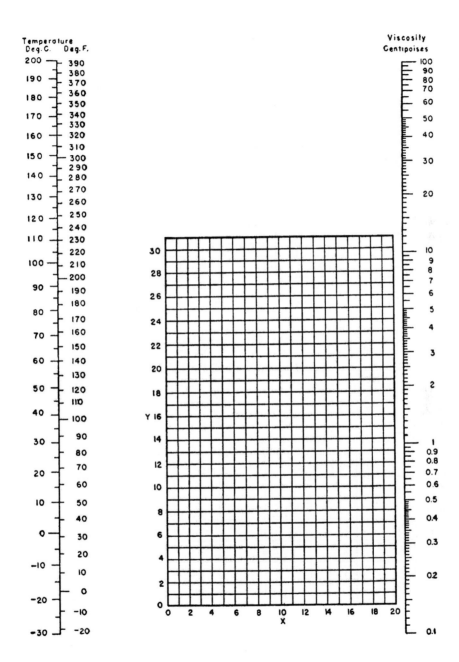

Figure E-2 Liquid Viscosity Nomograph, *Chemical Engineer's Handbook*, 5th ed, McGraw-Hill Co., N.Y., 1973, reproduced with permission of McGraw-Hill Companies

THERMAL CONDUCTIVITY DATA

Gases

Thermal Conductivity in w/m-K x 10^2

Gas	300 K	400	500	600	1000
Air	2.62	3.38	4.07	4.69	6.67
Carbon Dioxide	1.66	2.43	3.25	4.07	6.82
Hydrogen	18.3	22.5	26.6	30.5	44.8
Methane	3.42	4.93	6.68	8.52	
Water		2.66	4.12	4.63	9.78

Liquids

Thermal Conductivity in Btu/hr-ft-F x 10^2

Liquid	68 F	86	100	122	140	167	200	300
Acetone		10.2				9.5		
Benzene		9.2			8.7			
Gasoline		7.8						
Kerosene	8.6					8.1		
Methanol	12.4			11.4				
Water			36.3				39.3	39.5

APPENDIX F

Conversion Factors

Table F-1 Pressure, Shear Stresses, Momentum Flux - $\dfrac{F}{L^2}$, $\dfrac{m}{L\,t^2}$

{`Y`} units multiplied by table value gives { X } units

X Units

Y Units ↓↓↓	newtons m⁻² (pascals)	$lb_m\ ft^{-1}\ s^{-2}$	$lb_f\ ft^{-2}$	$lb_f\ in^{-2}$	atm	mm Hg	in H$_2$O
newtons m⁻² (pascals)	1	$6.72\ 10^{-1}$	$2.089\ 10^{-2}$	$1.45\ 10^{-4}$	$9.869\ 10^{-6}$	$7.50\ 10^{-3}$	$4.014\ 10^{-3}$
$lb_m\ ft^{-1}\ s^{-2}$	1.4882	1	$3.108\ 10^{-2}$	$2.158\ 10^{-4}$	$1.469\ 10^{-5}$	$1.116\ 10^{-2}$	$5.974\ 10^{-3}$
$lb_f\ ft^{-2}$	$4.788\ 10^{1}$	32.174	1	$6.944\ 10^{-3}$	$4.725\ 10^{-4}$	$3.591\ 10^{-1}$	$1.922\ 10^{-1}$
$lb_f\ in^{-2}$	$6.895\ 10^{3}$	$4.633\ 10^{3}$	144	1	$6.805\ 10^{-2}$	$5.172\ 10^{1}$	$2.768\ 10^{1}$
atm	$1.013\ 10^{5}$	$6.809\ 10^{4}$	$2.116\ 10^{3}$	14.7	1	760	$4.068\ 10^{2}$
mm Hg	$1.333\ 10^{2}$	$8.959\ 10^{1}$	2.785	$1.934\ 10^{-2}$	$1.316\ 10^{-3}$	1	$5.353\ 10^{-1}$
in H$_2$O	$2.49\ 10^{2}$	$1.674\ 10^{2}$	5.203	$3.613\ 10^{-2}$	$2.459\ 10^{-3}$	1.868	1

Table F-2 Energy $F \text{-} L, \dfrac{m}{L \text{-} t^2}$

{ Y } units multiplied by Table Value gives { X } units

X Units

Y Units ↓↓↓	$kg\ m^2\ s^{-2}$ (joules)	$lb_m\ ft^2\ s^{-2}$	$ft\ lb_f$	calories	BTU	kw-hr
$kg\ m^2\ s^{-2}$ (Joules)	1	$2.373\ 10^1$	$7.376\ 10^{-1}$	$2.39\ 10^{-1}$	$9.478\ 10^{-4}$	$2.778\ 10^{-7}$
$lb_m\ ft^2\ s^{-2}$	$4.214\ 10^{-2}$	1	$3.108\ 10^{-2}$	$1.007\ 10^{-2}$	$3.994\ 10^{-5}$	$1.171\ 10^{-8}$
$ft\ lb_f$	1.356	32.174	1	$3.241\ 10^{-1}$	$1.285\ 10^{-3}$	$3.766\ 10^{-7}$
calories	4.184	$9.929\ 10^1$	3.086	1	$3.966\ 10^{-3}$	$1.162\ 10^{-6}$
BTU	$1.055\ 10^3$	$2.504\ 10^4$	$7.782\ 10^2$	$2.522\ 10^2$	1	$2.931\ 10^{-4}$
kw-hr	$3.6\ 10^6$	$8.543\ 10^7$	$2.655\ 10^6$	$8.604\ 10^5$	$3.412\ 10^3$	1

Table F-3 Viscosity $\dfrac{F \text{-} t}{L^2}$

{ Y } units multiplied by Table Value gives { X } units

X Units

Y Units ↓↓↓	$g\ cm^{-1}\ s^{-1}$ (poises)	$kg\ m^{-1}\ s^{-1}$ (Pa sec)	$lb_m\ ft^{-1}\ s^{-1}$	centipoise
$g\ cm^{-1}\ s^{-1}$ (poises)	1	10^{-1}	$6.72\ 10^{-2}$	10^2
$kg\ m^{-1}\ s^{-1}$ (Pa sec)	10	1	$6.72\ 10^{-1}$	10^3
$lb_m\ ft^{-1}\ s^{-1}$	$1.488\ 10^1$	1.488	1	$1.488\ 10^3$
centipoise	10^{-2}	10^{-3}	$6.72\ 10^{-4}$	1

Table F-4 Thermal Conductivity $\dfrac{E}{L\text{-}t\text{-}T}$

{ Y } units multiplied by Table Value gives { X } units

X UNITS

Y Units ↓ ↓ ↓	kg m s^{-3} K^{-1} (Watts m^{-1} K^{-1}) (Joules m^{-1} s^{-1} K^{-1})	cal s^{-1} cm^{-1} K^{-1}	BTU hr^{-1} ft^{-1} F^{-1}
kg m s^{-3} K^{-1} (Watts m^{-1} K^{-1}) (Joules m^{-1} s^{-1} K^{-1})	1	2.39 10^{-3}	5.778 10^{-1}
cal s^{-1} cm^{-1} K^{-1}	4.184 10^{2}	1	2.418 10^{2}
BTU hr^{-1} ft^{-1} F^{-1}	1.731	4.137 10^{-3}	1

Table F-5 Heat Transfer Coefficients $\dfrac{E}{L^{2}\text{-}t\text{-}T}$

{ Y } units multiplied by Table Value gives { X } units

X UNITS

Y Units ↓ ↓ ↓	kg s^{-3} K^{-1} (Watts m^{-2} K^{-1}) (Joules m^{-2} s^{-1} K^{-1})	cal s^{-1} cm^{-2} K^{-1}	BTU hr^{-1} ft^{-2} F^{-1}
kg s^{-3} K^{-1} (Watts m^{-21} K^{-1}) (Joules m^{-2} s^{-1} K^{-1})	1	2.39 10^{-1}	0.176
cal s^{-1} cm^{-2} K^{-1}	4.184 10^{4}	1	7.369 10^{3}
BTU hr^{-1} ft^{-2} F^{-1}	5.678	1.357 10^{-4}	1

INDEX

W

Y

ABOUT THE AUTHOR

William J. Thomson is currently Professor of Chemical Engineering at Washington State University, Pullman, Washington, a position he has held since 1981. Prior to then he was a professor of Chemical Engineering at the University of Idaho for 11 years. Dr. Thomson received a BChE degree from Pratt Institute in 1960, an MS in chemical engineering from Stanford University in 1962 and a PhD in chemical engineering from the University of Idaho in 1969. During his career he has worked for AVCO Research, the National Security Agency, and UNOCAL Research. He has taught extensively in the areas of transport phenomena and kinetics and catalysis at both the undergraduate and graduate levels. Dr. Thomson's research interests are primarily focused on applied catalysis, including catalytic membranes, fuel reforming and synthesis gas conversion. He has authored or coauthored over 80 papers and presented seminars at professional meetings and at various research and industrial institutions. He has served as a consultant to over 30 industrial organizations, is an AIChE evaluator for the Accreditation Board for Engineering and Technology and serves on the advisory board of the Environmental Technology Division of Battelle Pacific Northwest National Laboratories.